2
Bioanalytical Reviews

Series Editor
Frank-Michael Matysik, Regensburg, Germany
Joachim Wegener, Regensburg, Germany

Aims and Scope

Bioanalytical Reviews is the successor of the former review journal with the same name, and it complements Springer's successful and reputed review book series program in the flourishing and exciting area of the Bioanalytical Sciences.

Bioanalytical Reviews (BAR) publishes reviews covering all aspects of bioanalytical sciences. It therefore is a unique source of quick and authoritative information for anybody using bioanalytical methods in areas such as medicine, biology, biochemistry, genetics, pharmacology, biotechnology, and the like.

Reviews of methods include all modern tools applied, including mass spectrometry, HPLC (in its various forms), capillary electrophoresis, biosensors, bioelectroanalysis, fluorescence, IR/Raman, and other optical spectroscopies, NMR radiometry, and methods related to bioimaging. In particular the series volumes provide reviews on perspective new instrumental approaches as they apply to bioanalysis, and on the use of micro-/nano-materials such as micro- and nanoparticles. Articles on μ-total analytical systems (μ-TAS) and on labs-on-a-chip also fall into this category.

In terms of applications, reviews on novel bioanalytical methods based on the use of enzymes, DNAzymes, antibodies, cell slices, to mention the more typical ones, are highly welcome. Articles on subjects related to the areas including genomics, proteomics, metabolomics, high-throughput screening, but also bioinformatics and statistics as they relate to bioanalytical methods are of course also welcome. Reviews cover both fundamental aspects and practical applications.

Reviews published in BAR are (a) of wider scope and authoratively written (rather than a record of the research of single authors), (b) critical, but balanced and unbiased; (c) timely, with the latest references. BAR does not publish (a) reviews describing established methods of bioanalysis; (b) reviews that lack wider scope, (c) reviews of mainly theoretical nature.

More information about this series at http://www.springer.com/series/11663

Joachim Wegener
Editor

Label-Free Monitoring of Cells in vitro

With contributions by

F. Alexander Jr. · A. Bauwens · D. Bettenworth · M. Brischwein ·
S. Eggert · Y. Fang · M. Götte · B. Greve · F. Hempel · S. Ingebrandt ·
D. Johannsmann · L. Kastl · B. Kemper · S. Ketelhut · J. K. Y. Law ·
P. Lenz · S. Mues · D. Price · J. Schnekenburger · J. A. Stolwijk ·
A. Vollmer · J. Wegener · J. Wiest

 Springer

Editor
Joachim Wegener
Institute of Analytical Chemistry,
Chemo- and Biosensors
University of Regensburg
Regensburg, Germany

Fraunhofer Research Institution for Microsystems
and Solid State Technologies EMFT
Munich, Germany

ISSN 1867-2086   ISSN 1867-2094   (electronic)
Bioanalytical Reviews
ISBN 978-3-030-32435-3   ISBN 978-3-030-32433-9   (eBook)
https://doi.org/10.1007/978-3-030-32433-9

This Springer imprint is published by the registered company Springer Nature Switzerland AG.
The registered company address is: Gewerbestrasse 11, 6330 Cham, Switzerland

This volume of Bioanalytical Reviews is dedicated to
Ivar Giaever (Nobel Laureate in Physics 1973)
on the occasion of his 90th birthday.
To an outstanding scientist and an inspiring academic teacher.
Thanks for all. J.

Preface

Across all areas of biomedical research, experiments with isolated animal (human) cells in a laboratory in vitro environment—also referred to as cell-based assays (CBA)—are considered an important intermediate between complex studies in living animals and simple biomolecular assays in well-defined binary or ternary systems. Accordingly, CBAs have emerged into an indispensable tool for fundamental cell biology aiming to unravel the yet unknown secrets of life, for drug development as well as for screening xenobiotics for their impact on living organisms. Even though cultured cells cannot represent all features of a multicellular organism as they miss the systemic level, they have tremendously contributed to our current understanding of cell structure and function just as much as they have proven their suitability to test potential drugs or harmful compounds on the cellular level. Nowadays, cell cultures from almost any mammalian tissue are available as primary cultures, isolated directly from the donor organism, or as established cell lines from commercial cell depositories. However, a proper cell culture model is just one prerequisite to extract useful information from CBAs. It also takes sensitive indicators and readout approaches to monitor the cell response to experimental manipulations or stimuli. Moreover, the response of the cells to a given stimulus needs to be measured quantitatively in order to determine threshold concentrations of bioactive compounds, establish structure-activity relationships, or to compare different classes of compounds with respect to their impact on living cells. Microscopic inspection of stained tissue specimens has been used in the early days of cell analysis and is still heavily applied with more advanced stains, functional probes, and image acquisition modalities. However, staining of cells typically requires fixation of the cells, sometimes even permeabilization of the membranes unless genetically encoded probes or membrane-permeable dyes are used. Accordingly, these techniques provide an endpoint analysis for the most part but do not allow continuous monitoring of the cells. This limitation also applies to non-imaging but label-based assays reading metabolic or enzyme activity, DNA content, or the activation of signaling cascades. In microscopy it has been Zernike's discovery of phase-contrast microscopy in the 1930s that allowed for time-resolved microscopic observation of unstained living cells as they migrate or proliferate without any interference by the measurement. In this sense phase-contrast microscopy has been the first label-free technique for cell analysis. Ever since its invention it has turned into a standard experimental approach that is used around the globe in any lab working with living cells as it enables routine inspections of the cell cultures at any time point along their life cycle. Other physical means have been developed over the years that provide label-free access to cell behavior. They share the advantages of phase-contrast microscopy in terms of being noninvasive and enabling time-resolved cell monitoring due to the absence of interfering labels or probes. This book is dedicated to those techniques.

The chapter "Impedance-Based Assays Along the Life Span of Adherent Mammalian Cells In Vitro: From Initial Adhesion to Cell Death" of the book reviews the use of impedance measurements to study adherent cells grown on thin-film electrodes. This technique, referred to as *electric cell-substrate impedance sensing* (ECIS), has been invented about 35 years ago by Charles R. Keese and Ivar Giaever to whom this book is dedicated. It is based on measuring the impedance of the electrodes with noninvasive AC currents and voltages. As the impedance readout is sensitive to the gradual coverage of the electrode with cells, ECIS provides a quantitative analysis of all cell behaviors that result in changes of surface coverage like cell spreading, cell proliferation, or cell migration. When the electrode is completely occupied by cells, impedance becomes sensitive to changes in cell morphology which paves the way to a huge number of additional applications like, for instance, drug or cytotoxicity testing. Stolwijk and Wegener describe the theoretical basis of the technique in extended detail in this chapter and demonstrate its versatility by presenting various impedance-based phenotypic assays along the life span of adherent cells in vitro.

The chapter "Transistor-Based Impedimetric Monitoring of Single Cells", written by Hempel and coauthors, addresses a variant of the original ECIS technique, which is no longer based on thin-film metal electrodes but on *field-effect transistors* (FET). The cells under study adhere to the gate of the transistor, and they modulate the AC voltage applied between the gate and a reference electrode in the bath according to their dielectric properties. The major advantage of using FETs is the size of the gate electrode that is tuned to be below the size of a single cell and thereby provides single cell measurements whereas regular ECIS readings report on cell ensembles that typically include more than 100 cells. Using transistors instead of metal film electrodes allows for tailored amplification of the signal so that attachment and detachment of single cells is recorded sensitively.

The chapter "Label-Free Monitoring of 3D Tissue Models via Electrical Impedance Spectroscopy" broadens the perspective of electrical impedance measurements to 3D cell culture models with special emphasis on spheroids but also other scaffold-free and scaffold-based systems. Due to the three-dimensional geometry of the biological object, the electrode placement or design inside the measurement device is of critical importance for the readout. Alexander and coauthors review the available literature on impedance measurements on 3D culture models and describe special devices for this purpose like microfluidic channels that match the size of spheroids to be studied or microneedle electrodes to pierce the cell aggregates.

The chapter "On the Use of the Quartz Crystal Microbalance for Whole-Cell-Based Biosensing" introduces acoustic transducers to monitor cell behavior. When the cells are grown on piezoelectric shear wave resonators as they are used in *quartz crystal microbalances* (QCM), the readout of the resonance parameters like resonance frequency and bandwidth provides noninvasive information on the viscoelastic properties of the cells under study. It is noteworthy that the amplitude of the mechanical shear oscillation is in the nanometer range and at MHz frequencies so that these QCM-like measurements are considered noninvasive to adherent cells. Johannsmann focuses in his article on technical aspects of this approach with respect to acoustic detection and data processing in order to highlight potential avenues for future applications not without clearly expressing current limitations. He particularly addresses the experimental options that are associated by systematically changing the shear amplitude.

Metabolic analysis of living cells in terms of extracellular acidification rate (EAR) or oxygen consumption rates (OCR) is provided by *microphysiometry* that is addressed in the chapter "Microphysiometry" by Brischwein and Wiest. In this approach electrodes for potentiometric/amperometric detection of pH and dissolved oxygen are placed in a cell culture dish in order to follow the rate of aerobic and anaerobic metabolism along an experiment. The chapter reviews the history of microphysiometry, introduces the technology, and describes applications. Besides the determination of EAR and OCR, the authors also shortly address alternative sensor concepts to determine carbon dioxide, glucose, or lactate by tailored biosensors.

In chapter "Optical Waveguide-Based Cellular Assays" the physical nature of the readout changes from electrochemical/piezoelectric measurements to optical signals. But in the line of this book, the approach does not rely on any labels but uses optical waveguides with internal diffractive nanograting structures (*resonant waveguide grating*) as the growth substrate for the cells. This device reads changes of the refractive index within a distance of approximately 200 nm from the surface. When cells attach and spread on the waveguide surface or when attached cells undergo morphological changes associated with mass redistribution, the device reports on it sensitively. Accordingly, the number of different phenotypic assays that are readily monitored by RWG-based sensing is huge. Fang reviews the history of the technique, explains the physical basis, and reports on its particular applications in drug discovery.

The chapter "Label-Free Quantitative In Vitro Live Cell Imaging with Digital Holographic Microscopy" introduces an emerging type of microscopy that operates without any labels but allows for 3D imaging of cells and biological structures: *digital holographic microscopy* (DHM). DHM is based on numerical reconstruction of quantitative phase images from digital holograms providing information from different focal planes and, thus, the 3D structure of the specimen. The approach uses low intensity incident light so that noninvasive long-term observations of cultured cells are accessible. As Kemper and coauthors point out after laying out the physical basis of this emerging technique, detailed DHM analysis provides more than just the 3D structure of biological specimens. It also reports on parameters as the refractive index of the cells, which is related to both their volume and intracellular solute concentrations.

The collection of label-free approaches to monitor living cells in vitro as they are covered in this book is not complete, but it includes the most prominent ones. All these techniques are integral in nature, reporting on the cells' response to a given stimulus in a holistic way without any specific information on the mode of action. This is different from assays that are based on specific molecular probes that report specifically on the concentration of one chemical species or a physical parameter. However, successful use of such specific label-based assays requires a priori knowledge on which functionality might change in the course of an experiment. Thus, label-based assays are biased and the researcher will only see what has been planned to see prior to the experiment. Label-free assays do not have such a bias. They report on anything, but it takes additional measures to pinpoint the detailed mode of action. So both strategies of cell monitoring, label-free and label-based, should be considered as complementing techniques rather than as competing ones. When they are used in parallel on a given problem to be solved, they unfold full synergy. This book will hopefully contribute to the appreciation of label-free holistic means of cell analysis.

This book would not have been possible without the fine work of all authors and coauthors to whom I would like to express my deepest gratitude. I hope you all agree that this big effort and the long hours of work have been worthwhile. Moreover, I would like to thank Simone Ruckdäschel and Christian Kade for the editorial preparation of some chapters of this volume.

Regensburg, Germany   Joachim Wegener
Summer 2019

Contents

BIOREV (2019) 2: 1–76
https://doi.org/10.1007/11663_2019_7
© Springer Nature Switzerland AG 2019
Published online: 15 September 2019

Impedance-Based Assays Along the Life Span of Adherent Mammalian Cells In Vitro: From Initial Adhesion to Cell Death

Judith A. Stolwijk and Joachim Wegener

Contents

J. A. Stolwijk
University of Regensburg, Regensburg, Germany

J. Wegener (✉)
University of Regensburg, Regensburg, Germany

Fraunhofer Research Institution for Microsystems and Solid State Technologies EMFT, Munich, Germany
e-mail: Joachim.Wegener@ur.de

Abstract Impedance-based monitoring of adherent cells has gained increasing acceptance in many areas of biomedical research and drug discovery, as it provides a noninvasive and label-free experimental tool to continuously monitor cell behavior in response to biological, chemical, or physical stimuli. It is based on growing cells on thin-film electrodes that are deposited on the bottom of regular cell cultureware and monitoring the electrical impedance of these electrodes as a function of frequency and time. Tailored hardware, data acquisition modes, electrode designs, and assay procedures have been developed for a fully automated and time-resolved observation of various cell phenotypes and phenotypic changes disclosed in cell-matrix adhesion, proliferation, differentiation, cell migration, signal transduction, or cell death. Impedance-based cell analysis is performed under regular cell culture conditions and applied for short-term and long-term observation of cell behaviour.

This review will recapitulate the physical principles of the measurement and highlight its applications not without critical assessment of its limitations. The most prominent impedance-based assays will be explained and discussed with reference to an extended literature survey of the field.

Keywords Cell adhesion · Cell migration · Cell-based assays · ECIS · Impedance · Label-free · Proliferation · Wholistic

Abbreviations

AC	Alternating current
BBB	Blood-brain barrier
CBA	Cell-based assays
DC	Direct current
ECIS	Electric cell-substrate impedance sensing
TEER	Transendothelial/transepithelial epithelial electrical resistance

1 Introduction

The body as a whole is composed of organs, the organs are built up from different tissues, and the tissues are a collection of cells – the latter are the fundamental building block of life. Dependent on their differentiation and phenotype, these cells perform highly complex functions to sustain our survival and pass on our genes to the next generation. Understanding the behavior of cells within their physiological context and the regulatory programs behind these processes is an important prerequisite to prevent and cure diseases. The most universal cell functions such as proliferation, respiration, and metabolism are often used as general biomarkers, as they provide fundamental information about the activity and metabolic state of a cell.

Depending on the cell type and its location in the body, other more distinctive structural or behavioral features become important to understand their full functionality such as cell-cell interactions, cell-matrix adhesion, cell migration, production and secretion of signaling molecules, as well as their correct response to signaling molecules. Accordingly, it is not surprising that cellular dysfunction may have harmful impacts on the functionality of a tissue, an organ, or the whole organism and may cause severe diseases. Aside from inherited mutations, environmental noxes play an important role in the acquisition of cellular dysfunctions and dysregulations. These environmental challenges may be of biological, chemical, or physical origin, including pathogens, toxins, radiation, and mechanical damage. With the advent of nanomaterials, a new class of industrial materials gained special attention in public health risk assessment. Even drugs, originally designed to cure human diseases, may have harmful impacts. Although it is the ultimate goal to find highly effective and safe drugs, it is almost impossible to design a drug that only shows the intended activity without having any side effects. Therefore, proper risk assessment by cytotoxicity testing and monitoring of phenotypic changes within the exposed tissues is a central aspect within the drug development process. Most of these studies start at the cell level.

Cell-based assays (CBA) are a valuable compromise between molecular interaction studies and animal experiments. Multiple assay types exist that are designed to study on different levels how the cell responds to a biological, chemical, or physical stimulus. However, most assays only address one feature of cell behavior. As one example out of many, the very popular *MTT assay* reads the metabolic activity of a cell ensemble with the help of a chemical probe that changes color when it is metabolized by the cells under study. A central caveat of this and other label-based assays is that they are endpoint assays and do not allow following the same sample over time, as the assay procedure typically leads to destruction of the cells by removal from their substrate, fixation, permeabilization, and/or staining. Thus, these types of assays provide only a snapshot of the cell behavior under investigation at a pre-defined time point.

In 1984 Giaever and Keese invented *electric cell-substrate impedance sensing* (ECIS) as a label-free and noninvasive technique to study cell behavior electrically [1]. ECIS is a widely applicable technique that allows analyzing any adherent cell type with respect to changes in fractional electrode coverage and cell shape [2–4]. This sensitivity for two rather unspecific and general descriptors of a cell population, namely, cell number and cell morphology, provides a broad applicability for many different types of cell-based assays. Typical applications include measurement of cell attachment and spreading [5], proliferation [6], barrier function of endothelia and epithelia [7], cellular micromotility [2], lateral migration and wound healing [8], receptor activation and signal transduction [9], cell differentiation [10], and cytotoxicity studies [11]. This plethora of accessible assays makes ECIS one of only few techniques that can be used to study cell behavior along the entire life span of adherent cells in vitro (Fig. 1). It provides the option to perform various phenotypic assays of a given cell population either in form of a single assay formats or by sequences of impedance-based assays. The latter become possible

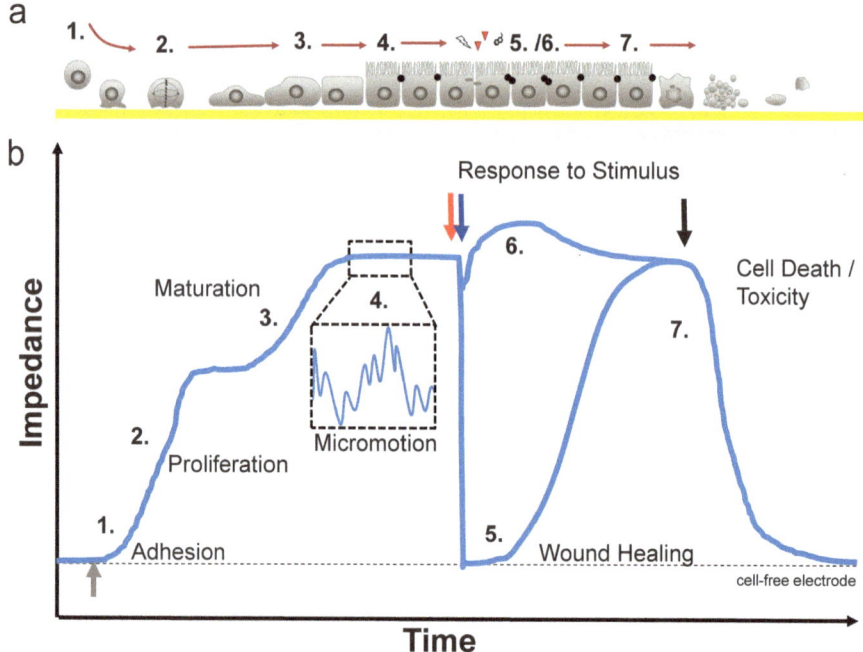

Fig. 1 Overview of consecutive activities of adherent cells in vitro monitored by impedance-based assays after seeding the cells upon substrate-integrated thin-film electrodes. (**a**) Schematic illustrating cell density and morphology changes during different phenotypic changes. (**b**) Schematic showing time-dependent changes in impedance of the electrodes along these processes. *(1)* Cell adhesion and spreading after inoculation of suspended cells (gray arrow); *(2)* proliferation; *(3)* cell layer maturation and differentiation; *(4)* micromotility (micromotion) within a confluent cell layer; *(5)* cell migration after electrical wounding of the cell layer (red arrow); *(6)* response of a confluent cell layer to a biological, chemical, or physical stimulus (blue arrow); *(7)* cell death induced by biological, chemical, or physical threats (black arrow)

since the technique is entirely noninvasive and does not rely on chemical additives or probes.

The in vitro life cycle of an adherently growing cell (Fig. 1) usually starts with the inoculation of a culture dish with the suspended cells. Adhesion and spreading of the cell bodies to the bottom of the dish are the first activities of the cells (1) followed by proliferation (2), provided that there is sufficient space for additional cells on the culture substrate. In many cases a certain time is required for the cells to fully mature and differentiate to a physiologically functional state (3). For example, epithelial and endothelial cells that are known to form interfacial cell layers separating two fluid compartments in the body build a two-dimensional diffusion barrier by expressing special cell-cell junctions. Formation of these junctions requires the cells to be in close proximity in a dense cell layer. Without the fully matured junctions, the cell layer will not show similar barrier and transport functions as a cell layer whose cell junctions are fully established. Many in vitro assays are performed with confluent, mature, and differentiated cell layers that reflect the in vivo situation as close as possible. For certain physiological but also pathological processes, cell motility and migration represent key phenotypes as these are involved in wound healing, angiogenesis, or metastatic progression. Several impedance-based assays exist to study cell motility (4), i.e., the dynamics of the cell body without any lateral movement, and cell migration (5) which describes the lateral movement of the cell bodies along the culture substrate. Lateral migration requires open spaces on the culture substrate which are experimentally introduced by mechanical or electrical wounding of an established cell layer (5). The time course of wound closure, provided by cells migrating from the wound edge to the center, is used to quantify the migration velocity. Moreover, during their in vitro lifetime, cells may get exposed to various biological, chemical, and physical stimuli (6) that cause measurable cell shape changes. Finally cell death, often induced experimentally by drugs or toxins, terminates the in vitro life cycle (7).

In the following chapters, we will address each of these phenotypic assays along the life span of cells in vitro and how they are monitored by electrochemical impedance measurements using tailored electrode layouts and data acquisition protocols. First, we will discuss the physical background necessary to understand the behavior of cells exposed to electric fields and the basics of impedance-based cell monitoring.

2 Physical Background of Impedance-Based Assays

2.1 Behavior of Cells in Electric Fields

Following cell behavior by means of electrical impedance measurements inevitably requires exposing the cells under study to external electric fields. This chapter discusses the fundamental behavior of cells in electric fields and why impedance-based assays are considered to be noninvasive.

Cells brought into a homogeneous electric field between two electrodes essentially behave like insulating particles. The insulating properties are due to the cell membranes. Their hydrophobic core of approx. 5 nm thickness insulates the conducting cytoplasm from the surrounding extracellular electrolyte [12, 13] (Fig. 2a). Despite the existence of some ion channels in the membrane, the overall membrane conductivity is low compared to the conductivity of the surrounding physiological buffers and cytoplasm. The membrane resistance is typically in the order of 10^3–10^5 Ω cm^2, depending on ion channel density and conduction state. Among other mechanisms, the ion channel-based selective permeability of the cell membrane to certain ions induces a natural resting potential difference across the membrane ($\Delta\Psi_0$) of about -40 to -70 mV, which is metabolically maintained by active membrane transport [14, 15]. When cells are exposed to external electric fields with field strength E, the *induced* membrane potential superimposes the natural resting potential of the membrane (Fig. 2a). The external electric field induced transmembrane potential $\Delta\Psi_E$ is best described by the formula:

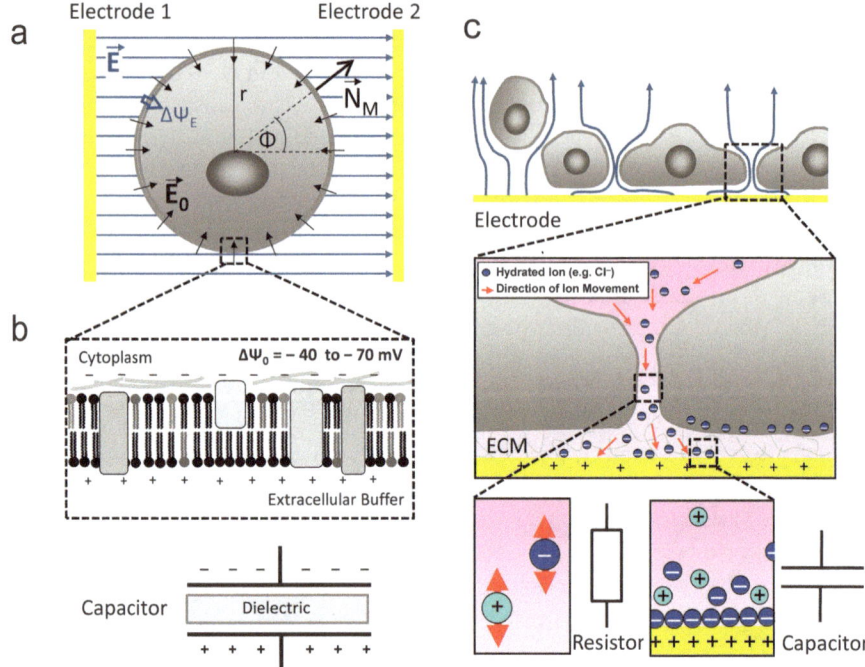

Fig. 2 Schematic illustration of cells in electric fields. (**a**) Suspended cell placed in an electric field (*E*) between two electrodes. Electric field lines are depicted by blue arrows. Note that in AC the orientation of *E* periodically changes direction. Cells can be electrically viewed as insulating shells filled with conducting electrolyte solution. The membrane potential ($\Delta\Psi_0$) generates an intrinsic electric field E_0 across the cell membrane that is superimposed by the external electric field (*E*). The magnitude of the net membrane potential ($\Delta\Psi_E$) depends on the angle between *E* and the normal to the cell membrane surface N_M (ϕ) as well as the cell radius *r*. (**b**) The cell membrane acts as a capacitor due to its hydrophobic dielectric core. Electric charges can accumulate at either side of the membrane and generate a transmembrane potential: $\Delta\Psi_{\text{total}} = \Delta\Psi_0 + \Delta\Psi_E$. (**c**) Cells adherently grown on a conducting surface. Cells adhere to extracellular matrix components (ECM) deposited on the electrode surface. In the culture medium, electric current flow is mediated by hydrated ions, e.g., Cl^- (blue) and Na^+ (green). The ions move along the electric field according to their charge. The insulating properties of the cell membrane force the ions to flow around the cell bodies, which increases the resistance especially in the intercellular and subcellular clefts. As the membrane is essentially impermeable to the hydrated ions, they accumulate at either side of the lipid bilayer. The ions close to the polarized electrode form a so-called electric double layer (Stern layer), with a first layer of oppositely charged hydrated ions in close vicinity to the electrode surface and a more distant diffuse layer. Formation of electric double layers explains the capacitor-like behavior of the electrode-electrolyte interface

$$\Delta\Psi_E = f \cdot g(\lambda) \cdot r \cdot E \cdot \cos\theta(M) \qquad (1)$$

It includes the cell radius *r* and a shape factor f ($f = 1.5$ for spherical cells). The term $E \cdot \cos\theta(M)$ describes the dependency of the locally induced transmembrane potential on the angle between *E* and the normal to the membrane at a position *M*. The factor $g(\lambda)$ stands for an expression that includes the different conductivities of the cell membrane, the cytoplasm, and the extracellular medium as well as the cell radius and the membrane thickness. It is discussed in more detail elsewhere [16–18]. The expression illustrates that the influence of the external electric field on cellular membranes is strongly dependent on the cell size and the relative position with respect to the electric field. At the cell poles facing the electrodes, the external field impact on the cell is strongest. The net membrane potential difference of a cell in an electric field is calculated by $\Delta\Psi_{total} = \Delta\Psi_0 + \Delta\Psi_E$. Many phenomena observed for suspended cells exposed to invasive or noninvasive electric field strengths are based on the polarization effect described above. Subcritical polarization of the cell body leads to phenomena like electrodeformation [19] and dielectrophoresis [20]. Most studies discuss the situation for suspended cells. But also in adherent cells, sublethal electric fields provoke cellular responses such as reorientation of the cellular axis, synchronized alignment of the cleavage plane during cell division, directed outgrowth of neurites, or electric field-induced migration [21–26]. All these electric field phenomena are attributed to electrophoretic or electroosmotic effects that change the lateral distribution of mobile membrane receptors, carriers, or channels within the lipid matrix causing intracellular concentration gradients [22]. However, these weak and mostly DC fields have to be applied to the cells over hours in order to evoke a cellular response.

Stability of the membrane is only impaired when the total transmembrane potential difference is raised above threshold values between 200 mV and 1 V dependent on cell type. Above these voltages the membrane lipids reorganize and form pores in the membrane [27–29]. Thus, the external electric field strength must be higher than a cell type-specific threshold (E_{perm}): $\Delta\Psi_E = f \cdot g(\lambda) \cdot r \cdot E_{perm}$. For most mammalian cells the critical transmembrane voltage is achieved when the external electric field strength is bigger than ~0.3 kV/cm [17]. Overcritical field strengths have been used to induce reversible pore formation in cell membranes for loading of the cells with membrane-impermeable substances such as DNA, most efficiently achieved for field strengths between 1 and 10 kV/cm applied for a few μs [17, 30]. Electroporation is generally considered being a reversible process as long as the length of overcritical field exposure is not too long and the field strength is not too high. Irreversible membrane poration will lead to cell lysis and cell death. Below the critical transmembrane voltage, as in regular impedance-based cell monitoring, the cell membrane behaves like an almost ideal capacitor, with the intracellular and extracellular side of the membrane acting as chargeable surfaces. The charging time of the membrane was estimated to be in the range of about 1 μs for a regular plasma membrane. It depends on the dielectric properties of the membrane, determined by the protein content and ion channel density, and the conductivities of the cytoplasm and the external medium [31].

The above equations have been derived for direct current (DC) field effects on cells. To include the influence of sinusoidal alternating current (AC) electric fields, a frequency-dependent term has to be included [32]:

$$\Delta\Psi_E = 1.5 \cdot \frac{g(\lambda) \cdot E \cdot r \cdot \cos\Phi(M)}{\sqrt{1 + (\omega\tau)^2}} \quad \text{with} \quad E = E_0 \cdot \sin\omega t \qquad (2)$$

The above equation indicates that the induced membrane potential difference in an AC field will be smaller than for a DC field of the same strength if the period of the applied AC signal is smaller than the charging time of the membrane. Only for frequencies smaller than 100 kHz AC signals can induce the same transmembrane potential as a DC pulse of the same field strength [33]. Moreover, the field strength of AC signals changes periodically with changing direction. Thus, potentially critical electric field strengths are only present for a fraction of the total exposure time.

The impedance-based assays described here are limited to anchorage-dependent cells grown as monolayers on solid planar substrates that are either permeable (e.g., Transwell cell culture inserts) or impermeable (e.g., thin-film electrodes). Excitation signals used to monitor living cells by impedance assays are typically in the order of a few mV or μA. For example, in a study provided by Gosh et al., the current amplitude used for impedance measurements was set to 1 μA, and it induced a voltage across the cell layer that was about 5 mV when culture medium was used as the electrolyte in a two-electrode system with an electrode area of 10^{-3} cm^2 [34]. Solly et al. used an excitation signal of 10 mV for impedance measurements [35]. They also tested the impact of 200 mV applied for 30 min on cell proliferation but did not detect any adverse effect [35]. Also Wegener et al. did not detect any long-term effects on MDCKII cells when using a voltage amplitude of 60 mV for the measurement [36]. Comparing these voltages applied in regular impedance-based monitoring with the natural resting potential explains the noninvasiveness of the measurement in particular as AC signals are non-polarizing in nature [34–39].

2.2 Physical Basics of Impedance Sensing

Impedance spectroscopy (IS) is a nondestructive and noninvasive method to characterize the electrical properties of a sample under study. IS has been widely used to characterize a variety of materials and electrochemical systems [40]. Today it is increasingly appreciated as label-free means to investigate biological samples as well [41–43]. Impedance spectroscopy provides information on how electric currents flow through a sample. The current pathway depends on the sample's composition and on the frequency of the alternating electric current that is injected for measurement. Electrical impedance is the ability of an electrochemical system to resist current flow in an alternating current (AC) circuit. According to Ohm's law, impedance Z is defined by the ratio of the time-dependent voltage $V(t)$ and current $I(t)$ at time t (Fig. 3a):

$$Z = \frac{V(t)}{I(t)} = \frac{V_0 \sin(\omega t)}{I_0 \sin(\omega t - \phi)} \tag{3}$$

V_0 and I_0 describe the amplitudes of voltage and current; ω is the angular frequency with $\omega = 2\pi f$, with the frequency f in Hz; φ is the phase shift between voltage $V(t)$

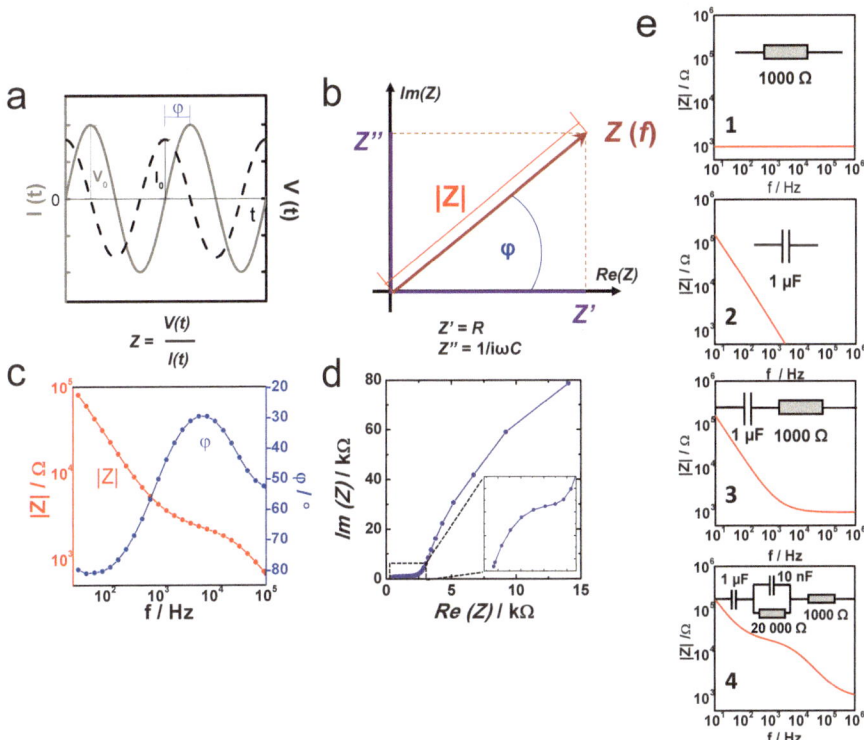

Fig. 3 Basic concepts of electrical impedance. (**a**) Impedance is the ratio of an applied alternating voltage and the resulting current through a system under study. (**b**) Illustration of complex impedance as a vector in a two-dimensional Cartesian plane, with a real (Re(Z)) and imaginary (Im(Z)) axis. Complex impedance is described by Cartesian coordinates Z' and Z'' or by polar coordinates $|Z|$ and φ; the latter represents the phase shift between voltage and current. (**c**) Bode plot presenting $|Z|(f)$ and $\varphi(f)$ of a cell-covered gold film electrode (5×10^{-4} cm^2) as a function of AC frequency. (**d**) Nyquist plot of the same cell-covered electrode presenting Im(Z) as a function of Re (Z). (**e**) Bode plots for *(1)* 1 kΩ resistor, *(2)* 1 µF capacitor, *(3)* 1 kΩ resistor in series with a 1 µF capacitor, *(4)* combination of a 20 kΩ resistor in parallel to a 10 nF capacitor in series to a 1 kΩ resistor and 1 µF capacitor

and current $I(t)$. Only if the circuit is perfectly resistive in nature, there is no phase shift between current and voltage ($\varphi = 0°$). Electrochemical systems that behave like ideal capacitors or coils cause a (+/−) 90° phase shift between voltage and current. Composite electrochemical systems with resistive and capacitive/inductive properties show frequency-dependent phase angles between +90° and −90°. Therefore, impedance becomes a vector in a two-dimensional complex plane with in-phase and 90° out-of-phase components (Fig. 3b). Complex impedance is typically expressed by its Cartesian coordinates that separate impedance contributions arising from current in-phase with the voltage (real part of impedance Z') from impedance contributions arising from 90° out-of-phase currents (imaginary part of impedance Z''). In-phase impedance contributions (Z') are termed resistance. Out-of-phase impedance contributions (Z'') are termed reactance. Thus, in Cartesian coordinates the complex impedance is expressed as:

$$Z = Z' + iZ'' \tag{4}$$

with the imaginary unit i, for which $i^2 = -1$. Alternatively, complex impedance, current and voltage are expressed in exponential form leading to a notation of impedance in polar coordinates:

$$Z = \frac{V(t)}{I(t)} = \frac{V_0 \cdot e^{i\omega t}}{I_0 \cdot e^{i(\omega t - \varphi)}} = \frac{V_0}{I_0} \cdot e^{i\varphi} = |Z| \cdot e^{i\varphi} \tag{5}$$

The impedance magnitude $|Z|$ denotes the length of the impedance vector, and the phase shift φ describes the angle between the vector and the x-axis. Cartesian coordinates (Z', Z'') and polar coordinates ($|Z|$, φ) are interconvertible as follows:

$$Z' = |Z| \cdot \cos\varphi \quad Z'' = |Z| \cdot \sin\varphi \tag{6}$$

$$|Z| = \sqrt{Z'^2 + Z''^2} \quad \varphi = arctan\left(\frac{Z''}{Z'}\right) \tag{7}$$

Note that real and imaginary contributions to the overall impedance may depend on the applied frequency. The impedance of purely resistive systems is frequency-independent (Fig. 3e, 1). The impedance of a capacitor decreases with increasing frequency ($Z_C = 1/(i\omega C)$ (Fig. 3e, 2), whereas the impedance of an inductor increases with increasing frequency ($Z_L = i\omega L$).

A so-called *Bode diagram* presents the frequency-dependent complex impedance of a system under test (Fig. 3c) by plotting the impedance magnitude $|Z|$ and the phase shift φ as a function of frequency in double logarithmic [$|Z|(f)$] or semilogarithmic [$\varphi(f)$] presentation. Alternatively, the imaginary impedance contribution is plotted versus the real impedance contribution resulting in a so-called *Cole-Cole plot* or *Nyquist diagram* (Fig. 3d). The most common way to illustrate the frequency-dependent behavior of cells o n dectrodes is the Bode diagram. Bode plots for several different electronic components and circuits are presented in Fig. 3e.

2.3   *Experimental Setup for Impedance Measurements of Adherent Cells In Vitro*

For studying adherent cells in vitro, two main impedance measurement setups have been developed: (1) Cells are cultured in cell culture inserts with a porous polymer membrane as growth substrate. The insert is sandwiched between two electrodes reaching into the liquid compartments above or below the cell-covered porous support (Fig. 4a). This setup is often called the Transwell™ setup. (2) Cells are

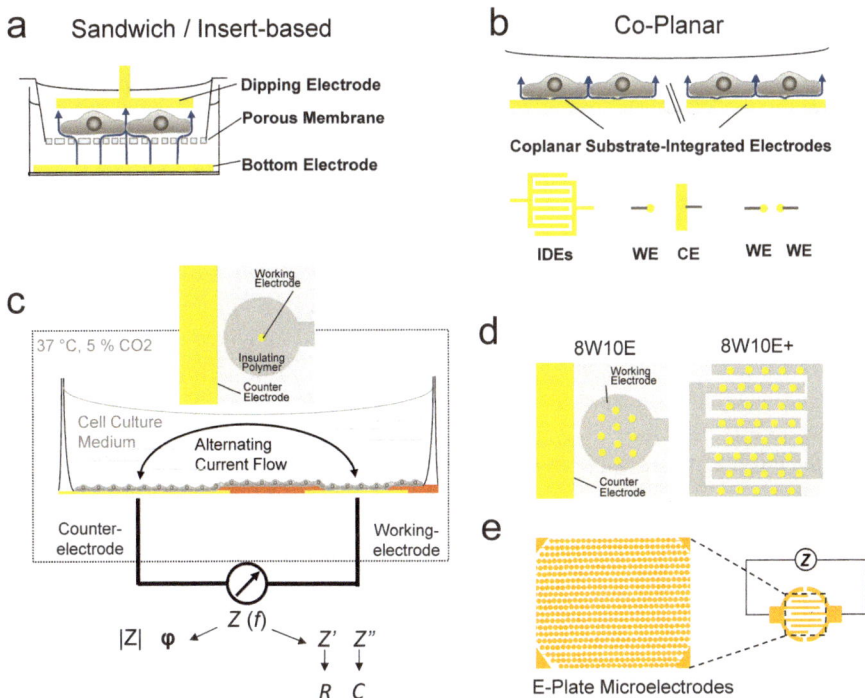

Fig. 4 Illustration of the most common setups and electrode arrangements for impedance-based cellular assays. (**a**) Insert-based *sandwich* setup. The cells are grown on permeable supports (porous polymer membrane) which are sandwiched between a set of electrodes. (**b**) Substrate-integrated, *coplanar* setups. The conductive electrode material (gold, ITO or others) is deposited as a thin film directly upon the cell culture substrate. The electrodes are often patterned (i) as interdigitated comb-like structures (IDE), (ii) in a working electrode/counter electrode (WE/CE) layout, in which the much smaller electrode dominates the impedance, or (iii) the two electrodes have the same surface area and contribute equally to the overall impedance (WE/WE). (**c**) Schematic illustrating the classical ECIS 8W1E electrode layout consisting of a small circular working electrode with an electrode area of 5×10^{-4} cm^2 and a significantly larger (~500×) counter electrode, both patterned on inert but transparent polymer substrate. An alternating current is injected into the two-electrode setup, and the impedance is calculated from readings of the in-phase (Z') and out-of-phase (Z'') voltage. (**d**) Schematic illustrating 8W10E and 8W10E+ type electrodes from Applied BioPhysics Inc. (**e**) Schematic illustrating the E-Plate electrode layout from ACEA Bioscience Inc. (adapted from [35])

directly grown on the surface of substrate-integrated, coplanar thin-film electrodes (Fig. 4b). This is the basis of the ECIS technology.

Insert-based setups have proven especially useful in the study of epithelia and endothelia that form tight diffusion barriers, like endothelial cells of the blood-brain barrier [44–47], or epithelial cells from the kidney or colon [48, 49]. The great advantage of insert-based setups is that the cells are grown as an interfacial cell layer between two separate fluid compartments – similar to the in vivo situation. The cells have access to nutrients from both compartments, and they become accessible from both sides for experimental manipulation. The presence of an apical and a basolateral compartment with potentially different chemical composition on the different sides of the cell layer is assumed to promote polarization and barrier properties of epithelia. Moreover, this setup facilitates co-culturing of two cell types that are expected to influence each other on either side of the porous support. Insert-based impedance measurements essentially provide information on the electrical tightness of the cell layer which is the most prominent functional differentiation marker of endothelial and epithelial cells [50]. However, for leaky cell layers, the contribution of the cell layer on top of the porous support to the overall resistance/impedance becomes marginal, and insert-based measurements become incapable of resolving changes in barrier function with sufficient sensitivity. Moreover, due to the extended surface area, insert-based measurements easily suffer from defects or leaks in the cell layer that are difficult to detect by microscopic inspection. In addition, changes in electrode position with respect to the cell layer along an experiment may lead to significant data scattering – in particular when handheld chopstick electrodes are used.

Due to the limited sensitivity of the insert-based approach for non barrier-forming cell layers that arises from their rather big surface area (mostly >0.33 cm^2), impedance-based techniques working with coplanar thin-film electrodes have been developed in which the cells under study are directly grown on the electrode surface (Fig. 4b). These electrodes are thin layers of conducting material on an inert base, and they do not change the topography of the culture substrate on cellular scales. The electrodes can be miniaturized down to the single cell level and therefore overcome the sensitivity issues associated with insert-based approaches. Accordingly, impedance measurements with substrate-integrated electrodes enable researchers to monitor any adherent cell type and characterize their behavior with respect to coverage of the electrode (e.g., adhesion and spreading, migration, proliferation, apoptosis), cell shape, or cell layer architecture (e.g., barrier function, transdifferentiation, epithelial-mesenchymal transition). It is noteworthy that most analytical techniques to study cell cultures such as microscopy, protein isolation and quantification, nucleic acid analysis, gene expression profiling, and others rely on harvesting cells from solid, impermeable substrates. Thus, performing in situ label-free impedance measurements with cells grown on impermeable electrodes is just straightforward and avoids different conditions for different readouts. This is particularly important as it is well known that impermeable or permeable growth substrates may cause rather significant differences in cell differentiation.

Thin layers of gold or indium tin oxide (ITO) are popular electrode materials, as they are easy to pattern by standard photolithography. Thin layers of both materials, patterned on transparent polymeric or glas substrates, are compatible with routine phase-contrast microscopy for quality control of the cell layer grown on the electrodes. Moreover they allow for microscopic correlation studies to support interpretation of impedance-based assays. ITO is significantly more transparent than gold. It which is also compatible with fluorescence microscopy using inverted microscopes. Miniaturization and well-established photolithographic patterning enable integration of the electrodes in multiwell plates making the technology suitable for high-throughput screening campaigns. Alternatively, multiple electrodes may be placed beneath a single cell layer to allow position-dependent evaluation of cell layer properties.

Insert-based impedance approaches and those relying on cell-covered thin-film electrodes make use of two-electrode arrangements, in which the same electrodes are used for current injection and voltage sensing. Using two instead of four electrodes (two for current injection and two for voltage sensing) is justified as long as no faradaic currents cross the electrode/electrolyte interface. In typical cell culture medium as bulk electrolyte, the small amplitude signals do not induce such processes. However, on the downside two electrode arrangements will make the electrode impedance contribute to the overall impedance readout so that electrode behavior has to be included in any analysis of recorded data and its interpretation.

Thin-film electrodes for cell-based assays are mostly arranged as either interdigitated comb-like structures (IDEs) or as a combination of a small working electrode (WE) and a significantly larger counter electrode (CE) (Fig. 4b). Both strategies are based on the fundamental concept of making impedance changes at the electrode-electrolyte interface dominate over impedance contributions from the bulk solution. It provides impedance measurements the required sensitivity to follow cell coverage of the electrodes and to detect subtle changes in cell morphology within a confluent cell layer. In some cases both electrodes are made small to fit in small dishes or wells, but they are not layouted as interdigitated electrodes. We refer to these electrodes as WE/WE arrangements indicating that both electrodes are small but of the same surface area. The small size of the second electrode makes it different from WE/CE layouts.

In the WE/CE setup, the serial bulk impedance contains a contribution that is called *constriction resistance* which originates from the constriction of the electric field to the dimensions of the small electrode. The constriction resistance is inversely proportional to the radius of a circular electrode ($1/r$), while the interface and cell layer impedance are inversely proportional to the area or $1/r^2$ [2]. Making the electrode small will therefore make the interface and cell layer impedance dominate over the impedance of the bulk. The size ratio of the two electrodes – WE and CE – determines their individual contribution to the overall signal. When the two electrodes are about the same size in surface area, they contribute equally to the overall signal. Is one of the two electrodes significantly smaller than the other, the electrode with the smaller surface area (typically diameter of commercial working electrodes are 250 µm) will dominate the impedance readout. With the WE/CE concept, using a surface area ratio of less than 1/100, as suggested by Giaever and Keese [2], the impedance signal is clearly dominated by changes at the small working electrode. This enables an easy correlation of impedance data with microscopic observation of the cell population on the small working electrode. In IDE and WE/WE layouts, both electrodes ideally have the same surface area and contribute equally to the overall impedance. In IDE setups the influence of the solution impedance is reduced by reducing the distance between the individual fingers of the comb so that it loses impact on the overall impedance in favor of the impedance of the electrode-electrolyte interface and the cell layer.

All three electrode arrangements have their advantages and limitations. Interdigitated electrodes allow for maximum coverage of the available growth surface. A well coverage of about 80% may be reached with IDE layouts, which means that the signal is averaged over a significant proportion of the cell population in a well. On small electrodes the cell population that is represented by the measurement is limited to a few hundred cells, depending on the size of the working electrode surface area and the cell size. The small fraction of cells under study may lead to data scattering in repetition experiments, but parallel microscopic observation is easily realized. Moreover, the small working electrodes provide a significantly improved sensitivity for subtle changes in cell morphology and allow monitoring cells that only loosely adhere with a rather small dielectric signature. Ultimately, different impedance-based assays require different electrode layouts to provide best sensitivity and data interpretation.

2.4   Popular Electrode Designs and Commercial Systems

The first system for measuring impedance of adherent cells grown on thin film electrodes has been reported in 1984 by Giaever and Keese [1]. It has been referred to as *electric cell-substrate impedance sensing* or short ECIS (Fig. 4c) and was commercialized by the authors soon after through *Applied BioPhysics Inc*. The classical ECIS electrode layout consists of a small working electrode with a diameter of 250 μm and an about 500 to 1,000 times larger counter electrode (Fig. 4c). The electrodes are created as openings in an insulating polymer layer that was deposited on a much bigger area of a sputter-coated gold film. The opening in the insulating overlayer determines the electrode size. An array of such electrodes consists of eight working electrodes and a common counter electrode. A corresponding multiwell top is glued to the electrode array providing a chip-like structure allowing for eight individual experiments. This basic electrode design has been modified for different applications. For instance, the single opening in the insulating polymer overlayer has been replaced by ten such openings which are several cell dimensions apart. With these electrode designs, the number of cells that are monitored in one experiment is tenfold higher, and the readout is averaged over this larger cell population (Fig. 4d). This strategy has been extended by using mm sized IDE gold film structures covered by an insulating passivation layer with 40 openings of the same size (250 μm diameter) to increase the number of cells that contribute to the measurement (Fig. 4d). In addition, different 96-well electrode formats have been developed. A variety of electrode designs have been specifically tailored to the needs of different impedance-based assays as they will be discussed below. The electrode arrays are connected to the impedance hardware via array holders that are compatible with the environment of typical cell culture incubators. Data acquisition and visualization are controlled by proprietary software run on a regular personal computer. The most advanced ECIS hardware measures complex impedance in a frequency range between 10 and 10^5 Hz. Impedance magnitude and phase shift are converted to real (R) and imaginary (X) components which are accessible for analysis. The system provides a high degree of freedom in selecting frequency and monitoring parameter ($|Z|$, φ, R, C).

The *xCELLigence RTCA* system is based on the same physical principles as ECIS but uses a different technical implementation and has been commercialized through *ACEA Biosciences Inc.* The RTCA system uses "pearl-chain" interdigitated electrode layouts, which result in a high bottom-of-well coverage of about 80% (Fig. 4d). The electrode arrays come in 16-well or 96-well format. The measurement is performed at three fixed frequencies using sinusoidal AC voltages of 10 mV amplitude [35]. The recorded impedance data is reported as the dimensionless *cell index* (CI). Originally, the CI has been calculated from the resistive portion of the impedance for all three frequencies to $CI_i = [R_{cell}(f_i)/R_0(f_i)] - 1$ [35]. $R_{cell}(f_i)$ and $R_0(f_i)$ represent the measured resistances for cell-covered (R_{cell}) and cell-free (R_0) electrodes, respectively. The subscript i denotes the frequencies $f_1 = 10$ kHz, $f_2 = 25$ kHz, and $f_3 = 50$ kHz. Typically, the frequency with the largest change of CI along the time course of the experiment is selected for data presentation, and the corresponding CI is plotted as the outcome of the experiment. For cell-free wells, CI is 0, while CI increases when cells grow on the electrodes. Inversely, when cells on the electrodes die and detach from the surface, the CI will decrease below CI values of an intact cell layer. In later publications CI has been redefined as $[Z_n(f_i)-Z_b(f_i)] - 1$ [51, 52], using the impedance magnitude as the readout parameter. Z_n denotes the impedance for a cell-containing well at a given time point during the experiment. Z_b is the background impedance of cell-free wells loaded with medium only. Z_n and Z_b are frequency-dependent. Default values for Z_b for the three frequencies are $f_1 = 10$ kHz $\Rightarrow Z_b(10\text{ kHz}) = 15\ \Omega$, $f_2 = 25$ kHz $\Rightarrow Z_b(25\text{ kHz}) = 12\ \Omega$, and $f_3 = 50$ kHz $\Rightarrow Z_b(50\text{ kHz}) = 10\ \Omega$, which are sometimes simply denoted as frequency factors [53]. The xCELLigence setup has been used in various phenotypic assay formats. Due to its very fast sampling rate, it has been applied extensively to study the periodic beating of cardiomyocytes in assays addressing cardiac safety. This particular application has been supported by hardware developments that enable simultaneous extracellular field potential (EFP) measurements so that action potentials and mechanical contraction are recorded side-by-side.

The *CardioExcyte* system produced and marketed by *NanJi[on Technologies GmbH* has been tailored to this application as well. Similar to xCELLigence RTCA, it provides simultaneous impedance and extracellular field potential measurements in 96-well formats and is, thus, compatible with high-throughput applications in cardiac safety and drug development campaigns.

Instrumentation suitable for impedance analysis of cells grown on permeable supports is available from *nanoAnalytics GmbH* under the brand name *cellZscope*. The device is tailored for barrier-forming cell types, reads the frequency-dependent impedance of cell-covered inserts, and returns the transepithelial or transendothelial electrical resistance (TEER) and the capacitance of the cell layer.

2.5   The Measurement Principle

The principle behind impedance-based cell monitoring when cells are grown on the surface of thin-film electrodes is illustrated in more detail in Fig. 5. The cell-free working electrode submerged in cell culture medium behaves electrically in good approximation like a capacitor in series with an ohmic resistor (Fig. 5a). Deviations from capacitive behavior is obvious as it shows phase angles different from $-90°$ but the capacitor is used here as a reasonable approximation. The Bode diagram for a cell-free electrode submerged in cell culture medium is shown in Fig. 5b (black open circles). At low frequencies the impedance of the electrode/electrolyte interface dominates the overall impedance and is represented by the straight descending line with a slope of approximately (-1) in a log-log plot of the frequency-dependent impedance magnitude $|Z|(f)$. At high frequencies the frequency-independent ohmic resistance of the bulk electrolyte plus the constriction resistance (cf. Sect. 2.3) dominate the signal. When adherent cells are grown upon the electrode surface, they contribute a cell type-specific impedance (Z_{cl}). Cells significantly alter the overall impedance of the system within a frequency range between ~100 Hz and 100 kHz for most cell types (Fig. 5b, red filled circles), which is attributed to the insulating properties of the cell membranes, the cell-cell and the cell surface junctions. The latter two determine the geometry of the extracellular current pathways.

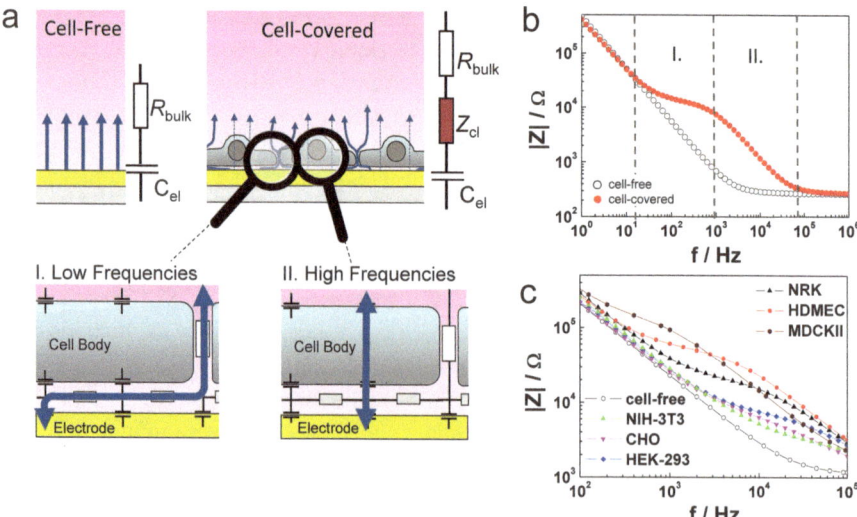

Fig. 5 Schematic illustrating the ECIS measurement principle. (**a**) The impedance of a cell-free electrode submersed in buffer or medium is modelled by a capacitor (C_{el}) and a resistor (R_{bulk}) in series. Cell coverage of the electrode is accounted for by the cell layer impedance (Z_{cl}). At low frequencies the current predominantly flows around the cell bodies (I); at high frequencies the current couples capacitatively across the plasma membranes (II). (**b**) Bode diagram of a cell-free (black open circles) and a cell-covered (red filled circles) electrode. (**c**) Impedance spectra for different confluent cell monolayers relative to the same cell-free electrode (open circles)

At low frequencies (10 Hz to 1 kHz), the current is forced to flow around the insulating membranes of the cell bodies, squeezing through the narrow clefts beneath cells and between neighboring cells (cf. arrow in Fig. 5a, panel I). At higher frequencies (>1 kHz), the current couples capacitatively across the cell membranes reducing the cell layer impedance towards higher frequencies (Fig. 5b panel II). Eventually the overall impedance is dominated by the bulk electrolyte at the high frequency end of the spectrum.

Because of this frequency dependency of the impedance, a proper selection of the monitoring frequency is crucial in single-frequency measurements to achieve sufficient sensitivity and to allow for correct data interpretation. As lower frequencies are predominantly sensitive for the paracellular current pathway, lower frequencies are typically chosen to perform measurements sensitive for cell shape fluctuations or changes in barrier properties of a cell layer. Monitoring cell behavior at high frequencies is useful to assess processes that are associated with changes in the fractional electrode coverage, such as cell attachment and spreading upon the electrode surface, proliferation, migration, or cell layer disintegration or detachment as a consequence of cell death. At intermediate frequencies, the impedance represents both current pathways with frequency-dependent contributions of the trans- and paracellular current flow.

The statements above are rather generalized but useful estimates. In detail, the frequency-dependent impedance spectrum $|Z|(f)$ depends on (1) electrode layout (WE/CE versus IDE), (2) electrode material (gold, ITO), and (3) of course the cell type under study. Because of their unique morphology and dielectric composition, each cell type provides individual impedance contributions (cf. Fig. 5c). Accordingly, frequencies well suited for monitoring either paracellular or transcellular current flows are unique for different cell types and electrode layouts. Taking a commonly used electrode of the WE/CE type with a surface area of 5×10^{-4} cm^2 as an example, changes in electrode coverage are typically monitored at 32, 40, or 64 kHz. Barrier function of epithelial or endothelial cells is typically observed at lower frequencies around 400 Hz. Selecting resistance R or capacitance C rather than the impedance magnitude $|Z|$ as monitoring parameter provides another way to focus the measurement on either paracellular or transcellular current pathways (Fig. 3b). Resistance readings report only on ohmic properties of the system (current and voltage are in phase), whereas capacitance readings solely mirror capacitive changes of the system (current and voltage are 90° out of phase). The impedance magnitude $|Z|$ always represents a more or less complex mixture of both contributions.

To visualize the rather theoretical statements from above, Fig. 6 shows the frequency-dependent impedance, resistance, and capacitance spectra of an IDE-type electrode (cf. Fig 4d, right) with an active electrode area of 0.01 cm^2 for each comb of the IDE without cells and at different time points after seeding a suspension of human dermal microvascular endothelial cells (HDMEC) to confluence. The spectra illustrate how the different parameters $|Z|$, R, and C change during cell layer formation and maturation (Fig. 6a–c). Typical monitoring frequencies of 400 Hz, 4 kHz, and 40 kHz are highlighted as vertical lines in these plots indicating the dynamic range of $|Z|$, R, and C at these frequencies

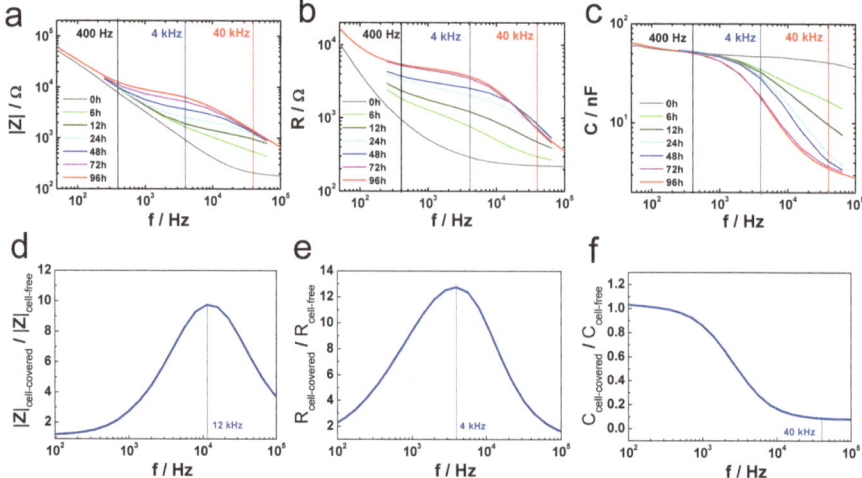

Fig. 6 Frequency-dependent dynamic range of impedance |Z|, resistance R, and capacitance C. (**a**)–(**c**) Frequency-dependent spectra of |Z| (**a**), R (**b**), and C (**c**) for a cell-free electrode (gray curve) compared to the same electrode covered with human dermal microvascular endothelial cells (HDMEC) at different time points after seeding the cell suspension (10^5 cells/cm^2). Data has been recorded for an IDE-type electrode (8W10E+) with an electrode area of 0.01 cm^2 for each comb. Popular monitoring frequencies of 400 Hz, 4 kHz, and 40 kHz are indicated by vertical lines. (**d**)–(**f**) Ratio of cell-covered (HDMEC, 96 h) to cell-free parameters: impedance magnitude (**d**), resistance (**e**), or capacitance (**f**)

during cell layer formation on the electrode. The most simple and straightforward approach to extract the monitoring frequency with the highest dynamic range for each quantity requires calculating the ratio of the frequency-dependent |Z|, R, or C of the electrode covered with a fully established mature cell layer and the corresponding values of the cell-free electrode: (1) $|Z|(f)_{cell-covered}/|Z|(f)_{cell-free}$, (2) R $(f)_{cell-covered}/R(f)_{cell-free}$, and (3) $C(f)_{cell-covered}/C(f)_{cell-free}$. Plotting cell-covered/cell-free values as a function of frequency provides bell-shaped curves for impedance and resistance (Fig. 6d, e). The frequency at which the ratio of this *normalized* |Z| or *normalized* R is maximum determines the frequency with maximal dynamic range between the cell-covered and cell-free electrode. It is considered as the most sensitive in single-frequency experiments for many applications. For capacitance as monitoring parameter, the most sensitive frequency is determined by the minimum of the *normalized* capacitance (Fig. 6f).

2.6 The ECIS Model: Correlating Impedance Readings with Cell Morphology

Impedance data acquired with two-electrode setups always represent the electrical properties of the entire circuit including cells, electrode, and bulk. To unravel their

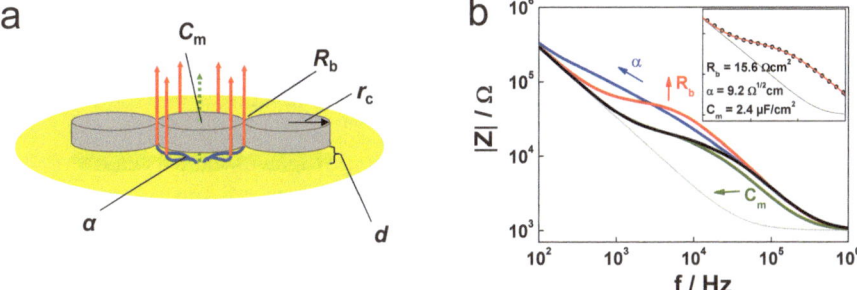

Fig. 7 Schematic of the ECIS model based on three morphology-related parameters α, R_b, and C_m. (**a**) Cells are modelled as circular discs with radius r_c hovering in a distance d above the electrode. (**b**) Frequency-dependent impedance magnitude |Z| as calculated from the ECIS model for a cell-covered electrode (black curve) with arbitrary values for α, R_b, and C_m. The blue, red, and green curves show the effect of distinct increases for α, R_b, and C_m, respectively, while the other parameters are kept constant. The spectrum of the cell-free electrode (gray) is included for comparison. The insert shows an experimental spectrum recorded from human dermal microvascular endothelial cells (HDMEC) grown on a gold film electrode. Fitting the ECIS model to the experimental data (red solid line) provides the parameter values as given in the figure. Adapted from [54]

individual impedance contributions, the system is typically modelled by equivalent circuits that ideally show the same frequency-dependent impedance as the experimental system under study. Such equivalent circuits are composed of circuit elements known from electric engineering (resistors and capacitors) and those specific to electrochemistry. These elements are arranged in series or in parallel such that the resulting network represents the electrochemical properties of the system. Using Ohm's and Kirchhoff's laws, a mathematical transfer function of this network is derived with the circuit elements as adjustable parameters. Fitting this transfer function to the experimental data by nonlinear least square methods returns values for the circuit elements that describe the electrochemical system.

To quantify the dielectric properties of cell layers grown on substrate-integrated film electrodes, Giaever and Keese derived a physical model and the corresponding transfer function for the impedance of such cell-covered electrodes from first principles. The transfer function is used for quantitative analysis of experimental data as described above [2]. This so-called ECIS model contains three adjustable parameters to describe the cells: α, R_b, and C_m (Fig. 7a). The parameter α (in $\Omega^{0.5}$·cm) is a measure for the constraint of current flow within the subcellular cleft underneath the cells, and it depends on (1) the cell radius r_c, (2) the average distance d between electrode surface and the basal plasma membrane, and (3) the specific resistivity ρ of the electrolyte in the subcellular space yielding $\alpha = r_c(\rho/d)^{1/2}$. The parameter R_b (in Ω cm^2) describes the electrical resistance arising from the restriction of current flow through the intercellular clefts which is highly influenced by the tightness of cell-cell contacts. C_m (in μF/cm^2) stands for the capacitance of the plasma membrane, and it essentially reflects the degree of membrane foldings. Each parameter causes characteristic features in the frequency-dependent impedance (Fig. 7b). Thus, fitting the transfer function of this model to experimental data of confluent cell layers grown on planar film electrodes provides the values for α, R_b, and C_m to describe the cell layers under study by their individual electrical properties and morphological features: (1) tightness of cell adhesion to the electrode (α), (2) tightness of cell-cell contacts (R_b), and (3) topography of the plasma membrane (C_m). Analysis of time course data reveals how these characteristics may change along an experiment. It is noteworthy that the model is only applicable to confluent cell layers as it cannot account for open spaces on the electrode.

3   Impedance-Based Assays to Study Cell Phenotypes and Behavior In Vitro

Impedance measurements with substrate-integrated electrodes sensitively mirror the degree of electrode coverage with cells and morphology changes within the cell layer. Major changes in electrode coverage occur during adhesion of initially suspended cells, during proliferation of sparse cultures, or when cells detach due to harmful insults. Once the cell layer is established, cell morphology and the strength of cell-cell and cell-matrix junctions are expressed in the frequency-dependent impedance. Accordingly, time-course impedance data may provide insight into their individual changes along an experimental observation. The cytoskeleton plays a fundamental role in orchestrating these changes in cell junctions or morphology making impedance measurements well suited to study changes in structure or dynamics of the cytoskeletal components. As many phenotypic changes involve adaptation of cell morphology, impedance-based cell observation is so widely applicable when the assays are designed and tailored to highlight a given feature of cell behavior.

In the following chapters, different phenotypic assays will be presented. The life cycle of cultured cells was used as a rough guideline to order the various assays for this manuscript. Some assays primarily address changes in electrode coverage. Others have a prime focus on more subtle changes of impedance resulting from changes in cell morphology within a confluent cell layer. However, a clear differentiation is not always possible, as different processes often intermix.

3.1   Cell-Matrix Adhesion

Cell adhesion is a central process in multicellular organisms. Anchorage to the extracellular matrix and to neighboring cells determines the cells' relative position, freedom of movement, and options of communication with the environment. The correct position of a cell relative to the surrounding tissues and fluids is crucial for its differentiation and functionality. Adhesion of cells to any noncellular surface is mediated by a complex interplay of extracellular matrix (ECM) molecules, transmembrane receptors, and the cytoskeleton. The ECM is a meshwork of proteins, glycoproteins, and proteoglycans that provide specific anchoring sites for transmembrane adhesion receptors like integrins which act as mechanical linkers between the ECM network and the intracellular cytoskeleton. In addition, integrins have essential signaling functions that are important in migration, proliferation, differentiation, and even for cell survival [55, 56]. Cell adhesion to endogenous extracellular matrices as well as to artificial materials became a central topic in cell biology, biomedical research, and biosensor technology. Especially the emerging role of artificial surfaces in medical implant technology and the use of cells as sensory elements in biosensor devices created a need for techniques to assess the cell-substrate contact. As cells do in fact not adhere directly to in vitro surfaces, but only to a pre-assembled adsorption layer of proteins, methods to study the attachment and spreading of cells to ECM proteins pre-deposited upon in vitro surfaces became extremely valuable [57]. However, experimental methods to quantitatively characterize cell-substrate interactions are limited. The most widely used approach involves simple counting of (stained) cells that adhere to a given surface under test. After the cells are allowed to adhere to a (coated) surface for a specified length of time, insufficiently adhered cells are removed by gentle washing or by applying more defined forces such as centrifugal acceleration [58, 59] or laminar shear flow [60]. The remaining cells on the test substrate are counted under the microscope, or, alternatively, the cell number is indirectly assessed by fluorescence or absorbance measurements after cell labeling. Widely used and popular assays of this type make use of crystal violet or carboxyfluorescein diacetate for cell staining. A central caveat of these classical assays is that they are endpoint assays and do not allow following the same sample along the adhesion process. Cells not yet adhered to the surface are discarded, while attached cells are sacrificed by the staining or even fixing procedures. Moreover, sticking to the surface alone is often not enough to properly evaluate cell adhesion. More advanced techniques are often microscopy-based, provide single-cell analysis, but require laborious experimental procedures and data analysis [60].

Impedance measurements, due to their sensitivity for changes occurring at the electrode-electrolyte interface, are well-suited to investigate both, the amount of cells adhering to the electrode and the quality of the adhesion [5, 62]. In addition, the time course of adhesion is accessible down to a time resolution of seconds, if required. The important features of an impedance-based adhesion assay are summarized in Fig. 8. It typically begins by recording the impedance of the cell-free electrode that may have been pre-coated with a specific protein or a defined mixture of ECM components. At this point of the protocol, it is important that the electrode impedance is stable before cells are inoculated, as drifting impedances due to equilibration processes occurring at the electrode-electrolyte interface will interfere with later data interpretation. Therefore, it is recommended to treat the gold surface with the amino acid cysteine before protein coating and cell inoculation. Cysteine binds to the gold surface via its thiol group forming a monomolecular layer that makes the electrode surface hydrophilic for later protein deposition, and it stabilizes

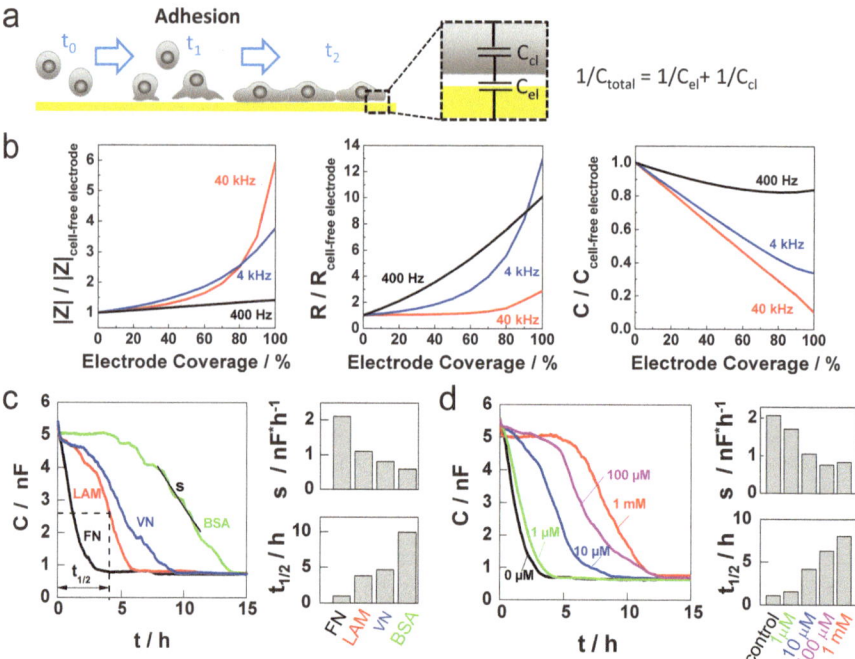

Fig. 8 Cell adhesion monitored by impedance readings. (**a**) Schematic illustrating cell adhesion to a substrate-integrated thin-film electrode upon inoculation with suspended cells at t_0. The capacitance of a cell-covered electrode (C_{total}) is a serial combination of the capacitances of the electrode (C_{el}) and the cell layer (C_{cl}). Since $C_{cl} < C_{el}$ and $1/C_{total} = 1/C_{el} + 1/C_{cl}$, the total capacitance decreases with increasing coverage by cells. (**b**) Changes of $|Z|$, R, and C when a cell-free electrode is gradually covered by cells at three frequencies (400 Hz, 4 kHz, 40 kHz). The calculation was performed using the ECIS model with $\alpha = 15\ \Omega^{1/2} cm$, $R_b = 0.5\ \Omega\ cm^2$, and $C_m = 1\ \mu F/cm^2$ for an electrode with a surface area of $5 \times 10^{-4}\ cm^2$. $|Z|$, R, and C are normalized to the respective values for the cell-free electrode. (**c**) Time course of capacitance (40 kHz) when suspended MDCKII cells are seeded on gold film electrodes ($5 \times 10^{-4}\ cm^2$) pre-coated with different proteins (*FN* fibronectin, *LAM* laminin, *VN* vitronectin, *BSA* bovine serum albumin). Time course data was analyzed for the apparent spreading rates s (slope of the curves between 4 and 2 nF) and the half-times required for adhesion ($t_{1/2}$). (**d**) Time course of capacitance (40 kHz) when suspended MDCKII cells are seeded on fibronectin-coated electrodes of the same type as in (*c*) in presence of increasing concentrations of the synthetic tetrapeptide RGDS (1 µM, 10 µM, 100 µM, 1 mM) compared to a control (0 µM). Apparent spreading rates (*s*) and half-times ($t_{1/2}$) are plotted as bar graphs. Graphs shown in (**c**) and (**d**) are adapted from [61]

the electrode impedance. Typically, cells are inoculated in high density to assure that the seeded cells cover the entire well surface without any further proliferation. Otherwise the recorded impedance data reflects both processes, and it may become difficult to differentiate the individual signal contributions. Often serum-free medium is used in adhesion assays in order to prevent deposition of serum proteins from the culture medium on top of the ECM preparation under study. After inoculation the suspended cells sediment to the bottom of the well and initiate the formation of cell-matrix contacts. During cell spreading free current flow is increasingly impeded.

The process of cell adhesion can be subdivided into roughly three stages [60]: (1) initial attachment of the cell body to the electrode driven by gravity, unspecific and specific interactions; (2) adhesion receptor-driven spreading of the cell body; and (3) enforcement of cell-substrate junctions by cytoskeletal rearrangements and focal adhesion complex formation. The rate of cell spreading was found to be proportional to the ratio of the adhesion energy and the cortical membrane tension [63–65]. Accordingly, cell spreading kinetics are a measure for the quality of the interaction between cell surface receptors and the ECM.

Readings of the capacitance at high frequencies (>10 kHz) have been proven to be especially useful to monitor the time course of cell spreading (Fig. 8b). Wegener et al. demonstrated that the capacitance of the system at high frequencies (>10 kHz) provides a direct and linear measure for electrode coverage [5]. The linear correlation between capacitance change and coverage of the electrode makes the analysis straightforward giving easy access to the apparent spreading rate and the time that is required for half-maximum cell spreading. Cell spreading is also expressed in readings of |Z| or R, but the nonlinear correlation makes interpretation more complex, and it does not provide a constant sensitivity of the readout along the entire spreading process. Whereas capacitance readings at lower frequencies become insensitive at high coverages, impedance and resistance are poorly sensitive for low electrode coverage (Fig. 8b). At higher frequencies the current preferentially couples capacitively through the membranes, while only a minor fraction uses the paracellular pathway. Along the transcellular route, the capacitance of the two cell membranes in series to the capacitance of the electrode-electrolyte interface provides additional capacitance contributions that scale linearly with the area that is covered by cells. As serial capacitances are added reciprocally, the overall capacitance of the system decreases when cell coverage of the electrode increases (Fig. 8a). Thus, a cell-free electrode exhibits maximum capacitance, whereas an electrode completely covered with cells provides the minimum capacitance. This relationship has been confirmed by studies using lipid vesicles and their spreading upon protein-coated electrodes. Sapper et al. prepared giant unilamellar lipid vesicles that were doped with increasing molar fractions of lipids carrying a biotin attached to their headgroups. The liposomes were allowed to spread upon avidin-coated gold electrodes [66]. The rate of the associated capacitance decrease at 40 kHz was shown to be depended on the biotin concentration in the liposome shell, a mimetic of the adhesion receptor. Accompanying Monte Carlo simulations of vesicle adsorption showed a good correlation with the time course of the normalized capacitance, whereas correlation of surface coverage with the resistance data was complicated.

Wegener et al. compared the spreading kinetics of suspended MDCKII cells on different ECM proteins fibronectin, laminin, vitronectin, and bovine serum albumin (BSA) and identified clear differences in the time course of capacitance decrease for the four protein coatings [5] (Fig. 8c). To characterize the adhesion process, two parameters were extracted from the capacitance traces: the slope of the curves and the time required for half-maximum capacitance change $t_{1/2}$ (Fig. 8c). Slopes of the curves were determined by linear regression between 4 and 2 nF to yield the apparent *spreading rate* of the cells. $t_{1/2}$ represents the time required to cover half of the available electrode area with cells. This approach paves the way for in-depth studies of cell spreading disclosing molecular details that are otherwise hard to get. For example, the specific interaction of integrins with highly conserved Arg-Gly-Asp-Ser (RGDS) sequences, present in various ECM proteins, was titrated by preincubating cells with increasing concentrations of soluble RGDS peptides before the cells were inoculated upon the protein-coated electrodes. While the cells did not show any RGDS-dependent spreading on laminin-coated electrodes [61], adhesion upon fibronectin-coated electrodes was significantly and concentration-dependently delayed (Fig. 8d). Using capacitance readings to monitor cell adhesion has been applied repeatedly in studies characterizing different aspects of cell-ECM interactions [67–70].

Despite the lack of linearity, the time courses of impedance and resistance have been used to study cell-ECM interactions [71, 72]. Luong et al. used ECIS to assess the $\alpha_2\beta_1$ integrin binding to collagen and laminin [72]. They compared the adhesion and spreading of human rhabdomyosarcoma (RD) cells recombinantly expressing the $\alpha_2\beta_1$ integrin type (RDX2C2) to mock transfected (RDpF) cells and to cells expressing a $\alpha_2\beta_1$ mutant lacking the α_2 cytoplasmic domain (RDX2CO). The increase in resistance (4 kHz) over time was used as a measurand when cells were inoculated upon collagen-, laminin- or fibronectin-coated gold electrodes. The observed resistance changes were normalized to the cell number and expressed as resistance change per cell $\Delta R/N$ (Ω/cell).

Sometimes both, capacitance and resistance readings, are used to monitor different aspects of the dynamic processes during cell layer formation on the electrodes. Since the resistance at lower frequencies is sensitive to cell-cell contact formation, spreading and junction formation may be observed simultaneously providing insight into the interdependence of both processes (cf. chapter "Cell Layer Maturation: Barrier Function as a Prominent Example"). For instance, Heink et al. monitored the capacitance at 40 kHz and the resistance at 400 Hz to compare spreading and junction formation of various human airway epithelial cells and primary bronchial epithelial cells (PBECs) [73].

Even though many of the initial impedance-based studies on cell adhesion have been performed using CE/WE electrode arrangements, interdigitated electrodes are equally useful. Ehret et al. chose capacitance readings for monitoring cell adhesion to interdigitated electrodes [74]. They observed that different cell types seeded in similar numbers showed clear differences in adhesion and spreading kinetics. Atienza et al. seeded NIH3T3 cells on interdigitated microelectrodes and compared adhesion and spreading kinetics for poly-L-lysine and fibronectin coating of the electrodes, respectively [75]. Anti-integrin antibodies and actin-disrupting latrunculin inhibited spreading of NIH3T3 fibroblasts, underlining the relevance of these molecular assemblies for cell adhesion. Also the role of the signaling molecule c-Src in coordinating the complex adhesion process was demonstrated by a c-Src inhibitor and by siRNA-mediated knockdown of c-Src in BxBC3 cells [75]. The authors compared the time course of cell index (CI) with standard adhesion assays [75].

The usefulness, reliability, and applicability of impedance measurements to monitor cell spreading have been validated many times by microtiter plate adhesion assays [72, 76, 77] or microscopic approaches including morphology inspection [67, 75, 78, 79] and cell counts [80–83]. As the above examples demonstrate, the applicability of impedance-based adhesion studies goes beyond analysis of different ECM compositions [5, 70, 75, 78, 84–86] and includes in-depth studies on adhesion receptors [68, 76, 81, 87, 88], intracellular protein expression, as well as regulation and signaling that play important roles in the adhesion process [69, 79, 80, 82, 89–95].

3.2 Cell Proliferation

Cell proliferation is probably the most central and unique characteristic of living organisms, as it is the basis for the generation of new tissue required for embryonic development, growth during adolescence, and repair mechanisms in adult organisms. Uncontrolled cell proliferation, however, leads to severe often fatal pathologies such as cancer. Therefore, understanding the regulatory mechanisms controlling cell cycle and cell proliferation is indispensable to identify drugs for efficient therapies so that proliferation assays are equally important in basic research as in drug development.

Classical proliferation assays quantify the increase in cell number over time. This typically involves direct counting of cells or indirect colorimetric/fluorimetric quantification of stained cells or their constituents at defined time points after inoculation of initially low cell numbers [11]. The dyes bind either to cellular proteins or to DNA and reflect gross increase in biomass. More sophisticated methods measure the integration of (radio-)labeled nucleotides and amino acids into proliferating cells as a descriptor of cell growth, or they use antibodies directed against cell proliferation markers. Another indirect approach to assess the amount of cells at a given point in time relies on quantifying their metabolic activity. The most popular assays determine (1) the cells' ability to reduce chromogenic tetrazolium salts (e.g., MTT, XTT, MTS) or resazurin, (2) the enzymatic activity of cellular esterases (calcein AM, fluorescein diacetate), (3) intracellular ATP levels (luciferase, luciferin), or (4) pH gradients in cellular organelles (neutral red).

Impedance-based analysis of cell proliferation offers many distinct advantages. For a typical assay, cells are seeded in low density, and impedance (or capacitance) of the electrodes is monitored as a function of time. After initial adhesion and spreading of the inoculated cell population leading to a sparse coverage of the well bottom, establishing a confluent cell layer requires cell division and cell growth (Fig. 9a). The lower the initial number of cells, the more time is required to fully cover the electrode (Fig. 9b). Very similar to the impedance-based adhesion and spreading assay, it is advantageous to use the capacitance at high frequencies (Fig. 9c) for cell proliferation monitoring to directly mirror the increasing electrode coverage in the raw data [39, 98–101].

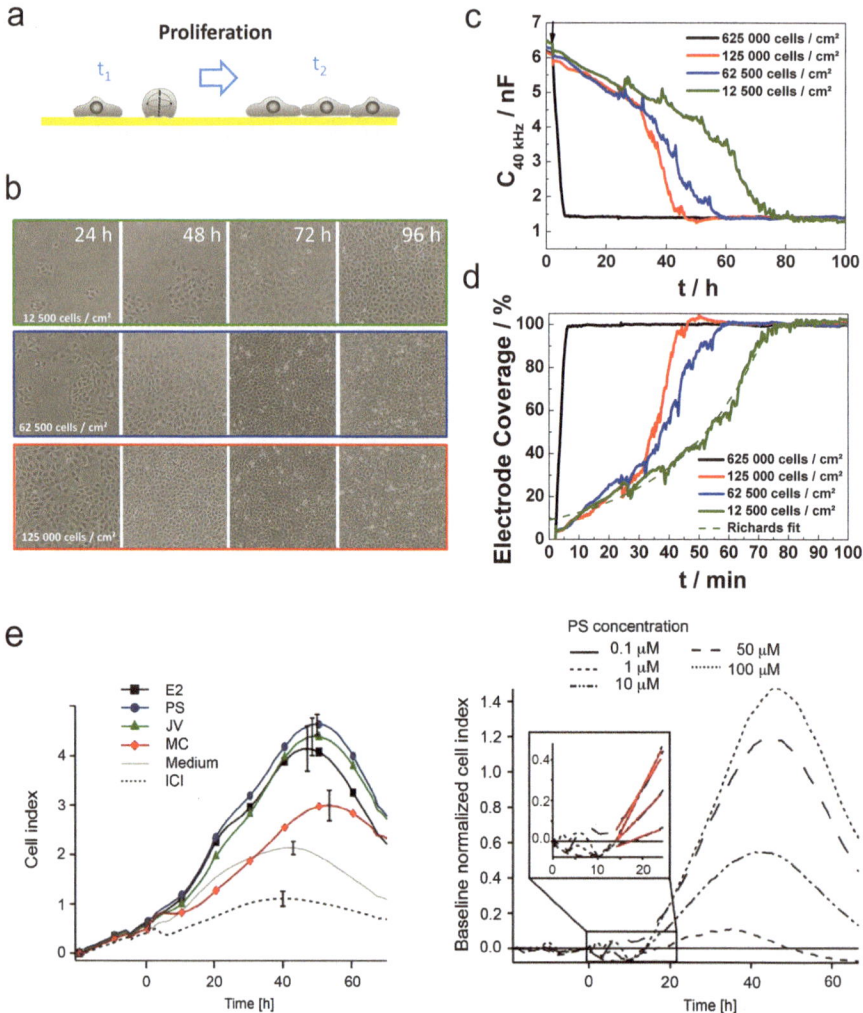

Fig. 9 Cell proliferation monitored by impedance measurements. (**a**) Schematic illustrating cell proliferation on a substrate-integrated thin-film electrode. (**b**) Phase-contrast micrographs of NRK cells 24, 48, 72, and 96 h after seeding three different cell densities as indicated in the micrographs. (**c**) Time course of capacitance at 40 kHz when electrodes of 5×10^{-4} cm^2 are inoculated with the same numbers of NRK cells as in (**b**). (**d**) Time course of electrode coverage calculated from capacitance data at 40 kHz shown in (**c**). The growth curve of cells seeded at an initial density of 12,500 cells/cm^2 was fitted with the Richards model as described in [96]. (**e**) Influence of different hormonelike odorants on MCF7 proliferation measured by impedance. Left: The initial 20 h worth of data ($t < 0$ h) reflect the phase of initial cell attachment and spreading. Then electrical impedance measurements were performed for another 70 h after adding different odorants: Polysantol (PS, 50 μM), Javanol (JV, 50 μM), Mousse Cristal (MC, 100 μM), and estrogen (E2, 0.1 nM) at time zero. Control measurements show the effect of adding cell medium or blocking the response of E2 by co-application of ICI 182,780 (1 μM). Data represent means of at least three independent experiments (±SEM). Right: Representative dose-dependent impedance measurements of MCF7 cell proliferation stimulated by Polysantol (PS) at the indicated concentrations. Inset enlargement of the first ascending phase (10–24 h) post-stimulation. Reprinted from [97] with permission from Elsevier

Duchateau et al. studied the proliferation of three different cell lines, CHO, HEK, and BV2 cells, for five different seeding densities using 96-well E-plates (Roche) in combination with an in-house eight-channel impedance analyzer [39, 102]. For each cell type, they extracted the optimal frequency for proliferation monitoring by plotting the impedance change ($\Delta|Z|$) over 100 h versus frequency. The frequency of maximum impedance increase was selected for studies on the respective cell type (CHO, 15 kHz; HEK and BV2, 20 kHz). Time course and magnitude of impedance change were found to be dependent on cell size and doubling time of the respective cell type (BV2, 15 h; HEK, 24 h; CHO, 14–17 h). Witzel et al. investigated the relationship between cell index (CI) from impedance measurements at 10 kHz and actual electrode coverage during proliferation of HT29 or RKO cells [96]. Once a day they took phase-contrast micrographs of the cells growing in electrode-free zones of the wells. Image analysis was used to calculate confluency. Plotting CI from impedance recordings of both cell types versus their relative confluency was used to test for any linear correlation of these quantities, yielding a Pearson's correlation coefficient of 0.82 and 0.93, respectively. Limame et al. compared the performance of an impedance-based proliferation assay (xCELLigence) with a classical method, a colorimetric assay that is based on the binding of the dye sulforhodamine B (SRB) to cell protein [103]. In this study proliferation of the breast cancer cell line MDA-MB-231 and the lung cancer cell line A549 was monitored over 10 days using four different plating densities (100, 500, 1,000, and 2,000 cells/cm^2). From the resulting time courses of CI, the cellular doubling times were extracted, which corresponded well with the doubling times obtained from the SRB assay (MDA-MB-231, 27.79 h SRB/29.93 h xCELLigence; A549, 27.94 h SRB/29.18 h xCELLigence).

Growth curves from impedance-based proliferation assays have been analyzed using mathematical models in order to estimate growth rates and describe contact inhibition in quantitative terms. Different mathematical models have been described to quantify growth behavior, and many models are based on exponential functions [6, 96, 104]. Three different mathematical models, namely, (1) a simple exponential growth, (2) a logistic function including a factor for growth limitation by the petri dish, and (3) a generalized logistic growth model (Richards model), have been tested for their usefulness in modelling impedance growth curves by Witzel et al. [96]. The Richards model, a flexible growth model applicable to cells with constant growth rate and contact inhibition, provided the most accurate description of CI time courses and returned the cells' characteristic cell cycle time (t_d). The cell cycle times of four different colon cancer cell lines were quantified and compared: RKO ($t_d = 20.8 \pm 0.6$ h), HT29 ($t_d = 22.2 \pm 2$ h), HCT116 ($t_d = 9.6 \pm 2.6$ h), and Lim1215 ($t_d = 21.1 \pm 3.9$ h). This model was also applied to the growth curve shown in Fig. 9 after transforming the time course of capacitance into the time-dependent change in fractional electrode coverage (Fig. 9d). Fitting the Richards model to the time course of fractional electrode coverage provided a doubling time of $t_d = 21.4$ h for NRK cells seeded at a density of 12,500 cells/cm^2.

Impedance-based assays have proven their usefulness to unravel the impact of various biological [39, 77, 84, 101, 105, 106] and chemical [6, 97, 107] stimuli on proliferation of different cell types, including the effect of different electrode coatings [52, 108]. For example, Masanetz et al. studied the effect of different pine pollen extracts on the proliferation of porcine ileal cells by monitoring the normalized impedance at 30 kHz over 50 h [101]. They nicely demonstrated that water/ethanol extracts (50:50) of pine pollen significantly delayed proliferation, while pure ethanol extracts only induced a transient delay, whereas water/methanol (20:80) and water/hexane extracts (20:80) had no influence. Dowling et al. used impedance-based proliferation assays to investigate the effect of different fibroblast-derived culture media on colon cells [106]. The human colon cancer cell line HCT116 proliferated fastest in human dermal fibroblast-conditioned medium compared to standard DMEM medium or HCT116-conditioned medium. Kim et al. studied the effects of ECM protein mimetics on adhesion and proliferation of chorion-derived mesenchymal stem cells [108]. The latter showed the highest adhesion and proliferation on electrodes coated with the fibronectin peptide mimetic GRGDSP. Voltan et al. studied the effect of human sera from young, old, and heart failure patients on endothelial cells [105]. They demonstrated that sera from older but healthy donors promoted cell proliferation when compared to sera from younger donors.

A more detailed example for the impact of various chemicals on cell proliferation is summarized in Fig. 9e. Pick et al. studied the effect of different hormone-like odorants on the proliferation of MCF-7 breast cancer cells in comparison to the effect of estrogen [97]. The compounds were added to the proliferating cells 20 h after seeding. In this study the maximum cell index change over a time period of 70 h upon odorant addition to the proliferating cell culture was used as a quantitative descriptor and compared to the alternative use of the initial slope of the CI time course. Dose-response curves were established for the compounds under study that revealed similar EC_{50} values for both methods of quantification. An in silico study addressing the question what curve parameter is best suited to quantify the influence of small-molecule inhibitors of proliferation was performed by Witzel et al. [96]. They analyzed the CI at a given time point after inoculation, the area under the curve from time zero to an arbitrary endpoint, and the normalized CI value $CI(t)/CI(t = 0)$ at an arbitrary time t or its logarithm. They found that most consistent data were obtained for the logarithmic approach.

The role of different cellular proteins in proliferation has been investigated using different genetic manipulations before running impedance-based assays [79, 98, 109, 110]. For example, Cai et al. investigated the role of a lentiviral accessory protein (Nef) in downregulation of the CXCR4 chemokine receptor and the associated influence on proliferation of tumor cells in response to the chemokine *stromal-derived factor 1α* (SDF-1 α) [109]. HeLa cells were transfected with vectors encoding either Nef or a Nef mutant (Nef-M8), and the inhibitory effect of both proteins on cell proliferation was studied relative to a mock control. In line with their results that expression of Nef led to a reduction of CXCR4 on the cell surface causing reduced proliferation, anti-CXCR4 siRNA-treated cells showed reduced proliferation compared to the control.

3.3 Cell Layer Maturation: Barrier Function as a Prominent Example

After cells reach confluency and make contact to neighboring cells, most cell types start forming cell-to-cell junctions accompanied by rearrangements of the cytoskeleton until a polarized cell morphology and a fully functional cell layer are attained. This process requires time – hours to days depending on cell type – during which cell morphology may change significantly (Fig. 10).

Barrier-forming cells, i.e., epithelia and endothelia, are good examples to highlight the use of impedance measurements to monitor cell maturation. Epithelial and endothelial cells line all inner and outer surfaces of the body efficiently separating chemically distinct areas within or adjacent to the body. By virtue of their position at the interface between two compartments, these cells perform important barrier, transport, and signaling functions. Structurally, cell-cell junctions such as *adherens junctions*, *desmosomes*, and *tight junctions* are fundamental for epithelial and endothelial functionality. *Adherens junctions* and *desmosomes* predominantly provide mechanical cohesion between individual cells within the cell layer. *Tight junctions* establish a highly selective and tightly regulated diffusion barrier between the two chemical compartments by sealing the intercellular clefts. This occlusion of the

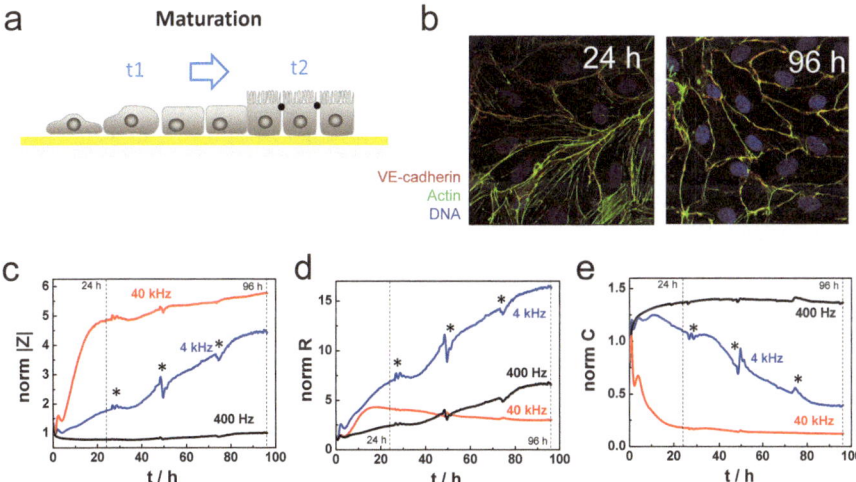

Fig. 10 Cell layer maturation monitored by impedance-based assays. (**a**) Schematic illustrating cell maturation on a substrate-integrated thin-film electrode using epithelial cells with apical microvilli as example. (**b**) Fluorescence micrographs of confluent layers of human dermal microvascular endothelial cells (HDMEC) stained for VE-cadherin (red), actin (green), and the nucleus (blue) after 24 and 96 h in culture. Regions of co-localization of actin and VE-cadherin appear yellow. (**c**)–(**e**) Time course of normalized impedance magnitude (**c**), resistance (**d**), and capacitance (**e**) at 400 Hz, 4 kHz, and 40 kHz recorded with electrodes of 5×10^{-3} cm^2 surface area inoculated with HDMEC (10^5 cells/cm^2) at time $t = 0$. Culture medium was exchanged every 24 h (∗). Micrographs in (**b**) are adapted from [54]

intercellular cleft is the structural basis to establish chemical gradients across the cell layer which is indispensable for many physiological processes. Moreover, *tight junctions* form a molecular fence within the plasma membrane which prevents lipids and proteins from diffusing within the membrane from apical to basolateral or vice versa. This enables a different lipid and protein composition in both membrane domains which is crucial for cell polarization and directed transport. Barrier function of epithelia and endothelia cultured in vitro is experimentally assessed by quantifying the permeation rate of tracer molecules (P_E) or by measuring the electrical resistance across the cell layer (TEER) [111–113].

Since impedance analysis is sensitive to changes in barrier-forming cell-cell junctions, the approach became a popular tool for in vitro studies of epithelia and endothelia [7, 54, 114]. When the sampling frequency is properly selected, single-frequency impedance measurements provide a much better time resolution and, thus, enable detection of fast changes in barrier properties. Single-frequency barrier function assays are typically performed using the in-phase part of the impedance (resistance). Giaever and Keese used an AC frequency of 4,000 Hz for the measurements in their first paper on the subject [1], and this has been widely accepted throughout the literature. Measurements at 4,000 Hz are, however, prone to artifacts when the cell layers become rather tight as, for example, endothelial cells forming the blood-brain barrier or epithelial cells derived from the kidney, colon, or skin. In these cases the use of lower frequencies (e.g., 400 Hz) and impedance instead of resistance as readout parameter should be considered. It is noteworthy in this context that the type of electrode and in particular the surface area available for current flow have an additional impact that makes general recommendations difficult.

A full comparison of the time courses of $|Z|$, R, and C at AC frequencies of 400 Hz, 4 kHz, and 40 kHz (all normalized to the values of the cell-free electrode) during spreading and maturation of a HDMEC cell layer is plotted in Fig. 10c–e. The corresponding changes in the spectra have been presented in Fig. 6 of chapter "The Measurement Principle". Data in Fig. 10 show that 24 h after inoculation, the normalized capacitance at 40 kHz reaches its minimum, indicating complete coverage of the electrode (Fig. 10e). Also the normalized impedance magnitude at 40 kHz has reached about 80% of its maximal value (Fig. 10c). In contrast, the impedance magnitude at 4 kHz has just reached about 20% of its maximal signal change within the first 24 h, which is finally stationary at 96 h. Resistance at 4 kHz levels off to its final plateau phase not before 96 h after inoculation (Fig. 10d). This indicates that even though a confluent cell layer has established within 24 h, it takes 3 more days to establish a mature cell layer with functional cell-cell junctions – an indicator for proper endothelial differentiation. This interpretation of impedance data was supported by immunocytochemical stainings of the actin cytoskeleton and the vascular endothelial (VE)-cadherin. After 96 h in culture HDMEC show more pronounced co-localization of VE-cadherin and actin at the cell borders as compared to the situation after 24 h in culture with the actin fibers irregularly distributed within the cytoplasm (Fig. 10b).

Some epithelial and endothelial cell layers may form extremely tight cellular barriers in certain parts of the body, for example, at the blood-brain barrier.

To reproduce this terminal differentiation in vitro often requires growth of the cells on permeable supports since active transport of solutes across the cell layer and the accompanying chemical gradients are considered as indispensable differentiation clues. As has been shown in Fig. 4a, barrier function of cell layers grown on permeable supports (inserts) is accessible when the insert is sandwiched between two electrodes integrated into each of the liquid compartments above and below the cell layer [44, 45, 47–50]. Moreover, when epithelial or endothelial cells are grown on permeable supports, a second cell layer may be cultured on the back side of the insert allowing for cell communication between both cell populations [115]. It is well known that co-culture of cells that are in close vicinity within the body may alter the barrier properties of a given interfacial epithelial or endothelial cell layer [116]. Using permeable growth substrates allows impedance-based studies to be combined with studies using molecular tracers to determine the cell layer's permeability for the tracer next to its impedance in the same setup. When cells are grown on porous supports, their integral electrical resistance is commonly referred to as TEER values (in Ω cm^2). TEER values are extracted from the impedance spectra by fitting a corresponding model to the experimental data that contains the resistance TEER and the cell layer capacitance C_{cl} as adjustable parameters. TEER values vary widely for different epithelia and endothelia and often mirror the cells' physiological function in their respective tissues. Human umbilical vein endothelial cells (HUVEC), for example, show TEER values in the order of 10–12 Ω cm^2 [117, 118], while endothelial cells from brain capillaries that form the highly selective blood-brain barrier exhibit TEER values between 1,000 and 2,000 Ω cm^2 when properly differentiated [119–121]. Unfortunately, impedance readings of cells grown on permeable inserts become increasingly insensitive for cell types with weak barrier function [7, 50, 54].

3.4   Micromotion: An Indicator for Viability and Motility

The first publication by Giaever and Keese on the use of impedance measurements to monitor cells grown on thin-film electrodes already discussed subtle fluctuations in time-course impedance data of cell-covered electrodes [1]. As these fluctuations were significantly reduced in amplitude by cytochalasin B treatment, a fungal toxin that interferes with the actin cytoskeleton, they concluded that this "noise" is biological in nature and a consequence of cell motion. Indeed, when a confluent mature cell layer is inspected by time-lapse video microscopy, it becomes evident that individual cells are not static but rather perform constant shape fluctuations with considerable translocation of organelles within the cell bodies (Fig. 11a). The sensitivity and temporal resolution of impedance measurements enable the detection of cell body dynamics and the accompanying geometric changes in the subcellular and paracellular shunts down to the level of several tens of nanometers. They have been referred to as *micromotion* in the literature. In this respect impedance measurements exceed the resolution of optical microscopes [2]. Impedance fluctuations

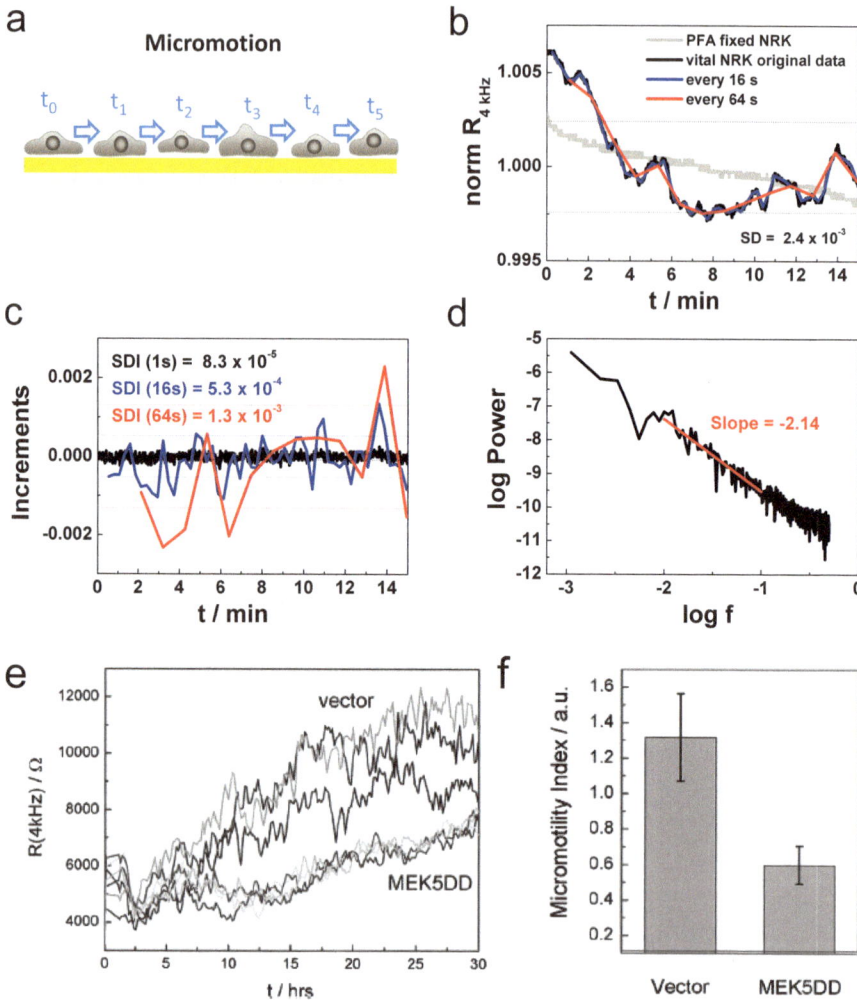

Fig. 11 Cellular micromotion monitored by impedance readings. (**a**) Schematic illustrating cellular micromotion on a substrate-integrated thin-film electrode. (**b**) Time course of normalized resistance of confluent NRK cell layers: black, vital cell layer; gray, cell layer fixed with paraformaldehyde (PFA). Resistance fluctuations are described by the standard deviation of increments (SDI) (**c**) or the corresponding power spectrum (**d**). (**c**) For SDI analysis the increments between successive (black), every 16th (blue), or every 64th (red) data points are calculated after detrending and plotted as a function of time. SDI is the standard deviation of the time course of increments. (**d**) Power spectral density of the incremental time course shown in (**c**). The slope of the power spectrum was determined by linear regression between 0.01 and 0.1 Hz. (**e, f**) Example for micromotion analysis. (**e**) Time series of the resistance at 4 kHz for HUVEC that were transfected with MEK5DD or a control plasmid. MEK5DD inhibits cell motility. (**f**) Quantitative analysis of HUVEC micromotion. The bars show the average SDI for consecutive data frames of 64 s in arbitrary units. Data shown in (**e**) and (**f**) are reprinted from [123] with permission

vanished when the cell layer was immobilized by fixation reagents like formalin [2] and decreased when the temperature or glucose levels in the medium were lowered [4]. The latter findings support the notion that impedance fluctuations mirror the metabolic activity of the cells on the electrode. Most likely, micromotions originate from metabolically driven dynamics within the cytoskeleton and the sub-membraneous actin cortex. As such, micromotions induce fluctuations in cell-substrate adhesion and cell-cell cohesion [84, 122–124] and became a promising indicator for the metastatic potential of cancer cells [125–127]. Moreover, as an attribute of metabolically active cells, micromotion signatures of cells were recognized as a sensitive indicator for cytotoxicity [128–132]. Micromotion recordings often reveal a cytotoxic insult much earlier than the gross change in impedance or the readouts of label-based assays [130]. It is important to note that micromotion recordings require small electrodes to ensure sufficient sensitivity and to avoid that the impedance signature of individual cells is averaged out when too many cells reside on the electrode.

To extract quantitative measures for the intensity of the signal fluctuations, most studies have used time course data of the resistance at a given frequency. Datasets are analyzed with respect to their variance, variance of increments, or Fast Fourier transformation (FFT) [4] (Fig. 11b–d). For a typical increment analysis, the time series is first normalized by dividing every data point by the time average. If necessary, the data is detrended if strong up- or downward drifts superimpose the subtle signal fluctuations. Then the difference between two successive data points is calculated representing $N-1$ increments from N original data points. Instead of calculating the increments for successive data points, this is also done for every 4th, 16th, 32nd, or 64th data point of a given time series. This way, fluctuations are analyzed on different time scales. For each time series, the standard deviation of increments (SDI) or the variance of increments (SD^2I) is computed providing a quantitative estimate for the amplitude of signal fluctuations for a given time resolution. The most sensitive reading frame to analyze micromotions has to be determined empirically [4, 130], and it may be cell type-dependent. The variance of increments for frames of 64 s is often used for analysis and denoted as the so-called *micromotility index* [123] (Fig. 11f). In a most recent study, Lang et al. found a direct correlation between the SDI of cancer cells and their metastatic potential as determined from animal experiments [133].

Alternatively, the resistance time series is subjected to Fast Fourier Transformation (FFT) providing the corresponding power density spectrum [4, 130, 131, 134] (Fig. 11d). Distinct peaks in the power spectrum indicate characteristic periodicities in micromotion. For instance, Lo et al. found a characteristic peak in the micromotion power spectrum at a frequency of 0.0015 Hz which corresponds to a period of 11 min. The regular CO_2 injections into the incubator occurred with the same period [134] leading to periodic changes of medium pH and corresponding changes in cell morphology. Linear regression analysis of the low frequency part of the power spectrum (log-log) provides slopes between -2.1 and -2.7 for viable cells, while slopes of power spectra recorded from chemically fixed, dead cells or an empty electrode are close to zero [131]. Opp et al. report on slopes of micromotion power spectra recorded for viable HUVEC cells of -2.9 under control conditions that decrease gradually down to -2.0 for increasing concentrations of cytochalasin B (1–10 μM) [130].

Additional measures used to estimate the impact of long-term correlations in micromotion time series are the Hurst exponent and the exponent of a procedure referred to as *detrended fluctuation analysis* [122, 125, 128–130]. These parameters quantify the existence of correlations within the time course of resistance at a certain time point $R(t_1)$ of the experiment and some later time point $R(t_2)$. They serve as the quantitative basis to classify a time series as Brownian motion (Hurst Coefficient ≈ 0.5) with no correlation between $R(t_1)$ and $R(t_2)$, an anti-persistent time series when values at t_2 tend to match a long-term mean, or a persistent time series in which an increase or decrease at t_1 will likely be followed by a further increase or decrease at t_2, respectively. Persistence is indicated by a Hurst exponent between 0.5 and 1, with a larger Hurst exponent indicating stronger correlations. Based on such a time series analysis (*Hurst's rescaled range analysis*), Giaever and Keese discovered a fractal character in micromotion data of human fibroblasts. According to their analysis, micromotion data recorded for transformed WI-38 VA13 cells were slightly different from micromotion data acquired for the non-transfected parental cell line WI-38. Accordingly, the Hurst exponent for WI-38 V13 was slightly higher than that of WI-38, both were in the range of $H = 0.65$ [128]. Lovelady et al. found Hurst exponents of 0.77 and 0.74, respectively, for two different non-cancerous (HOSE) and cancerous (SKOV) human ovarian epithelial cells [125]. Opp et al. reported Hurst exponents of 0.66 for HUVEC under control conditions [130]. When exposed to increasing concentrations of cytochalasin B, the Hurst exponent first increased to 0.97 and then decreased to 0.6. For NIH3T3 fibroblasts cytochalasin B exposure led to a decrease in Hurst exponent from 0.76 for untreated cells to 0.6 at the highest concentration of 10 μM [129]. A wavelet-based multiscale quantitative analysis technique has been applied by Das et al. to analyze micromotion of normal epithelial and breast cancer cells [135]. Impedance fluctuations were also dissected by *discrete wavelet transform* (DTW) techniques that provide quantitative descriptors for cell dynamics that have been named *cellular energy, cellular power dissipation,* and *cellular moments* according to their anticipated cellular correlate. Cancerous cells showed higher *cellular energy* and *cellular power dissipation* associated with stronger micromotion compared to normal cells. All the above examples indicate that cancerous and normal cells can be classified by the signature of their impedance fluctuations.

Another application of analyzing fast cell shape fluctuations has emerged in recent years enabled by access to human-induced pluripotent stem cell-derived cardiomyocytes (iPSC-CMs) that show a stable phenotype and develop a synchronized beating pattern in culture [136]. Beating rhythm and amplitude of cardiomyocytes has been recorded with high temporal resolution (12.9 ms) by impedance measurements and has evolved as a promising in vitro technique for cardiac toxicity testing in preclinical drug screens [136–139]. For instance, Guo et al. tested 28 different compounds with known cardiac effects and analyzed the changes in impedance signature of iPSC-CMs grown on interdigitated electrodes by comparison to extracellular field potential measurements with microelectrode arrays (MEA) [137]. Impedance measurements accurately assessed drug-induced abnormalities in the rhythmic, synchronous contractions of the cardiomyocytes. Some commercial instruments like the CardioExcyte (nanion GmbH, Germany) or xCELLigence (ACEA Bioscience Inc., USA) have been tailored for this particular application and allow measuring impedance and extracellular field potential simultaneously.

3.5   Cell Migration: The Electrical Wound-Healing Assay and 3D Approaches

Migration of cells plays an important role in processes such as development, wound healing, and cancer progression. For this reason a diverse set of assays has been developed to investigate the migratory activity of cells in vitro [140].

Widely used are the so-called *scratch assay* for two-dimensional migration on a culture dish and the *Boyden chamber assay* for migration across three-dimensional (3D) matrices. In the *scratch assay*, a confluent cell layer grown on conventional tissue culture plates is wounded by mechanically introducing a scratch into the cell layer. The initial wound and the migration of cells from the wound edges into the free space is documented by microscopy in regular time intervals. A major caveat of this technically simple approach is poor wound size reproducibility. Therefore, special chambers have been designed that allow for more defined wound sizes [140]. Similar to all microscopy-based migration assays, the scratch assay suffers from low temporal resolution and laborious image analysis.

Impedance-based wound-healing and migration assay provide a good alternative since they are completely automated, provide a time resolution down to minutes, and are performed in a regular cell culture incubator and – if needed – in a 96-well format. In the so-called *electric wound-healing assay*, the cell-covered thin-film electrodes are used both, for creating a defined wound in a confluent cell layer and for real-time documentation of wound closure [8]. Lethal voltages (or currents) are applied to the electrodes for several seconds, which leads to immediate and irreversible permeabilization of the cells that are residing on the electrode (Fig. 12a, upper panel). As the cells around the electrode are protected by the insulating passivation layer, a well-defined and very reproducible wound is created that is restricted to the surface area of the electrode (Fig. 12b, 2). Intact cells at the wound edge sense the lesion and start to migrate into the open spaces that have been created upon the electrode (Fig. 12a, b, 3). The cell debris remaining on the electrode is removed by the migrating cells in a snowplow-like fashion. Because impedance measurements are sensitive to the degree of electrode coverage with living intact cells, the repopulation of the electrode surface is sensitively followed with high time resolution.

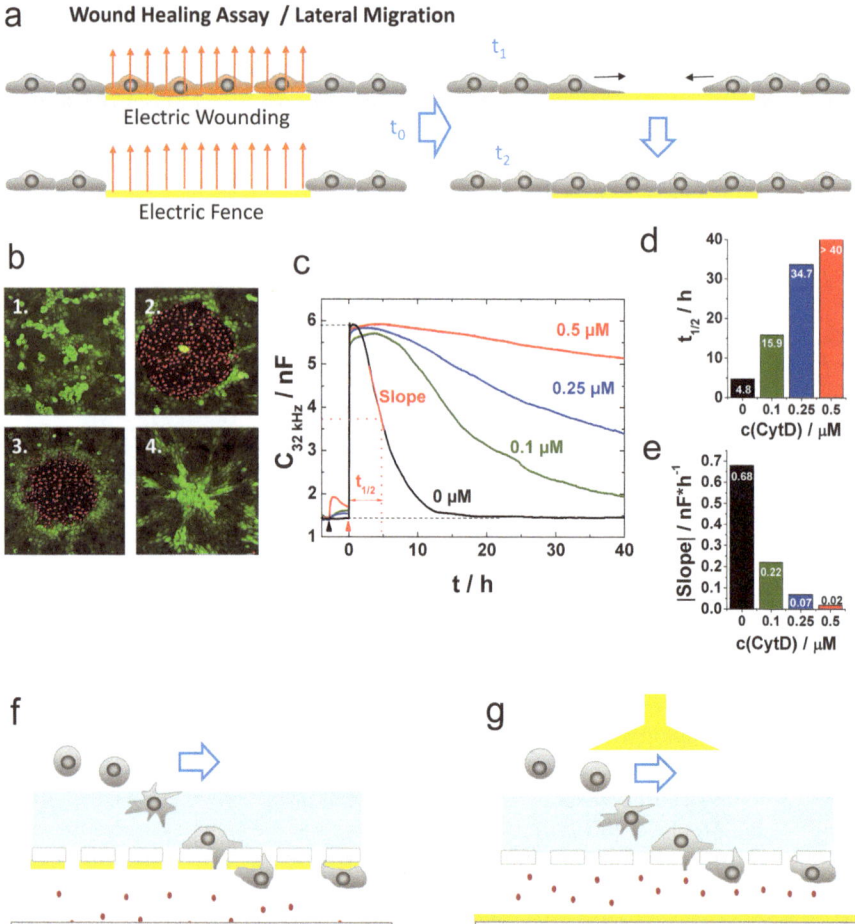

Fig. 12 Migration monitored by impedance-based assays. (**a**) Schematic illustration of the electrical wound-healing assay for cells grown on a substrate-integrated thin-film electrode. A cell-free lesion restricted to the electrode area is created by a lethal electric field pulse that kills the cells on the electrode (upper panel) or by the so-called "electric fence" technique which prevents cell adhesion during cell inoculation. Cell migration starts after the invasive electric field is turned off and cells from the periphery of the wound (re)populate the electrode. (**b**) NRK cells grown on circular gold film electrodes stained by a dual-color vital stain: *(1)* prior to wounding, *(2)* immediately after wounding, *(3)* at ~50% recovery, and *(4)* after full recovery. (**c**) Time course of capacitance at a sampling frequency of 32 kHz for NRK cells grown to confluence on circular electrodes with 250 µm diameter (cf. **b**) and wounded at time $t = 0$ by an AC pulse of 5 V, 40 kHz, and 30 s duration (red arrow). 4 h before wounding (black arrow), cells were exposed to different concentrations of cytochalasin D. (**d**) $t_{1/2}$ values for the experiment shown in (**c**). (**e**) Apparent migration speed determined from a linear fit to the capacitance curve between 3.5 and 5 nF (except for the highest concentration where the linear fit was applied to the data between 10 and 40 h). (**f**) Schematic illustration of the 3D impedance-based migration assay developed by ACEA Inc. (**g**) Schematic illustration of the 3D impedance-based migration assay developed by Applied BioPhysics Inc. Images in (**b**) are reprinted from [57] with permission

Instead of killing the cells that have already formed a confluent cell layer on the electrode, the *electrical fence assay* prevents cell adhesion to the electrode, while the cells populate the rest of the growth substrate (Fig. 12a, lower panel) [141]. Strong electric pulses applied in regular intervals to the electrode prevent cell adhesion during the attachment process right after inoculation of the electrode with cells. The electrode remains free of cells and ECM that is otherwise deposited by adhering cells. The area surrounding the electrode is regularly populated by cells. When the electric fence is turned off, the cells from the periphery of the electrode start to migrate into the free space of the cell-free electrode. The fractional coverage of the electrode surface with cells is then followed by impedance or capacitance readings at high frequencies (>10 kHz). Figure 12c shows a characteristic time course of the capacitance at 32 kHz along an electrical wound-healing assay. Confluent NRK cells were electrically wounded by a high-amplitude AC pulse (40 kHz, 5 V, 30 s) at time zero indicated by the red arrow. Immediately after wounding the capacitance approaches values of a cell-free electrode. Thereafter, due to non-wounded cells repopulating the electrode from the periphery, the capacitance decreases gradually until pre-wound capacitance values are re-established (Fig. 12c, black curve). In this example three other cell populations (green, blue, red) have been pre-treated with different concentrations of the actin-depolymerizing cytochalasin D. The impact of such a manipulation of the actin cytoskeleton on wound healing and cell migration has been studied by the impedance assay compared to a vehicle control (black). Quantitative analysis of the raw data provides the time required for half-maximum repopulation of the electrode $t_{1/2}$ or the apparent migration rate, which is the slope of the descending capacitance curve (Fig. 12d, e). The impedance-based wound-healing assays were applied in studies addressing the migratory potential of cancer cells [142–149] as well as wound repair/tissue recovery of epithelia [67, 150–160] and endothelia [73]. Also angiogenesis [161, 162] and smooth muscle cell migration [163, 164] have been addressed with this assay as both phenomena are important hallmarks of vascularization of tumor tissue and vascular remodeling upon injury.

The ability of cells to migrate in 3D environments representing the connective tissue is of obvious relevance in studies on cancer progression. Classical *Boyden chamber* assays use *matrigel*-coated permeable supports with a pore size of about 8 μm. Matrigel is a basement membrane-like mixture of different ECM proteins, primarily laminin, collagen IV, and nidogen, extracted from murine tumors [165]. If required, the porous supports are alternatively coated with other protein matrices of defined composition. Cells are then seeded to the upper compartment on top of the matrigel, and the lower chamber is filled with a chemoattractant to direct migration towards the bottom compartment. Actively migrating cells pass the 3D protein network and the pores of the permeable support. After a specified time, the matrigel on top of the permeable inserts is gently removed, and cells on the back side of the filter are stained with crystal violet or DNA-specific stains for quantification.

ACEA Biosciences Inc. developed a system that works in a similar way like the Boyden chamber assay but replaced the optical readout by impedance measurements (Fig. 12f). The back side of the porous membrane is coated with a thin gold film, which serves as sensing electrode for those cells that have successfully migrated through the hydrogel and reached the back side of the permeable support. The number of cells on the backside of the permeable support is quantified by impedance measurements. This setup has been used in various studies addressing the migratory potential of different cancer cell types [166–170], endothelial cells [171], or stem cells [51]. A slightly different setup developed by Applied BioPhysics Inc. was recently used to study chemokine-directed migration of bone marrow-derived stem cells [172] (Fig. 12g). Conventional matrigel-coated permeable inserts as typically used for Boyden chamber assays are inserted into an 8-well holder which contains a circular thin-film electrode on the bottom of each well directly below the insert. The lid of the chamber is equipped with eight dipping electrodes, each of which reaches into the upper chamber of a matrigel-coated insert. Jurkat T cells were inoculated into this 8-well setup, and the chemokine *stromal cell-derived factor-1* (SDF-1) was introduced into the lower compartment. Chemokine-dependent migration of Jurkat cells across the matrigel matrix *into* the pores of the insert resulted in an increased resistance of the matrigel-coated insert, which was less pronounced without chemokine addition. The increase in resistance occurred within 2 h and was therefore likely due to migration only, with minimal contribution of proliferation. Addition of a SDF-1 receptor antagonist inhibited the chemokine-driven migration and resistance increase. Moreover the migration of peripheral blood mononuclear cells (PB-MNC) in response to the chemokines SDF-1 and MIP-1 as well as exposure to radiation was studied with this setup. These measurements clearly demonstrated the inhibitory effect of radiation on PB-MNC migration.

3.6 Morphology Changes in Response to Biological, Chemical, or Physical Stimuli

This chapter will summarize current knowledge in impedance-based assays that monitor the response of established cell layers to experimental stimuli that are not expected to be cytotoxic. These assays include those studying receptor activation and signal transduction, differentiation, cell-cell and cell-pathogen interactions, as well as cell responses to physical stimuli such as electric fields or mechanical loads. Cell death and toxicity monitoring will be covered by a separate chapter.

3.6.1 Receptor Activation and Signal Transduction

Cells respond to a myriad of external stimuli with the activation of intracellular signaling cascades. Signal transduction describes the process of converting signals coming from the outside into functional changes within the cell. It generally requires molecular recognition, intracellular amplification, and integration. A signaling cascade is typically initiated by activation of a receptor located on the cell surface or inside the cell [173]. Receptors located on the cell surface are classified as ionotropic, metabotropic, or enzyme-linked receptors. For membrane-permeable hormones, the corresponding receptors are located in the cytoplasm (nuclear receptors). Agonist-induced conformational changes of the receptor are transduced into specific relaying mechanisms including ion transport across membranes, enzyme activation, protein-protein interactions, and modulation of gene transcription. Because of its central role in controlling cellular functions, measuring the activity of signal transduction events has two major motivations: (1) unravel signaling pathways within the cell for gaining a better understanding of the underlying principles that discriminate diseased from healthy phenotypes and (2) evaluate the efficiency and potency of drugs that interfere with signaling pathways involved in certain pathologies.

The first report on impedance sensing applied to signal transduction in cultured cells was published by Tiruppathi et al. in 1992 [9]. The impact of the inflammatory mediator thrombin on bovine aortic endothelial cells grown on gold film electrodes was followed with time. Thrombin cleaves the N-terminal end of the G protein-coupled receptor (GPCR) PAR-1, which is abundantly expressed in endothelial cells and involved in the regulation of vascular permeability. The impedance-based assay returned the same results as had been reported before in studies measuring the permeation rate of radio- or fluorescence-labeled tracers across endothelial cell layers grown on permeable supports [9]. A critical advantage of impedance-based monitoring over the conventional method was its time resolution that provided the kinetics of barrier breakdown and its recovery in detail. Since this initial publication, impedance-based assays have evolved as a valuable and accepted tool in measuring receptor-mediated signaling events.

After its potential in monitoring receptor signaling had been recognized, impedance measurements have been increasingly applied for signal transduction studies for all kinds of receptor classes (Fig. 13a) and cell types addressing a diversity of biological functions that are controlled by signaling events. Among the receptors that were successfully studied by impedance assays are many GPCRs including the adenosine A1 and A2A receptors [176, 177], the angiotensin receptor [178, 179], ß-adrenergic receptor [180, 181], cytokine receptors (e.g., CXC3) [182], histamine receptor H1R [183–185], kinin B1 receptor [186], lysophosphatidic acid (LPA) receptor [187], lysophosphatidylcholin (LPC) receptor [188], niacin receptor [189], neurokinin 1 receptor [190, 191], prostaglandin (PGE2) receptor [192–194], PAR-1 receptor [9], and sphingosine-1-phosphate (S1P) receptors [195–198]. Figure 13b shows the response of human dermal microvascular endothelial cells to three different GPCR agonists upon binding to their receptors: thrombin, histamine, and S1P (Fig. 13b). The effects of these three mediators on endothelial barrier and morphology are very different in terms of magnitude and kinetics. For each receptor-agonist pair, a very typical response profile is observed. Other important receptor classes that have been studied in terms of signal transduction pathways and physiological function are growth factor receptors (receptor tyrosine kinases) [91, 199–205], cytokine receptors [206–211], Toll-like receptors [212, 213], serine/threonine kinase receptors such as the TGF-beta receptor [214–216], nuclear hormone receptors [217–221], and even adhesion molecule-mediated signaling

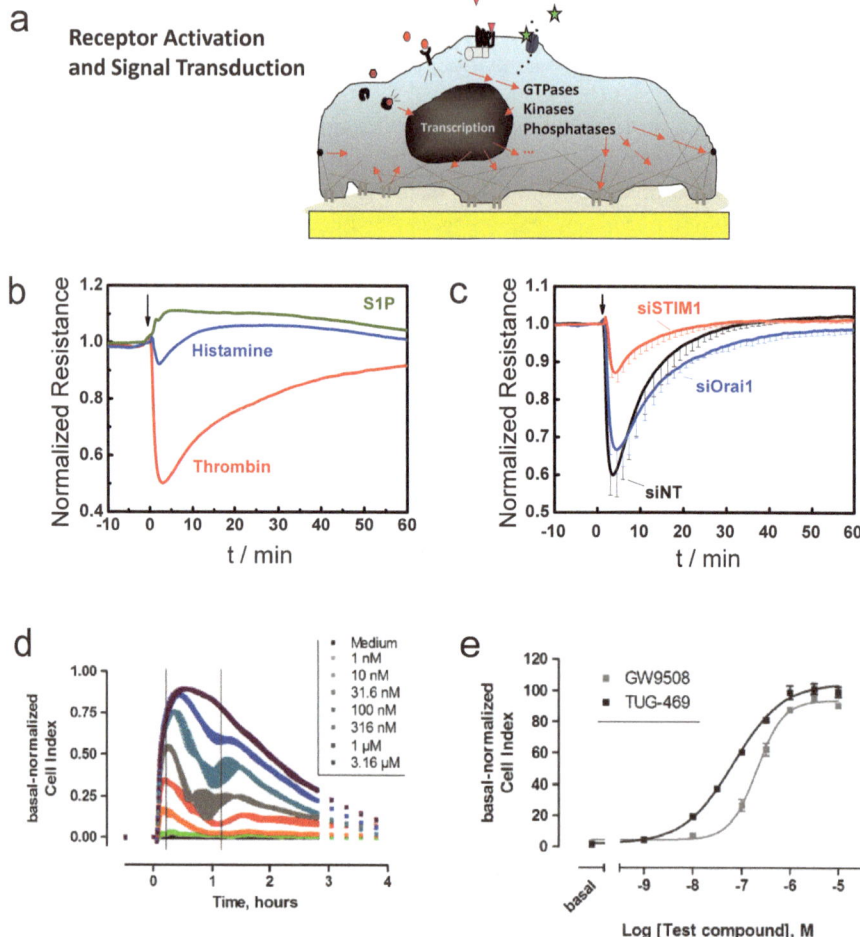

Fig. 13 Receptor activation and signal transduction monitored by impedance-based assays. (**a**) Schematic illustrating receptor-mediated signal transduction of a cell grown on a substrate-integrated thin-film electrode, including (from left to right) intracellular receptors, enzyme-linked receptors, G protein-coupled receptors (GPCR), ligand-gated ion channels, and adhesion receptors. (**b**) Response profiles of human dermal microvascular endothelial cells upon addition of three different endogenous GPCR agonists: thrombin, histamine, and sphingosine-1-phosphate (S1P). Adapted from [54] with permission. (**c**) Investigating the role of individual proteins in signaling cascades: knockdown of the calcium channel *Orai1* and the ER-resident calcium sensor *stromal interacting molecule-1* (STIM1) by siRNA affects thrombin-mediated HDMEC response. Adapted from [174] with permission. (**d**, **e**) Characterizing GPCR agonists: TUG-469, agonist to the G protein-coupled free fatty acid receptor 1 (FFA1), was compared to the standard FFA1 agonists GW9508 with respect to their individual dose-response relationship. (**d**) TUG-469-induced, concentration-dependent cell responses in impedance-based assays. (**e**) Dose-response analysis for both agonists based on impedance readout. $pEC_{50} \pm SE$: GW9508 (6.69 ± 0.03), TUG-469 (7.17 ± 0.04); $E_{max} \pm SE$: GW9508 (93.8 ± 1.5), TUG-469 (105.0 ± 2.3). From [175] with permission by Springer

[123, 222–224]. Impedance-based studies on ligand-gated ion channel-mediated signaling are limited, but it has been successfully demonstrated for the transient receptor potential cation channel subfamily V 1 (TRPV1) expressed in CHO cells and its agonist capsaicin [225].

Impedance-Based Assays in Receptor-Mediated Signaling Research

In basic research it is the major goal to unravel the downstream signaling pathways of cell surface receptors upon stimulation with endogenous agonists that induce a certain change in cell function and/or phenotype. Classically, impedance-based signal transduction assays have been performed on confluent cell layers, focusing on the immediate change in cell shape of a mature, well-established cell layer. Cellular responses to receptor stimulation detected by impedance most often occur within seconds or minutes but can also take several hours, depending on the receptor type and the downstream signaling. GPCR activation is most often processed very rapidly on the time scale of seconds to minutes (cf. Fig. 13 and examples given in [181, 183, 226–228]), whereas activation of growth factor and cytokine receptors usually leads to rather slow responses that require hours or even days as they may include alterations in gene expression [200, 204, 206, 207, 210, 211]. Unravelling cellular signaling patterns that lead to certain phenotypic changes is strongly supported by the availability of pharmacological inhibitors/modulators that specifically shut down/modulate individual pathways and thereby highlight their importance. These molecules have become extremely valuable and important tools in signal transduction research in general and to the same degree for impedance-based assays in this field. These include modulators of cytoskeletal proteins [196, 197, 228–230], kinases [183, 226, 229, 231–234], GTPases [235], phospholipases [190, 232, 236], phosphodiesterases [181, 237], ion channel blockers [174, 238], and calcium chelators [174, 232]. Furthermore, intracellular signaling cascades may be triggered directly by membrane permeable derivatives of second messengers such as CPT-cAMP, bypassing the corresponding cell surface receptors [181, 192, 226, 239, 240]. Unfortunately, pharmacological modulators do not always inhibit or stimulate a certain target protein in the signaling branch exclusively. To overcome possible crosstalk and off-target effects, modern molecular biology has introduced tools to specifically alter the expression level of individual target proteins, like protein knockdown using interfering RNAs (siRNA, shRNA) [174, 177, 188, 206, 222, 233, 239, 241–243] or ectopic expression of cDNA encoding protein variants [123, 187, 221, 244–246] (Fig. 13c). These techniques require chemical/physical transfection or viral transduction but are compatible with an impedance-based readout. Cells derived from knockout animals may serve as an alternative [238, 247]. Strategies such as "erase and replace" experiments combine both approaches, starting with an siRNA- or shRNA-mediated knockdown of the protein of interest, followed by expression of either a wild-type or a mutated/truncated form of the same protein [222, 238, 247, 248].

Signal transduction is the basis of converting environmental cues to cellular adaptation. So it is not surprising that it is involved in virtually any aspect of regulating cell function. Accordingly, all of the assays described in the preceding chapters monitoring cell adhesion [95, 249], proliferation [250–252] or migration [180, 195, 252, 253] have been used to study certain aspects of signal transduction. For example, Peizman et al. studied the effect of different ß-adrenergic receptor (ß-AR) agonists and antagonists on the wound-healing capacity of normal human bronchial epithelial (NHBE) cells [180]. Well-known ß-AR agonists like nonselective epinephrine, ß1-selective dobutamine, and ß2-selective salbutamol were shown to slow down wound healing of NHBE. The effect of the ß-AR agonists was fully blocked by the nonselective ß-AR antagonist propranolol. Also, inhibition of protein phosphatase PP2A was able to block the delay in wound closure produced by the ß-agonists. Xu et al. used a calcium switch protocol to first open an endothelial barrier by chelation of extracellular calcium followed by addition of S1P to study its impact on subsequent barrier recovery [230].

Impedance-Based Assays in Drug Discovery Targeting Cell Surface Receptors

Because of their key role in regulating cell functions and their accessibility from the extracellular side, cell surface receptors belong to the most widely addressed and promising drug targets [254]. Thus, assays reporting on ligand binding to a particular receptor are of utmost importance in the drug development process and the basis of many screening campaigns. Such assays include ligand binding studies, e.g., radioligand binding assays [255, 256], as well as quantification of a ligand's efficiency and potency to activate cellular downstream signals. Several different experimental strategies exist to measure receptor-induced signal transduction pathway: (1) measuring the activity of G proteins directly coupled to the receptor by using, for example, the *GTPγS assay* when GPCRs are the receptors under study; (2) detecting intracellular *second messenger* levels, i.e., calcium, inositol-phosphates, or cAMP; (3) quantifying the activity of GTPases, kinases, phosphatases, and other downstream proteins; (4) testing of transcription activity via reporter gene constructs with specific responsive promotor elements; and (5) assessing receptor internalization or dimerization events [257–259]. Many of these approaches are endpoint assays that require labeling, exposure to harmful reagents, or transfection with a reporter gene. The situation has gotten more complicated ever since it was discovered that the established "one-receptor-one-pathway" paradigm may not be generally valid. Some receptors change their coupling scheme dependent on the chemical nature of the agonist. This so-called *functional selectivity* requires studying several if not all potential signal transduction pathways when a new receptor-ligand couple is under investigation. In this context unbiased, label-free assays based on measurement of the refractive index or the electrical impedance are increasingly appreciated [229, 257–267] as they do not require any a priori selection of potential signaling pathways but integrate over all signaling events in the cell that contribute to the final cell shape change. This integrative nature of label-free physical approaches is the reason to call them *(w)holistic* – similar to whole organ studies. In return, the identity of the actual pathway(s) activated by a receptor under study remains obscure and has to be unraveled using pharmacological tools with molecular specificity [229, 234]. Initially it has been claimed that time-resolved impedance recordings upon receptor activation show pathway-specific and distinguishable profiles, suggesting that the signaling pattern is deducible from the shape of the impedance curve [266, 268]. However, a broader perspective shows that impedance profiles are also strongly cell type-dependent [264] so that a clear-cut assignment of the signal transduction involved in receptor activation is still difficult and prone to error when only impedance curves are used for analysis.

Activation of all receptor classes is assessable by impedance-based readout techniques as long as the cellular response to receptor activation leads to some kind of morphology change. GPCRs and to a lesser extent *receptor tyrosine kinases* (RTK) have been the prime targets in the literature. One of the biggest advantages of impedance-based approaches in receptor research is their sensitivity which allows studying cells with endogenous receptor levels like primary cells, instead of relying on heterologous expression of the receptor of interest [268, 269]. Impedance assays have proven useful to characterize both agonists [175, 176, 227, 266, 269–272] (Fig. 13d, e) and antagonists [266, 269, 271, 273] in terms of dose-response relationships that provide information about efficacy and potency of a ligand of interest (Fig. 13e). Partial agonists, partial antagonists, inverse agonists, and allosteric modulators have been addressed and characterized by such assays [266, 268, 271, 274]. Impedance-based assays were also applied to study functional selectivity of cell surface receptors as described above [232, 264]. Taken together, impedance-based assays have found extensive appreciation in cell signaling studies and are powerful readouts for drug screening projects targeting cell surface receptors and beyond [83, 266].

3.6.2 Cell Differentiation

Differentiation is defined as the complete and full expression of a certain cell phenotype. As such it needs to be distinguished from simple maturation of a cell layer. Differentiation plays a central role in stem cell development and cell lineage determination, with relevance to regenerative medicine and the transformation of somatic cells into malignant cancer cells. Cell differentiation oftentimes involves complex cellular reorganizations that are associated with major changes in cell morphology that are easily detected by impedance readings (Fig. 14a). Differentiation of stem cells has been studied by impedance-based assays [10, 275, 276], as well as developmental differentiation [277] and epithelial to mesenchymal transition [122, 278]. For instance, Bagnaninchi et al. monitored differentiation of stem cells derived from adipose tissue for 17 days and followed their osteogenic or adipogenic paths upon exposure to the respective differentiation cues [10]. The two cell lineages had individual dielectric properties which were reflected by distinct time

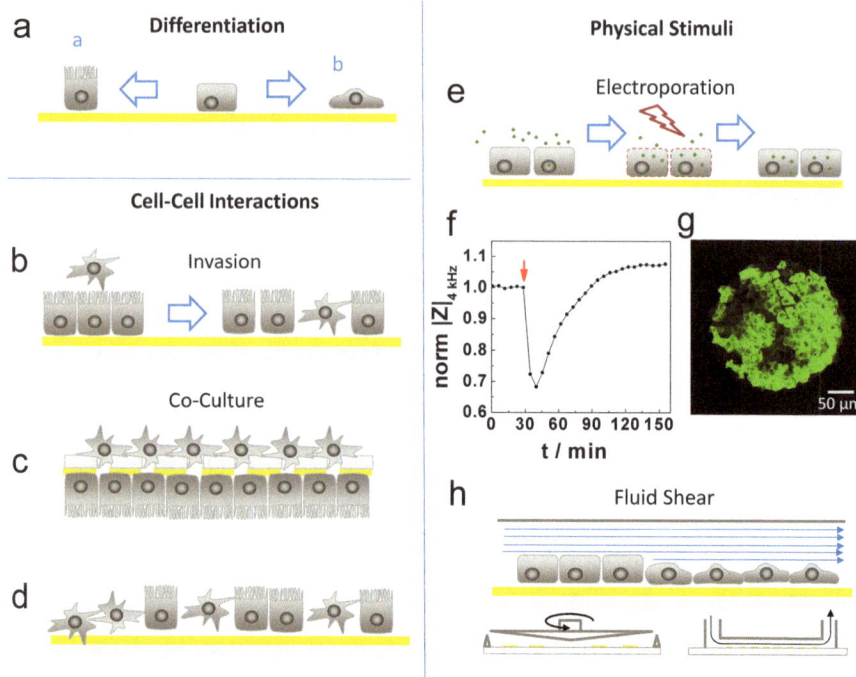

Fig. 14 Differentiation, cell-cell interactions, and cell response to physical stimuli studied by impedance-based assays. (**a**) Schematic illustrating cell differentiation on a thin-film electrode. (**b**) Schematic illustrating the invasion of cells (cancer cells, blood cells, pathogens) across a confluent layer of barrier-forming cells grown on a thin-film electrode. (**c**) Schematic illustrating the co-culture of two different adherent cell types on opposite sides of a permeable cell culture insert with an integrated electrode on the underside of the insert. (**d**) Schematic illustrating the co-culture of two different adherent cell types grown on planar electrodes. (**e**) Schematic illustrating in situ electroporation of cells grown on thin-film electrodes. (**f**) Response profile of NRK cells to an electroporation pulse of 4 V, 200 ms, and 40 kHz. (**g**) Confocal fluorescence micrograph of NRK cells loaded with FITC-labeled dextran by in situ electroporation. (**h**) Schematic illustrating cells on thin-film electrodes exposed to fluid shear stress by a rotating cone in a cone-plate setup or by pumping fluid through a narrow channel above the cells. The micrograph shown in (**g**) is adapted from [279] with permission

courses of the complex impedance with the first significant difference occurring as early as about 12 h after induction. Multifrequency impedance readings allowed for a later analysis of the cell-related parameters R_b, α, and C_m throughout the experiment (cf. chapter "The ECIS Model: Correlating Impedance Readings with Cell Morphology"). The membrane capacitance C_m was identified as an early marker for the two differentiation pathways.

3.6.3 Cell-Cell and Cell-Pathogen Interactions

Cell-cell and cell-pathogen interactions have been studied particularly for epithelia and endothelia. Functioning as a selective and well-regulated barrier, endothelial and epithelial cell layers often encounter other cell types that are not supposed to pass the barrier. Various pathological implications induce impairment of the barrier function so that it becomes permeable and allows other cells or pathogens to migrate across. Barrier opening is either induced by the cell-cell contact itself or by the local secretion of chemical mediators. Prominent examples for such scenarios are extravasation of circulating cancer cells from the bloodstream into the surrounding tissue during metastasis, immune cells extravasation under inflammatory conditions, or pathogen-induced barrier breakdown.

Impedance-based assays have been established to model such scenarios in vitro (Fig. 14b). Typically, the barrier-forming cell type is grown to confluence on the gold electrode (or filter), and the pathogens [280–284], cancer [285–292], or blood cells [224, 228, 293–296] are suspended in the supernatant. Keese et al. established confluent human umbilical vein endothelial cell (HUVEC) layers on planar gold film electrodes and exposed them to suspensions of different sublines of the Dunning murine prostate adenocarcinoma (G, AT1, AT2, AT3, MLL) that differ with respect to their metastatic potential [288]. The highly invasive cancer cell lines (MLL, AT2, AT3) induced an immediate and drastic decrease of the electrical impedance indicating barrier breakdown, whereas the weakly metastatic sublines showed a less pronounced impact on HUVEC barrier properties. A similarly gradual transmigratory behavior was observed for four human prostate cancer cell lines (DU145, PC3, TSU, PPC1) indicating that the metastatic potential in vivo is correctly mirrored in these in vitro experiments.

On the other hand, it has been shown that many barrier-forming cells enhance their barrier properties when they are co-cultured with other cell types that are found in close proximity in vivo (Fig. 14c, d). A prominent example is the endothelium of the blood-brain barrier (BBB). Insert-based systems are typically used for this kind of co-cultures with the endothelial cells grown on one side of the insert, while the interacting cell type, e.g., astrocytes or pericytes, is grown on the opposite side. Such inserts with gold electrodes prepared on the backside of the permeable support (CIM plates by ACEA Bioscience Inc.) were used to establish an inverted BBB model by growing brain microvessel endothelial cells (BMEC) on these electrodes and adding astrocytes to the upper compartment of the insert [297]. With this setup, changes of the endothelial barrier were monitored by time-resolved impedance readings. Rother et al. studied the cardiomyocyte-fibroblast crosstalk by co-culturing cardiomyocytes and fibroblasts in different ratios on gold film microelectrodes (Applied BioPhysics Inc.) [298]. The different cell populations were characterized with respect to their spreading kinetics, the regular cardiomyocyte beating, and their dielectric characteristics as it has been extracted from frequency-dependent impedance measurements.

3.6.4 Response to Physical Challenges: Invasive Electric Fields, Fluid Shear, and More

In addition to biological and chemical stimuli, impedance-based assays have been used to study the impact of different sorts of physical challenges such as radiation [299], gravity [300], electric fields [8, 34, 37, 61, 279], and fluid shear [301–303].

When cells are exposed to invasive electric fields, they face membrane permeabilization and Joule heating. As cells are capable of recovering from short-term membrane permeabilization, short electric pulses are used to load cells with extracellular, membrane-impermeable probes and molecules such as DNA (electroporation). Indeed, reversible membrane electroporation became a widely used technique for transfection. During the permeabilized state of the membrane, small molecules in the external buffer may diffuse into the cytoplasm via *electropores*, or they are accumulated on the plasma membrane by the electric field for later uptake by membrane-mediated mechanisms (e.g., endocytosis). The electrodes that are used for impedance sensing may be used for delivery of the electroporation pulses as well (Fig. 14e) [37]. This has been shown by delivering membrane-impermeable fluorescent probes into the cytoplasm of adherent cells that were grown on the surface of gold film electrodes and subjected to AC electric fields of about 1 kV/cm for 200 ms at 40 kHz [61]. Loading efficiency was quantified by fluorescence microscopy (Fig. 14g). Since the same electrodes are used for pulse application and impedance sensing, cell recovery from the electroporation pulse was monitored by impedance readings providing detailed insight into the invasiveness of the electroporation event (Fig. 14f). When the pulse parameters were properly selected for the cell line under study, loading efficiency was higher than 90%, while the cells completely recovered from the pulse within less than 45 min [61]. Recovery of the impedance to pre-pulse values does, however, not report on the time course of membrane resealing which occurs on the time scale of seconds. Instead it mirrors the morphological recovery of the cells after pulse application and disturbance of cytoplasmic homeostasis [37, 279]. In addition to recording the mere response to the electroporation pulse, the combined impedance sensing/electroporation setup allows studying the impact of membrane-impermeable molecules that get introduced into the cells by this technique [279]. For example, the anticancer drug bleomycin shows a rather low cytotoxicity when applied to a cell population in the extracellular buffer. However, it efficiently induces apoptosis at about 1,000-fold lower concentrations when it is delivered to the cytoplasm by means of reversible membrane permeabilization [279]. When the applied electric fields are too invasive, the cell population on the electrode surface is irreversibly damaged and killed (irreversible membrane electroporation). This is the basic principle of the ECIS-based wound-healing assay [8]. By selectively killing the cell population directly grown upon the electrode, a defined microlesion is introduced into the cell layer, which is subsequently repopulated by cells migrating in from the periphery of the electrode (cf. chapter "Cell Migration: The Electrical Wound-Healing Assay and 3D Approaches").

Some cells within the body are exposed to hydrodynamic shear forces, in particular endothelial cells lining the blood vessels. Such mechanical impacts due to shear forces or hydrostatic pressure have also been studied by means of impedance-based cell monitoring [7]. Well-defined experimental shear forces are typically applied by one of the two major setups (Fig. 14h):

1. Parallel-plate, microfluidic flow channels with the cells growing on the bottom of the plate. Fluid is pumped through the channel producing highly reproducible mechanical shear forces [302].
2. Cone-plate rheological setups create laminar flow with the help of a rotating cone that is placed above the cell layer grown on a planar substrate [303].

In both cases thin-film electrodes are deposited on the bottom of the chambers on which the cells are grown enabling in situ impedance-based cell observation. Based on such measurements, various molecular mechanisms have been described by which cells sense shear forces [304, 305] that lead to morphological and functional changes assessable by impedance measurements.

The first report about the effect of shear stress on adherent endothelial cells cultured on thin-film electrodes in a flow channel has been provided by Phelps and DePaola [302]. They investigated the effect of laminar shear on the barrier function of bovine aortic endothelial cells (BAEC). During exposure to flow, the electrical resistance of the cell layer decreased. This finding was consistent with supporting experiments revealing an increased permeability of the BAEC layer for fluorescence-labeled dextrans. The flow channel contained a step-shaped obstacle to induce turbulence and locally break the laminar flow profile. This way the cell response to laminar and turbulent flow was studied at different sites of the channel with individual electrodes. The cell response to the onset of flow was studied as a function of time for the different flow profiles. About the same time, Seebach et al. reported on the response of porcine pulmonary trunk endothelial cells (PSEC) to laminar shear stress of different magnitudes ($0–200$ dyn/cm^2) using a rotating cone-plate setup [303]. The onset of fluid shear-induced increases in electrical resistance depended on shear stress level which was then followed by a characteristic decrease. This characteristic response was fully reversible after turning off shear forces. Impedance data was found to correlate with immunocytochemical stainings for VE cadherin, actin, and catenin that changed their subcellular distribution pattern upon flow exposure [303, 306].

Besides studying the pure mechanical load on adherent cells, impedance measurements under flow have also been applied to identify flow-dependent differences in cell responses to cell surface receptor activation [307, 308], signaling pathway modulators [309], other cell types [310–313], nanoparticles [314], or toxicants [315, 316]. With respect to the current trend towards cell culture in microfluidic devices, including organ-on-a-chip or multi-organ-on-a-chip approaches, the number of studies combining microfluidics and impedance-based cell observation will likely increase in the near future.

3.7   Monitoring Cell Death and Cytotoxicity

Within multicellular organisms the number of cells is strictly controlled by cell division on the one hand and cell death on the other. Cell division and proliferation have been addressed above (cf. chapter "Cell Proliferation"). This paragraph will discuss the use of impedance-based assays to monitor cell death induced by toxins, environmental pollutants, or endogenous triggers.

Cell death is typically categorized according to two major pathways along which cells die: (1) *apoptosis* and (2) *necrosis*. Apoptosis is a genetically encoded, well-orchestrated cell death mechanism, which is mediated by special proteases (caspases) that enable a highly controlled degradation and clean removal of cell remnants from the body. Any dysregulation of the internal mechanisms to undergo apoptosis has dramatic consequences. Both cell accumulation due to an impaired eradication of defective cells and excessive cell loss due to faulty induction of apoptosis lead to severe problems. Pathologies associated with dysregulated apoptosis include diverse neurodegenerative, hematological, cardiovascular, metabolic, autoimmune- and development-associated disorders, as well as malignant and pre-malignant diseases [317]. Apoptosis has been particularly well studied with respect to its implications in neoplasia and cancer progression. Besides internal signaling, there are external triggers for apoptosis like pathogen infection, UV or ionizing radiation, certain pharmaceuticals such as cytostatics and hormones, lack of growth factors, or hypoxia [318]. *Necrosis* describes a more immediate form of cell death as a consequence of external impacts like heat, mechanically or electrically induced trauma, and intoxication – to mention just a few. It commonly occurs when a well-controlled elimination of the cell via apoptosis is not possible anymore. Caspases are not activated. Although we typically think of apoptosis and necrosis as the only and very distinct modes by which cells die, studies in recent years have revealed that the boundaries between these two are often blurred and that additional mechanisms of cell death exist [317, 319–321].

A number of experimental methods have been developed to distinguish apoptosis from necrosis [322]. Methods identifying cell death by apoptosis rely on the detection of caspase activity or other biochemical hallmarks, such as DNA condensation and fragmentation or flipping of phosphatidylserine to the outer membrane leaflet. Morphological indicators for apoptosis are cell shrinkage, membrane blebbing, and the formation of apoptotic bodies, i.e. vesicles containing cell fragments ready to be taken up by macrophages. In contrast, necrosis is characterized by cell swelling and membrane rupture, which is often induced by failure of ion pumps in the cell membrane and the associated osmotic disbalance. Intracellular material is spilled into the extracellular space which causes an activation of the immune system, triggers inflammation, and may be used to identify necrotic cell death.

There is a strong need to test chemicals or biologicals for their ability to induce cell death in vitro as this is a first hint for their potential harmfulness to humans, animals, and the ecosystem. Accordingly, toxicity monitoring has been addressed extensively in the scientific literature. Toxicity is a function of the toxin's concentration and the duration of exposure. It is subdivided into three categories: (1) acute, (2) subchronic, and (3) chronic toxicity [323]. Acute toxicity involves harmful effects in an organism through a single or short-term exposure. Subchronic toxicity is the ability of a toxic substance to cause harmful effects within more than 1 year but less than the lifetime of the exposed organism. Chronic toxicity is the ability of a substance or mixture of substances to cause harmful effects over an extended period of time, usually as a consequence of repeated or continuous exposure, sometimes lasting for the entire life of the exposed organism. Further classifications have been described, which are based on the nature of the toxic impact (chemical, biological, physical), the target tissue (hepatotoxicity, neurotoxicity), the mechanism of action (anticholinergic, inhibitor, uncoupler), special harmful effects (carcinogenic, mutagenic, endocrine disruptor), or the type of toxic response (local, systemic, immediate, delayed). Accordingly, the term cytotoxicity is not always clearly defined in the rather large pool of experimental toxicity studies. The obvious need for standardized toxicity testing is indicated by the establishment of international programs like the European program on Registration, Evaluation, Authorization and restriction of CHemicals (REACH) and the American "Toxicology in the 21st Century" Tox21 program [324, 325] that are aiming at improving and streamlining cytotoxicity test systems for a better risk assessment of potentially hazardous materials. In order to reduce animal testing to a minimum, the implementation of alternative in vitro assays for routine cytotoxicity testing is encouraged, which are typically faster, allow higher throughputs, are less expensive, and are more robust than animal tests. For in vitro cell cultures, a compound or treatment is considered to be cytotoxic if it interferes with cell adhesion, significantly alters cell morphology, adversely affects cell growth rate, or causes cell death [326].

Classical in vitro tests for toxicity testing in cell cultures are based on measuring (1) compromised plasma membrane integrity by dye inclusion or exclusion (e.g., neutral red uptake) or enzyme leakage (e.g., LDH release) or (2) a reduction in metabolic activity (e.g., concentration of reducing coenzymes, esterase activity, ATP levels, pH gradients) [11, 326]. Alternatively, cell number, total protein mass (e.g., sulforhodamine B assay), and the potential to form new cell colonies after replating are used as experimental indicators [11]. *Cytotoxicity assays* are distinguished from *viability assays* even though they are obviously inversely related. Precisely speaking, cytotoxicity assays measure parameters that are proportional to the fraction of dead cells in a population, while viability assays record parameters that are proportional to the number of viable cells. Complementary assays that measure both, viability and cytotoxicity, are considered as useful means to mitigate liabilities of single mode testing and prevent faulty data interpretation. The MTT assay is often regarded as the gold standard and used as a reference for the development of new viability/cytotoxicity assays [326]. Similar to MTT, most of the classical assays are endpoint assays that sacrifice the culture for the measurement. Thus, it is of critical importance to define the proper time point for the readout. This is particularly tricky for dye inclusion/exclusion assays as dying cells may retain an intact plasma membrane over a long period of time before permeabilization and lysis occur. The same problem applies to assays that detect the activity of enzymes released from dying cells to the culture medium. Moreover, marker enzymes like LDH released from the cytosol into the culture medium may undergo degradation before readout. Reductase-based assays with tetrazolium and resazurin reagents require incubation times of up to 4 h to generate measurable signals, and, thus, assay results represent an average over the entire incubation time [326]. Therefore, there is considerable interest in using time-resolved label-free technologies for cytotoxicity testing because of their striking advantages.

In 1992 Giaever and Keese suggested the use of impedance-based cell monitoring (ECIS) for toxicology studies [327]. Later, they showed that impedance measurements serve as a reliable predictor of cytotoxicity [328]. In this study WI-38/VA13 and MDCK cells were subjected to four different detergents at increasing concentrations. Dose-response curves were established from the impedance raw data that correctly ranked these toxicants according to their in vivo toxicity. The authors used the gross changes in overall impedance as well as the more subtle changes in impedance fluctuations (micromotion) as experimental indicators. It is noteworthy that cellular micromotion reported more sensitively on toxic insults than the overall impedance. Ever since this initial report, impedance measurements have been used in diverse cytotoxicity studies using a wide range of cell types and toxins. The applicability for toxicity testing has been demonstrated with various cell models and toxicants, such as heavy metal ions [6, 38, 315, 329–332], toxicants of biological origin [130, 333, 334], cytotoxic drugs [335], nanomaterials [131, 132, 336, 337], biologically induced cell death [338–340], photocytotoxicity [341], or other chemical stressors [6, 38, 340]. Comparing the outcome of impedance-based toxicity screening with conventional colorimetric testing (MTT, neutral red assay) provided a similar or better sensitivity for impedance-based techniques in almost all studies (compare Table 1). The mechanism of cell death, apoptosis, or necrosis has been addressed in some reports as well by measuring biochemical indicators of apoptosis (cf. above) at pre-defined times along the experiment in addition to time-course impedance data [86, 331, 342–344].

Common impedance-based toxicity assays apply the compound under test to a healthy, mature cell layer. Exposing a confluent, equilibrated cell layer to an acutely toxic compound induces a dramatic change in a cell layer's gross impedance compared to control. Typically, acute toxicity is associated with a decline in impedance (or resistance) to values of cell-free electrodes [130, 330, 335, 342]. Dying cells undergo shape changes, lose their plasma membrane integrity, and eventually detach from the electrode surface. These events cooperatively increase the available current pathways across the cell layer and decrease the impedance. Alternatively, the increase in capacitance is used as a measurand for cell death, specifically reporting on electrode coverage with cells [331].

Figure 15 shows the response of confluent normal rat kidney (NRK) cells to different concentrations of the respiratory chain inhibitor antimycin A. The capacitance at 40 kHz reports on electrode coverage with cells (Fig. 15b), whereas the resistance at 4 kHz indicates cell shape changes (Fig. 15c) along the same experiment. Surprisingly the capacitance remained constant for over 40 h even for the highest concentrations of antimycin A (Fig. 15b) before it increased to values of a cell-free electrode – indicating cell detachment or loss of membrane integrity due to

Table 1 EC$_{50}$ values obtained from impedance-based assays and compared to more classical label-based assays

Toxicant	EC$_{50}$ impedance assay	EC$_{50}$ classical assay (assay type)	Reference
CdCl$_2$	3.9 ± 0.4 μM	3.0 ± 0.4 μM (NRU)	[38]
Na$_2$HAsO$_4$	51.0 ± 6.7 μM	52.2 ± 7.7 μM (NRU)	
BAK	13.8 ± 0.5 μM	15.3 ± 0.9 μM (NRU)	
Quarz NP	0.06 mg/ml	0.08 mg/ml (AP)	[343]
ZnO NP	13 mg/ml	22 mg/ml (MTT)	[336]
Mn$_2$O$_3$ NP	33 mg/ml	18 mg/ml (MTT)	
Ag^0 NP	72 mg/ml	59 mg/ml (MTT)	
Sodium arsenite (III)	NIH 3T3: 8.22 ± 0.43 μM	NIH 3T3: 10.51 ± 0.72 μM (MTT) 8.96 ± 0.37 μM (NRU)	[332]
	BALB/c 3T3: 7.71 ± 0.27 μM	BALB/c 3T3: 24.6 ± 0.63 μM (MTT) 38.66 ± 0.87 μM (LDH) 7.96 ± 0.54 μM (NRU)	
	CHO-K1: 9.91 ± 0.54 μM	CHO-K1: 19.02 ± 0.58 μM (MTT) 56.9 ± 1.15 μM (LDH) 11.7 ± 0.43 μM (NRU)	
Mercury (II) chloride	NIH 3T3: 16.74 ± 0.35 μM	NIH 3T3: 20 ± 0.78 (MTT) μM 16.9 ± 0.24 (NRU) μM	
	BALB/c 3T3: 15.72 ± 0.24 μM	BALB/c 3T3: 16.39 ± 0.82 μM (MTT) 40 ± 0.76 μM (LDH) 13.11 ± 0.36 μM (NRU)	
	CHO-K1: 19.75 ± 0.56 μM	CHO-K1: 27.8 ± 0.45 μM (MTT) 43.8 ± 0.81 μM (LDH) 20.73 ± 0.66 μM (NRU)	
Pure CTAB	24 h: 6.4 μM 48 h: 5.5 μM	24 h: 111.2 μM (MTT) 48 h: 16.5 μM (MTT)	[131]
Pure Cd(Ac)$_2$	24 h: 4.3 μM 48 h: 0.3 μM	24 h: 73.9 μM (MTT) 48 h: Nontoxic up to 0.3 μM (MTT)	
Gold NP/CTAB	24 h: 1.2 × 10^{11} p/ml 48 h: 8.5 × 10^{10} p/ml	24 h: Nontoxic up to 2.3 × 10^{11} p/ml (MTT) 48 h: 8.7 × 10^{10} p/ml (MTT)	
Multishell QDs	24 h: 2 × 10^{14} p/ml 48 h: 1 × 10^{14} p/ml	24 h: Nontoxic up to 4 × 10^{14} p/ml (MTT) 48 h: 190 × 10^{14} p/ml (MTT)	
Ricin	0.11 ng/ml	1.5 ng/ml (MTT)	[334]
Agglutinin	30 ng/ml	50 ng/ml (MTT)	

Data are provided as single determinations or mean ± SD. *AP* alkaline phosphatase assay, *LDH* lactate dehydrogenase assay, *MTT* 3-(4,5-dimethylthiazol-2-yl)-2,5-diphenyltetrazolium bromide assay, *NP* nanoparticles, *NRU* neutral red uptake assay

cell death. The onset of the capacitance rise was concentration-dependent; the slope of the curve was not. Resistance readings at 4 kHz suggest a moderate impact of antimycin A on cell morphology already during the initial phase of exposure within the first hours (Fig. 15c) which is, however, not strictly concentration-dependent.

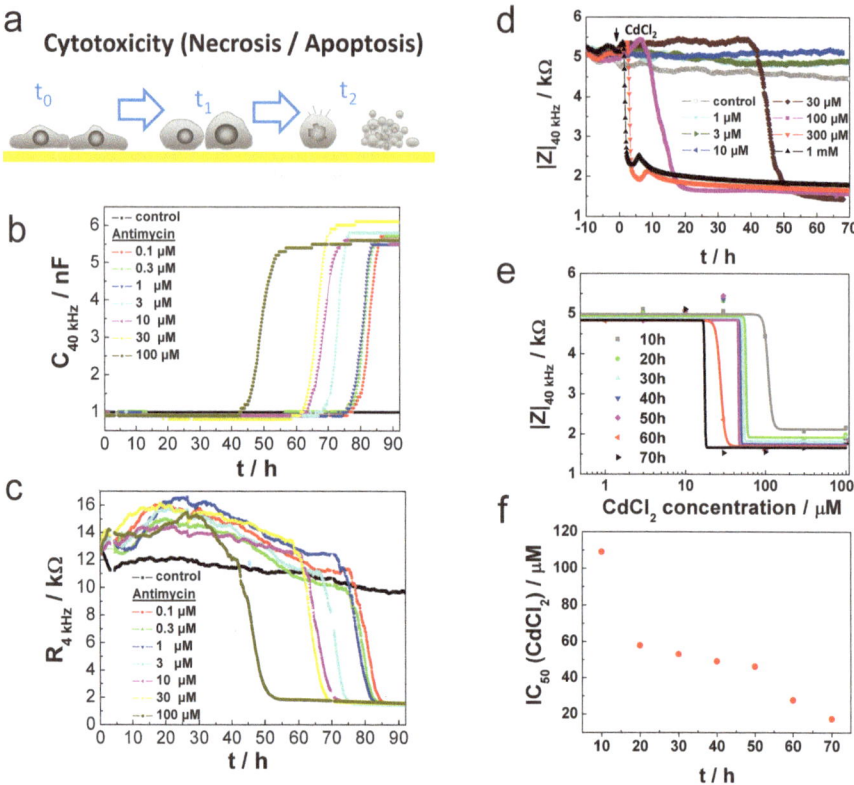

Fig. 15 Cell death and cytotoxicity monitored by impedance-based assays. (**a**) Schematic illustrating cell death on a substrate-integrated thin-film electrode. (**b**) Time course of capacitance at 40 kHz for a confluent layer of NRK cells grown on gold film electrodes of the WE/CE type ($\varnothing = 250$ μm) when exposed to increasing concentrations of the respiratory chain inhibitor antimycin A. (**c**) Time course of the resistance at 4 kHz for the same experiment as shown in (**b**). (**d**) Time course of the impedance magnitude |Z| at 40 kHz for a confluent layer of NRK cells grown on gold film electrodes of the WE/CE type ($\varnothing = 250$ μm) upon exposure (arrow) to different concentrations of cadmium chloride ($CdCl_2$). (**e**) Dose-response curves extracted from the experiment shown in (**d**). The impedance magnitude is plotted as a function of $CdCl_2$ concentration for different exposure times. A logistic function has been plotted to the data points. (**f**) Dynamic EC_{50} values. The EC_{50} concentrations extracted from the logistic fits in (**e**) are shown as a function of exposure time to $CdCl_2$

Gross changes in resistance are synchronized with the capacitance time course at later stages of the response profile. Such details of the cell response would be difficult if not impossible to observe with conventional toxicity endpoint assays.

Micromotion of adherent cells has been used as alternative indicator of cytotoxicity that is assessable by impedance readings [130–132, 328]. Even when there is no gross change in impedance, resistance, or capacitance, changes in the fluctuation pattern ("noise") of the readout parameter reveal even subtle impacts on cell physiology. Fluctuations of the resistance at intermediate frequencies (4 kHz) are considered especially sensitive for monitoring such cell shape dynamics that are characteristic for metabolically active and motile cells (see chapter "Micromotion: An Indicator for Viability and Motility"). A reduction of the fluctuation amplitude has been linked to reduced metabolic activity, cell motility, and viability. Opp et al. recorded micromotion data of cells challenged with different concentrations of cytochalasin B [130]. Whereas the overall resistance showed significant changes only for higher concentrations (2.5–10 μM cytochalasin B) within 20 h, micromotion analysis revealed an impact of cytochalasin B on cell shape dynamics already at concentrations of 0.1 μM. Recording frequency spectra of the complex impedance repeatedly along the time course of the experiment followed by modelling of the raw data (cf. chapter "The ECIS Model: Correlating Impedance Readings with Cell Morphology") allows assigning the observed impedance changes to the level of cell-cell junctions, cell-matrix junctions, or the plasma membrane [130, 342]. For instance, data reported by Opp et al. indicated that cytochalasin B treatment decreases the tightness of cell-cell junctions but increases the membrane capacitance in human umbilical vein endothelial cells [130].

In an alternative, frequently applied assay, the compounds under test are added to a proliferating cell population at different times of an experiment [6, 35, 38, 332, 333, 338, 343]. Cells are either inoculated in presence of the toxicant or the cells are allowed to settle and proliferate for several hours before they are exposed to the test substance. Whereas the former assay unravels any impact of the test compound on cell adhesion, the latter reports on its impact on cell proliferation. For example, Xiao et al. studied the cytotoxicity of cadmium chloride, sodium arsenate, and benzalkonium chloride using V79 fibroblasts as cell culture model [38]. Suspended cells were seeded upon fibronectin-coated electrodes in presence of the test compounds at different concentrations. Adhesion, spreading, and proliferation of cells in presence of these substances were followed for 24 h. The raw data was evaluated as resistance changes (4 kHz) normalized to the initial cell number (in Ω/cell). The concentration of half-maximum response EC_{50} as derived from the dose-response function agreed well with data from a neutral red assay. In general, impedance-based cytotoxicity studies agree well with reference data derived from label-based biochemical assays such as the MTT assay [332], neutral red uptake (NRU) assay [6, 38], LDH assay [332], or luminescence-based assays [35] for the same exposure times (cf. Table 1).

It is the strength of impedance-based approaches to provide time-resolved data so that dose-response relationships are accessible for any exposure time within the limits of the experiment. Accordingly, it is possible to extract the EC_{50} values of a given toxin as a function of exposure time, which has been referred to as the *dynamic EC_{50}* [6, 131, 345]. Figures 15d–f illustrate the extraction of the dynamic EC_{50} values from time-resolved impedance data (40 kHz) recorded for confluent NRK cells exposed to increasing concentrations of $CdCl_2$. Dose-response relationships are established from the raw data for different exposure times (Fig. 15e) providing a set of EC_{50} values. When these are plotted against the corresponding exposure time, the resulting graph (Fig. 15e) illustrates the detailed toxicity profile of the test compound. In this particular example, the EC_{50} decreases from about 110 μM for a 10 h exposure to 10 μM for a 70 h exposure. Using this approach Xiao and Luong found that the EC_{50} of V79 cells to trinitrobenzene (TNB) quickly dropped from 160 μM for 15 h of exposure to 40 μM for 20 h of exposure and eventually 6 μM when the cells were exposed to TNB for 40 h [345]. In contrast, the EC_{50} of $HgCl_2$ only slightly decreased from 77.5 to 54.8 μM within 25 h. While for $HgCl_2$ a toxic cell response was detected immediately for all concentrations under test, TNB toxicity was only observed delayed not before 15 h into the measurement which gave rise to the time-dependent EC_{50} [345]. Instead of using the observed net impedance change as measurand, the *time* of the initial impedance change has been used successfully as a descriptor of toxicity as well [335, 342].

In summary, the great advantage of impedance-based assays and other label-free techniques in general is the time-resolved nature of the raw data so that a single experiment provides the critical concentrations of the compound under test (EC_{50}) for any exposure time. Preselection of the exposure time at the beginning of the experiment becomes dispensable. Real-time impedimetric observation provides the entire time-resolved response profile for a compound under test from a single experiment. Moreover, analysis of the time course data may hold information on the mechanism of cell death or other subtle details not revealed by endpoint assays.

4 Concluding Remarks

Many phenotypes, phenotypic changes, and functional characteristics of adherent cells are assessable by impedance-based assays in vitro: cell-matrix adhesion, proliferation, migration, expression of epithelial/endothelial barriers, response to receptor stimulation, differentiation, interactions with other cells or pathogens, response to electrical and mechanical stimuli, and eventually cell death. This list is not complete but encloses those prominent applications of impedance-based cell monitoring that have been addressed and reviewed in the preceding chapters. The very simple and adaptable measurement principle of detecting changes (1) in cell morphology and (2) fractional cell coverage of the electrode is the technical basis for this broad applicability. This perfectly matches the fundamental observation in cell biology that cells respond very sensitively to a myriad of different stimuli by cell morphology changes. Moreover, fundamental cell phenotypes like adhesion, proliferation, migration, and cell death are inevitably associated with an increase or decrease in cell coverage of the growth surface. In this respect, the sensitivity of impedance-based assays is perfectly in line with a prominent fraction of adherent cell behaviors. Besides this broad applicability, the label-free, noninvasive, and time-resolved character of the measurement provides unique advantages over other means of cell analysis.

However, it is noteworthy that the method is blind for phenotypic changes that are not associated with changes in cell shape or electrode coverage. Moreover, its integral unbiased character, often considered as a big advantage in applications like cytotoxicity or signal transduction monitoring, does not provide any information with molecular specificity. In this respect, impedance-based cell monitoring should be considered as a complementary approach to assays that rely on highly specific molecular probes. Very different strategies have been applied to overcome this inherently limited specificity of impedance-based approaches, often giving up on broad applicability but focusing on certain cell phenotypes. These include adaptation of electrode layout, hardware, data acquisition modes, and last but not least data analysis. A special endeavor to increase information depth combines impedance-based cell monitoring with other techniques for cell analysis that are at least in part covered by individual chapters of this book. An obvious example is the combination of impedance monitoring with microscopic live cell imaging in particular when transparent electrodes made from indium-tin-oxide (ITO) are used. Combining impedance sensing with the *quartz crystal microbalance* (QCM) provides a hybrid approach that reports on changes in cellular micromechanics and morphology for one cell population [346–348]. Similarly, combining impedance readings with *surface plasmon resonance* spectroscopy (SPR) reveals information on cell morphology and mass redistribution (refractive index) close to the surface [349]. As described above, impedance monitoring is readily combined with cell manipulations by elevated electric fields like electroporation or wounding [8, 61, 279]. In analogy to other fields of analytical chemistry, we refer to these combinatorial approaches as *hyphenated* techniques with special reference to the acronym of the original impedance approach ECIS: ECIS-microscopy, ECIS-QCM, ECIS-SPR, or ECIS-ELPO. As highlighted in a designated chapter of this volume, impedance-based cell and tissue monitoring has also entered the third dimension following the most recent trend from 2D to 3D tissue models to better mimic in vivo conditions [350–352].

About 35 years after its invention, impedance-based cell monitoring is widely accepted in academic labs and increasingly appreciated by private enterprises and regulatory institutions. An important requirement for its involvement in drug development schemes is scalability. Accordingly, 96-well and 384-well formats have been developed to meet *high-throughput screening* (HTS) standards. With the amount of data provided by multifrequency recordings in multi-well formats, meaningful and efficient data analysis is required, for which a profound understanding of the measurement parameters is equally important as powerful software packages for data handling and automation. The coming years will show to what extend novel impedance-based assay formats and implementations of these in analytic/diagnostic devices will succeed in improving our understanding of cell behavior. To this point it seems that impedance-based monitoring may become as important for the analysis of adherent cells as flow cytometry already is for suspended cells.

Acknowledgments The authors gratefully acknowledge financial support by the Research Training Group 1910 "Medicinal chemistry of selective GPCR ligands" funded by the German Research Foundation (DFG) (grant number 222125149) as well as support by a regular DFG research grant (grant number 253182429). JS is particularly grateful for a scholarship granted by the *Bavarian Gender Equality Program*.

References

1. Giaever I, Keese CR (1984) Monitoring fibroblast behavior in tissue culture with an applied electric field. Proc Natl Acad Sci U S A 81(12):3761–3764
2. Giaever I, Keese CR (1991) Micromotion of mammalian cells measured electrically. Proc Natl Acad Sci U S A 88(17):7896–7900
3. Keese CR, Giaever I (1991) Substrate mechanics and cell spreading. Exp Cell Res 195 (2):528–532
4. Lo CM, Keese CR, Giaever I (1993) Monitoring motion of confluent cells in tissue culture. Exp Cell Res 204(1):102–109
5. Wegener J, Keese CR, Giaever I (2000) Electric cell-substrate impedance sensing (ECIS) as a noninvasive means to monitor the kinetics of cell spreading to artificial surfaces. Exp Cell Res 259(1):158–166
6. Xiao C, Luong JH (2003) On-line monitoring of cell growth and cytotoxicity using electric cell-substrate impedance sensing (ECIS). Biotechnol Prog 19(3):1000–1005
7. Wegener J, Seebach J (2014) Experimental tools to monitor the dynamics of endothelial barrier function: a survey of in vitro approaches. Cell Tissue Res 355(3):485–514
8. Keese CR, Wegener J, Walker SR, Giaever I (2004) Electrical wound-healing assay for cells in vitro. Proc Natl Acad Sci U S A 101(6):1554–1559
9. Tiruppathi C, Malik AB, Del Vecchio PJ, Keese CR, Giaever I (1992) Electrical method for detection of endothelial cell shape change in real time: assessment of endothelial barrier function. Proc Natl Acad Sci U S A 89(17):7919–7923
10. Bagnaninchi PO, Drummond N (2011) Real-time label-free monitoring of adipose-derived stem cell differentiation with electric cell-substrate impedance sensing. Proc Natl Acad Sci U S A 108(16):6462–6467
11. Stolwijk JA, Michaelis S, Wegener J (2012) Cell growth and cell death studied by electric cell-substrate impedance sensing. Electric cell-substrate impedance sensing and cancer metastasis. Springer, Heidelberg, pp 85–117
12. Dimitrov D (1995) Electroporation and electrofusion of membranes. Handb Biol Phys 1:851–901
13. Ramos C, Teissié J (2000) Electrofusion: a biophysical modification of cell membrane and a mechanism in exocytosis. Biochimie 82(5):511–518
14. Barrau C, Teissie J, Gabriel B (2004) Osmotically induced membrane tension facilitates the triggering of living cell electropermeabilization. Bioelectrochemistry 63(1-2):327–332
15. Tinoco I, Sauer K, Wang JC, Puglisi JD (1995) Physical chemistry: principles and applications in biological sciences, 4th edn. Prentice-Hall Inc., Upper Saddle River
16. Lojewska Z, Farkas DL, Ehrenberg B, Loew LM (1989) Analysis of the effect of medium and membrane conductance on the amplitude and kinetics of membrane potentials induced by externally applied electric fields. Biophys J 56(1):121–128
17. Teissie J, Rols MP (1993) An experimental evaluation of the critical potential difference inducing cell membrane electropermeabilization. Biophys J 65(1):409–413
18. Gross D, Loew LM, Webb WW (1986) Optical imaging of cell membrane potential changes induced by applied electric fields. Biophys J 50(2):339–348
19. Riske KA, Dimova R (2005) Electro-deformation and portion of giant vesicles viewed with high temporal resolution. Biophys J 88(2):1143–1155
20. Voldman J (2006) Electrical forces for microscale cell manipulation. Annu Rev Biomed Eng 8:425–454
21. McCaig CD, Rajnicek AM, Song B, Zhao M (2005) Controlling cell behavior electrically: current views and future potential. Physiol Rev 85(3):943–978
22. Zhao M, Forrester JV, McCaig CD (1999) A small, physiological electric field orients cell division. Proc Natl Acad Sci U S A 96(9):4942–4946

23. Rajnicek AM, Foubister LE, McCaig CD (2006) Growth cone steering by a physiological electric field requires dynamic microtubules, microfilaments and Rac-mediated filopodial asymmetry. J Cell Sci 119(Pt 9):1736–1745
24. Yao L, McCaig CD, Zhao M (2009) Electrical signals polarize neuronal organelles, direct neuron migration, and orient cell division. Hippocampus 19(9):855–868
25. Yan X, Han J, Zhang Z, Wang J, Cheng Q, Gao K, Ni Y, Wang Y (2009) Lung cancer A549 cells migrate directionally in DC electric fields with polarized and activated EGFRs. Bioelectromagnetics 30(1):29–35
26. Li J, Nandagopal S, Wu D, Romanuik SF, Paul K, Thomson DJ, Lin F (2011) Activated T lymphocytes migrate toward the cathode of DC electric fields in microfluidic devices. Lab Chip 11(7):1298–1304
27. Gabriel B, Teissie J (1997) Direct observation in the millisecond time range of fluorescent molecule asymmetrical interaction with the electropermeabilized cell membrane. Biophys J 73 (5):2630–2637
28. Rols MP (2006) Electropermeabilization, a physical method for the delivery of therapeutic molecules into cells. Biochim Biophys Acta 1758(3):423–428
29. Chen C, Smye SW, Robinson MP, Evans JA (2006) Membrane electroporation theories: a review. Med Biol Eng Comput 44(1-2):5–14
30. Neumann E, Schaefer-Ridder M, Wang Y, Hofschneider PH (1982) Gene transfer into mouse lyoma cells by electroporation in high electric fields. EMBO J 1(7):841–845
31. Bernhardt J, Pauly H (1973) On the generation of potential differences across the membranes of ellipsoidal cells in an alternating electrical field. Biophysik 10(3):89–98
32. Marszalek P, Liu DS, Tsong TY (1990) Schwan equation and transmembrane potential induced by alternating electric field. Biophys J 58(4):1053–1058
33. Tsong TY (1991) Electroporation of cell membranes. Biophys J 60(2):297–306
34. Ghosh PM, Keese CR, Giaever I (1994) Morphological response of mammalian cells to pulsed ac fields. Bioelectrochem Bioenerg 33(2):121–133
35. Solly K, Wang X, Xu X, Strulovici B, Zheng W (2004) Application of real-time cell electronic sensing (RT-CES) technology to cell-based assays. Assay Drug Dev Technol 2(4):363–372
36. Wegener J, Sieber M, Galla HJ (1996) Impedance analysis of epithelial and endothelial cell monolayers cultured on gold surfaces. J Biochem Biophys Methods 32(3):151–170
37. Ghosh PM, Keese CR, Giaever I (1993) Monitoring electropermeabilization in the plasma membrane of adherent mammalian cells. Biophys J 64(5):1602–1609
38. Xiao C, Lachance B, Sunahara G, Luong JH (2002) Assessment of cytotoxicity using electric cell-substrate impedance sensing: concentration and time response function approach. Anal Chem 74(22):5748–5753
39. Duchateau S, Broeders J, Croux D, Janssen D, Rigo JM, Wagner P, Thoelen R, De Ceuninck W (2013) Cell proliferation monitoring by multiplexed electrochemical impedance spectroscopy on microwell assays. Phys Status Solidi C 10(5):882–888
40. Ende D, Mangold K (1993) Impedance spectroscopy. Chem Unserer Zeit 27(3):134–140
41. Chang B-Y, Park S-M (2010) Electrochemical impedance spectroscopy. Annu Rev Anal Chem 3:207–229
42. Lukic S, Wegener J (2015) Impedimetric monitoring of cell-based assays. Encyclopedia of life sciences. Wiley, Chichester
43. Khalil SF, Mohktar MS, Ibrahim F (2014) The theory and fundamentals of bioimpedance analysis in clinical status monitoring and diagnosis of diseases. Sensors 14(6):10895–10928
44. Daniels BP, Cruz-Orengo L, Pasieka TJ, Couraud PO, Romero IA, Weksler B, Cooper JA, Doering TL, Klein RS (2013) Immortalized human cerebral microvascular endothelial cells maintain the properties of primary cells in an in vitro model of immune migration across the blood brain barrier. J Neurosci Methods 212(1):173–179
45. Nakagawa S, Deli MA, Kawaguchi H, Shimizudani T, Shimono T, Kittel A, Tanaka K, Niwa M (2009) A new blood-brain barrier model using primary rat brain endothelial cells, pericytes and astrocytes. Neurochem Int 54(3–4):253–263

46. Paolinelli R, Corada M, Ferrarini L, Devraj K, Artus C, Czupalla CJ, Rudini N, Maddaluno L, Papa E, Engelhardt B, Couraud PO, Liebner S, Dejana E (2013) Wnt activation of immortalized brain endothelial cells as a tool for generating a standardized model of the blood brain barrier in vitro. PLoS One 8(8):e70233

47. von Wedel-Parlow M, Schrot S, Lemmen J, Treeratanapiboon L, Wegener J, Galla HJ (2011) Neutrophils cross the BBB primarily on transcellular pathways: an in vitro study. Brain Res 1367:62–76

48. Karczewski J, Troost FJ, Konings I, Dekker J, Kleerebezem M, Brummer RJ, Wells JM (2010) Regulation of human epithelial tight junction proteins by Lactobacillus plantarum in vivo and protective effects on the epithelial barrier. Am J Physiol Gastrointest Liver Physiol 298(6): G851–G859

49. Rehder D, Iden S, Nasdala I, Wegener J, Brickwedde MK, Vestweber D, Ebnet K (2006) Junctional adhesion molecule-a participates in the formation of apico-basal polarity through different domains. Exp Cell Res 312(17):3389–3403

50. Wegener J, Abrams D, Willenbrink W, Galla HJ, Janshoff A (2004) Automated multi-well device to measure transepithelial electrical resistances under physiological conditions. Biotechniques 37(4):590, 592–594, 596–597

51. Hu G-W, Li Q, Niu X, Hu B, Liu J, Zhou S-M, Guo S-C, Lang H-l, Zhang C-Q, Wang Y (2015) Exosomes secreted by human-induced pluripotent stem cell-derived mesenchymal stem cells attenuate limb ischemia by promoting angiogenesis in mice. Stem Cell Res Ther 6(1):1

52. Martinez-Serra J, Gutierrez A, Muñoz-Capó S, Navarro-Palou M, Ros T, Amat JC, Lopez B, Marcus TF, Fueyo L, Suquia AG (2014) xCELLigence system for real-time label-free monitoring of growth and viability of cell lines from hematological malignancies. Onco Targets Ther 7:985

53. Volakis LI, Li R, Ackerman WE, Mihai C, Bechel M, Summerfield TL, Ahn CS, Powell HM, Zielinski R, Rosol TJ, Ghadiali SN, Kniss DA (2014) Loss of myoferlin redirects breast cancer cell motility towards collective migration. PLoS One 9(2):e86110

54. Stolwijk JA, Matrougui K, Renken CW, Trebak M (2015) Impedance analysis of GPCR-mediated changes in endothelial barrier function: overview and fundamental considerations for stable and reproducible measurements. Pflugers Arch 467(10):2193–2218

55. Bokel C, Brown NH (2002) Integrins in development: moving on, responding to, and sticking to the extracellular matrix. Dev Cell 3(3):311–321

56. Harburger DS, Calderwood DA (2009) Integrin signalling at a glance. J Cell Sci 122 (Pt 2):159–163

57. Michaelis S, Robelek R, Wegener J (2011) Studying cell–surface interactions in vitro: a survey of experimental approaches and techniques. Tissue engineering III: cell-surface interactions for tissue culture. Springer, Berlin, pp 33–66

58. Giacomello E, Neumayer J, Colombatti A, Perris R (1999) Centrifugal assay for fluorescence-based cell adhesion adapted to the analysis of ex vivo cells and capable of determining relative binding strengths. Biotechniques 26(4):758–762, 764–756

59. Reyes CD, Garcia AJ (2003) A centrifugation cell adhesion assay for high-throughput screening of biomaterial surfaces. J Biomed Mater Res A 67(1):328–333

60. Khalili AA, Ahmad MR (2015) A review of cell adhesion studies for biomedical and biological applications. Int J Mol Sci 16(8):18149–18184

61. Wegener J, Keese CR, Giaever I (2002) Recovery of adherent cells after in situ electroporation monitored electrically. Biotechniques 33(2):348, 350, 352

62. Mitra P, Keese CR, Giaever I (1991) Electric measurements can be used to monitor the attachment and spreading of cells in tissue culture. Biotechniques 11(4):504–510

63. Frisch T, Thoumine O (2002) Predicting the kinetics of cell spreading. J Biomech 35 (8):1137–1141

64. Dubin-Thaler BJ, Giannone G, Dobereiner HG, Sheetz MP (2004) Nanometer analysis of cell spreading on matrix-coated surfaces reveals two distinct cell states and STEPs. Biophys J 86 (3):1794–1806
65. Cuvelier D, Thery M, Chu YS, Dufour S, Thiery JP, Bornens M, Nassoy P, Mahadevan L (2007) The universal dynamics of cell spreading. Curr Biol 17(8):694–699
66. Sapper A, Reiss B, Janshoff A, Wegener J (2006) Adsorption and fluctuations of giant liposomes studied by electrochemical impedance measurements. Langmuir 22(2):676–680
67. Castaneda FE, Walia B, Vijay-Kumar M, Patel NR, Roser S, Kolachala VL, Rojas M, Wang L, Oprea G, Garg P, Gewirtz AT, Roman J, Merlin D, Sitaraman SV (2005) Targeted deletion of metalloproteinase 9 attenuates experimental colitis in mice: central role of epithelial-derived MMP. Gastroenterology 129(6):1991–2008
68. Driss A, Charrier L, Yan Y, Nduati V, Sitaraman S, Merlin D (2006) Dystroglycan receptor is involved in integrin activation in intestinal epithelia. Am J Physiol Gastrointest Liver Physiol 290(6):G1228–G1242
69. Nguyen HT, Dalmasso G, Yan Y, Laroui H, Dahan S, Mayer L, Sitaraman SV, Merlin D (2010) MicroRNA-7 modulates CD98 expression during intestinal epithelial cell differentiation. J Biol Chem 285(2):1479–1489
70. Navdaev A, Heitmann V, Desantana Evangelista K, Morgelin M, Wegener J, Eble JA (2008) The C-terminus of the gamma 2 chain but not of the beta 3 chain of laminin-332 is indirectly but indispensably necessary for integrin-mediated cell reactions. Exp Cell Res 314(3):489–497
71. Luong JH, Habibi-Rezaei M, Meghrous J, Xiao C, Male KB, Kamen A (2001) Monitoring motility, spreading, and mortality of adherent insect cells using an impedance sensor. Anal Chem 73(8):1844–1848
72. Luong JH, Xiao C, Lachance B, Leabu ŠM, Li X, Uniyal S, Chan BM (2004) Extended applications of electric cell-substrate impedance sensing for assessment of the structure–function of α2β1 integrin. Anal Chim Acta 501(1):61–69
73. Heijink IH, Brandenburg SM, Noordhoek JA, Postma DS, Slebos DJ, van Oosterhout AJ (2010) Characterisation of cell adhesion in airway epithelial cell types using electric cell-substrate impedance sensing. Eur Respir J 35(4):894–903
74. Ehret R, Baumann W, Brischwein M, Schwinde A, Wolf B (1998) On-line control of cellular adhesion with impedance measurements using interdigitated electrode structures. Med Biol Eng Comput 36(3):365–370
75. Atienza JM, Zhu J, Wang X, Xu X, Abassi Y (2005) Dynamic monitoring of cell adhesion and spreading on microelectronic sensor arrays. J Biomol Screen 10(8):795–805
76. Sharma KV, Koenigsberger C, Brimijoin S, Bigbee JW (2001) Direct evidence for an adhesive function in the noncholinergic role of acetylcholinesterase in neurite outgrowth. J Neurosci Res 63(2):165–175
77. van Gils JM, Stutterheim J, van Duijn TJ, Zwaginga JJ, Porcelijn L, de Haas M, Hordijk PL (2009) HPA-1a alloantibodies reduce endothelial cell spreading and monolayer integrity. Mol Immunol 46(3):406–415
78. Asphahani F, Thein M, Veiseh O, Edmondson D, Kosai R, Veiseh M, Xu J, Zhang M (2008) Influence of cell adhesion and spreading on impedance characteristics of cell-based sensors. Biosens Bioelectron 23(8):1307–1313
79. ten Klooster JP, Jaffer ZM, Chernoff J, Hordijk PL (2006) Targeting and activation of Rac1 are mediated by the exchange factor beta-Pix. J Cell Biol 172(5):759–769
80. Charboneau AL, Singh V, Yu T, Newsham IF (2002) Suppression of growth and increased cellular attachment after expression of DAL-1 in MCF-7 breast cancer cells. Int J Cancer 100 (2):181–188
81. Davies S, Jiang WG (2010) ALCAM, activated leukocyte cell adhesion molecule, influences the aggressive nature of breast cancer cells, a potential connection to bone metastasis. Anticancer Res 30(4):1163–1168

82. Jiang WG, Ye L, Sanders AJ, Ruge F, Kynaston HG, Ablin RJ, Mason MD (2013) Prostate transglutaminase (TGase-4, TGaseP) enhances the adhesion of prostate cancer cells to extracellular matrix, the potential role of TGase-core domain. J Transl Med 11:269
83. Abassi YA, Xi B, Zhang W, Ye P, Kirstein SL, Gaylord MR, Feinstein SC, Wang X, Xu X (2009) Kinetic cell-based morphological screening: prediction of mechanism of compound action and off-target effects. Chem Biol 16(7):712–723
84. Sawhney RS, Zhou GH, Humphrey LE, Ghosh P, Kreisberg JI, Brattain MG (2002) Differences in sensitivity of biological functions mediated by epidermal growth factor receptor activation with respect to endogenous and exogenous ligands. J Biol Chem 277(1):75–86
85. Okochi M, Nomura S, Kaga C, Honda H (2008) Peptide array-based screening of human mesenchymal stem cell-adhesive peptides derived from fibronectin type III domain. Biochem Biophys Res Commun 371(1):85–89
86. Yun Y, Dong Z, Tan Z, Schulz MJ (2010) Development of an electrode cell impedance method to measure osteoblast cell activity in magnesium-conditioned media. Anal Bioanal Chem 396(8):3009–3015. https://doi.org/10.1007/s00216-010-3521-2
87. Kundranda MN, Ray S, Saria M, Friedman D, Matrisian LM, Lukyanov P, Ochieng J (2004) Annexins expressed on the cell surface serve as receptors for adhesion to immobilized fetuin-A. Biochim Biophys Acta 1693(2):111–123
88. Aikio M, Alahuhta I, Nurmenniemi S, Suojanen J, Palovuori R, Teppo S, Sorsa T, Lopez-Otin C, Pihlajaniemi T, Salo T, Heljasvaara R, Nyberg P (2012) Arresten, a collagen-derived angiogenesis inhibitor, suppresses invasion of squamous cell carcinoma. PLoS One 7(12): e51044
89. Ablin RJ, Kynaston HG, Mason MD, Jiang WG (2011) Prostate transglutaminase (TGase-4) antagonizes the anti-tumour action of MDA-7/IL-24 in prostate cancer. J Transl Med 9:49
90. Du P, Ye L, Ruge F, Yang Y, Jiang WG (2011) Metastasis suppressor-1, MTSS1, acts as a putative tumour suppressor in human bladder cancer. Anticancer Res 31(10):3205–3212
91. Lee CC, Putnam AJ, Miranti CK, Gustafson M, Wang LM, Vande Woude GF, Gao CF (2004) Overexpression of sprouty 2 inhibits HGF/SF-mediated cell growth, invasion, migration, and cytokinesis. Oncogene 23(30):5193–5202
92. Negash S, Wang HS, Gao C, Ledee D, Zelenka P (2002) Cdk5 regulates cell-matrix and cell-cell adhesion in lens epithelial cells. J Cell Sci 115(Pt 10):2109–2117
93. Nethe M, Anthony EC, Fernandez-Borja M, Dee R, Geerts D, Hensbergen PJ, Deelder AM, Schmidt G, Hordijk PL (2010) Focal-adhesion targeting links caveolin-1 to a Rac1-degradation pathway. J Cell Sci 123(Pt 11):1948–1958
94. Rotundo RF, Curtis TM, Shah MD, Gao B, Mastrangelo A, LaFlamme SE, Saba TM (2002) TNF-alpha disruption of lung endothelial integrity: reduced integrin mediated adhesion to fibronectin. Am J Physiol Lung Cell Mol Physiol 282(2):L316–L329
95. Schmidt MH, Chen B, Randazzo LM, Bogler O (2003) SETA/CIN85/Ruk and its binding partner AIP1 associate with diverse cytoskeletal elements, including FAKs, and modulate cell adhesion. J Cell Sci 116(Pt 14):2845–2855
96. Witzel F, Fritsche-Guenther R, Lehmann N, Sieber A, Bluthgen N (2015) Analysis of impedance-based cellular growth assays. Bioinformatics 31(16):2705–2712
97. Pick H, Terrettaz S, Baud O, Laribi O, Brisken C, Vogel H (2013) Monitoring proliferative activities of hormone-like odorants in human breast cancer cells by gene transcription profiling and electrical impedance spectroscopy. Biosens Bioelectron 50:431–436
98. Chakraborty PK, Lee WK, Molitor M, Wolff NA, Thevenod F (2010) Cadmium induces Wnt signaling to upregulate proliferation and survival genes in sub-confluent kidney proximal tubule cells. Mol Cancer 9:102
99. Horimoto N, Kitamura S, Tsuji K, Makino H (2014) Mizoribine inhibits the proliferation of renal stem/progenitor cells by G1/S arrest during renal regeneration. Acta Med Okayama 68 (1):7–15

100. Wang L, Wang L, Yin H, Xing W, Yu Z, Guo M, Cheng J (2010) Real-time, label-free monitoring of the cell cycle with a cellular impedance sensing chip. Biosens Bioelectron 25 (5):990–995

101. Masanetz S, Kaufmann C, Letzel T, Pfaff M (2012) Effects of pine pollen extracts on the proliferation and mRNA expression of porcine ileal cell cultures. J Appl Bot Food Qual 83 (1):14–18

102. Broeders J, Duchateau S, Van Grinsven B, Vanaken W, Peeters M, Cleij T, Thoelen R, Wagner P, De Ceuninck W (2011) Miniaturised eight-channel impedance spectroscopy unit as sensor platform for biosensor applications. Phys Status Solidi A 208(6):1357–1363

103. Limame R, Wouters A, Pauwels B, Fransen E, Peeters M, Lardon F, De Wever O, Pauwels P (2012) Comparative analysis of dynamic cell viability, migration and invasion assessments by novel real-time technology and classic endpoint assays. PLoS One 7(10):e46536

104. Chen SW, Yang JM, Yang JH, Yang SJ, Wang JS (2012) A computational modeling and analysis in cell biological dynamics using electric cell-substrate impedance sensing (ECIS). Biosens Bioelectron 33(1):196–203

105. Voltan R, Zauli G, Rizzo P, Fucili A, Pannella M, Marci R, Tisato V, Ferrari R, Secchiero P (2014) In vitro endothelial cell proliferation assay reveals distinct levels of proangiogenic cytokines characterizing sera of healthy subjects and of patients with heart failure. Mediators Inflamm 2014:257081

106. Dowling CM, Ors CH, Kiely PA (2014) Using real-time impedance-based assays to monitor the effects of fibroblast-derived media on the adhesion, proliferation, migration and invasion of colon cancer cells. Biosci Rep 34(4):e00126

107. Horimoto N, Kitamura S, Tsuji K, Makino H (2013) Mizoribine inhibits the proliferation of renal stem/progenitor cells by G1/S arrest during renal regeneration. Acta Med Okayama 68 (1):7–15

108. Kim JH, Jekarl DW, Kim M, Oh EJ, Kim Y, Park IY, Shin JC (2014) Effects of ECM protein mimetics on adhesion and proliferation of chorion derived mesenchymal stem cells. Int J Med Sci 11(3):298–308

109. Cai C, Rodepeter FR, Rossmann A, Teymoortash A, Lee JS, Quint K, Pietro DIF, Ocker M, Werner JA, Mandic R (2012) SIVmac(2)(3)(9)-Nef down-regulates cell surface expression of CXCR4 in tumor cells and inhibits proliferation, migration and angiogenesis. Anticancer Res 32(7):2759–2768

110. Danussi C, Petrucco A, Wassermann B, Pivetta E, Modica TM, Del Bel BL, Colombatti A, Spessotto P (2011) EMILIN1-alpha4/alpha9 integrin interaction inhibits dermal fibroblast and keratinocyte proliferation. J Cell Biol 195(1):131–145

111. Yuan SY, Rigor RR (2010) Regulation of endothelial barrier function. Integrated systems physiology: from molecule to function to disease. Morgan & Claypool Life Sciences, San Rafael

112. Wang Y, Alexander JS (2011) Analysis of endothelial barrier function in vitro. Methods Mol Biol 763:253–264

113. Monaghan-Benson E, Wittchen ES (2011) In vitro analyses of endothelial cell permeability. Methods Mol Biol 763:281–290

114. Benson K, Cramer S, Galla HJ (2013) Impedance-based cell monitoring: barrier properties and beyond. Fluids Barriers CNS 10(1):5

115. Hajek K, Wegener J (2017) Independent impedimetric analysis of two cell populations co-cultured on opposite sides of a porous support. Exp Cell Res 351(1):121–126

116. Cecchelli R, Aday S, Sevin E, Almeida C, Culot M, Dehouck L, Coisne C, Engelhardt B, Dehouck MP, Ferreira L (2014) A stable and reproducible human blood-brain barrier model derived from hematopoietic stem cells. PLoS One 9(6):e99733

117. Dewi BE, Takasaki T, Kurane I (2004) In vitro assessment of human endothelial cell permeability: effects of inflammatory cytokines and dengue virus infection. J Virol Methods 121(2):171–180

118. Schnoor M, Lai FP, Zarbock A, Klaver R, Polaschegg C, Schulte D, Weich HA, Oelkers JM, Rottner K, Vestweber D (2011) Cortactin deficiency is associated with reduced neutrophil recruitment but increased vascular permeability in vivo. J Exp Med 208(8):1721–1735

119. Hoheisel D, Nitz T, Franke H, Wegener J, Hakvoort A, Tilling T, Galla HJ (1998) Hydrocortisone reinforces the blood-brain barrier properties in a serum free cell culture system. Biochem Biophys Res Commun 244(1):312–316

120. von Wedel-Parlow M, Wolte P, Galla HJ (2009) Regulation of major efflux transporters under inflammatory conditions at the blood-brain barrier in vitro. J Neurochem 111(1):111–118

121. Malina KC-K, Cooper I, Teichberg VI (2009) Closing the gap between the in-vivo and in-vitro blood–brain barrier tightness. Brain Res 1284:12–21

122. Schneider D, Tarantola M, Janshoff A (2011) Dynamics of TGF-beta induced epithelial-to-mesenchymal transition monitored by electric cell-substrate impedance sensing. Biochim Biophys Acta 1813(12):2099–2107

123. Spiering D, Schmolke M, Ohnesorge N, Schmidt M, Goebeler M, Wegener J, Wixler V, Ludwig S (2009) MEK5/ERK5 signaling modulates endothelial cell migration and focal contact turnover. J Biol Chem 284(37):24972–24980

124. Ko KS-C, Lo C-M, Ferrier J, Hannam P, Tamura M, McBride BC, Ellen RP (1998) Cell–substrate impedance analysis of epithelial cell shape and micromotion upon challenge with bacterial proteins that perturb extracellular matrix and cytoskeleton. J Microbiol Methods 34 (2):125–132

125. Lovelady DC, Richmond TC, Maggi AN, Lo CM, Rabson DA (2007) Distinguishing cancerous from noncancerous cells through analysis of electrical noise. Phys Rev E Stat Nonlinear Soft Matter Phys 76(4 Pt 1):041908

126. Tarantola M, Marel AK, Sunnick E, Adam H, Wegener J, Janshoff A (2010) Dynamics of human cancer cell lines monitored by electrical and acoustic fluctuation analysis. Integr Biol 2 (2–3):139–150

127. Sawhney RS, Sharma B, Humphrey LE, Brattain MG (2003) Integrin α2 and extracellular signal-regulated kinase are functionally linked in highly malignant autocrine transforming growth factor-α-driven colon cancer cells. J Biol Chem 278(22):19861–19869

128. Giaever I, Keese C (1989) Fractal motion of mammalian cells. Physica D 38(1–3):128–133

129. Lovelady DC, Friedman J, Patel S, Rabson DA, Lo CM (2009) Detecting effects of low levels of cytochalasin B in 3T3 fibroblast cultures by analysis of electrical noise obtained from cellular micromotion. Biosens Bioelectron 24(7):2250–2254

130. Opp D, Wafula B, Lim J, Huang E, Lo J-C, Lo C-M (2009) Use of electric cell–substrate impedance sensing to assess in vitro cytotoxicity. Biosens Bioelectron 24(8):2625–2629

131. Tarantola M, Schneider D, Sunnick E, Adam H, Pierrat S, Rosman C, Breus V, Sonnichsen C, Basché T, Wegener J (2008) Cytotoxicity of metal and semiconductor nanoparticles indicated by cellular micromotility. ACS Nano 3(1):213–222

132. Tarantola M, Sunnick E, Schneider D, Marel AK, Kunze A, Janshoff A (2011) Dynamic changes of acoustic load and complex impedance as reporters for the cytotoxicity of small molecule inhibitors. Chem Res Toxicol 24(9):1494–1506

133. Lang O, Kohidai L, Wegener J (2017) Label-free profiling of cell dynamics: a sequence of impedance-based assays to estimate tumor cell invasiveness in vitro. Exp Cell Res 359 (1):243–250

134. Lo CM, Keese CR, Giaever I (1994) pH changes in pulsed CO2 incubators cause periodic changes in cell morphology. Exp Cell Res 213(2):391–397

135. Das D, Shiladitya K, Biswas K, Dutta PK, Parekh A, Mandal M, Das S (2015) Wavelet-based multiscale analysis of bioimpedance data measured by electric cell-substrate impedance sensing for classification of cancerous and normal cells. Phys Rev E Stat Nonlinear Soft Matter Phys 92(6):062702

136. Jonsson MK, Wang QD, Becker B (2011) Impedance-based detection of beating rhythm and proarrhythmic effects of compounds on stem cell-derived cardiomyocytes. Assay Drug Dev Technol 9(6):589–599

137. Guo L, Abrams RM, Babiarz JE, Cohen JD, Kameoka S, Sanders MJ, Chiao E, Kolaja KL (2011) Estimating the risk of drug-induced proarrhythmia using human induced pluripotent stem cell-derived cardiomyocytes. Toxicol Sci 123(1):281–289

138. Lamore SD, Kamendi HW, Scott CW, Dragan YP, Peters MF (2013) Cellular impedance assays for predictive preclinical drug screening of kinase inhibitor cardiovascular toxicity. Toxicol Sci 135(2):402–413

139. Scott CW, Zhang X, Abi-Gerges N, Lamore SD, Abassi YA, Peters MF (2014) An impedance-based cellular assay using human iPSC-derived cardiomyocytes to quantify modulators of cardiac contractility. Toxicol Sci 142(2):331–338

140. Kramer N, Walzl A, Unger C, Rosner M, Krupitza G, Hengstschlager M, Dolznig H (2013) In vitro cell migration and invasion assays. Mutat Res 752(1):10–24

141. Renken C, Keese C, Giaever I (2010) Automated assays for quantifying cell migration. Biotechniques 49(5):844

142. Desai SD, Reed RE, Burks J, Wood LM, Pullikuth AK, Haas AL, Liu LF, Breslin JW, Meiners S, Sankar S (2012) ISG15 disrupts cytoskeletal architecture and promotes motility in human breast cancer cells. Exp Biol Med 237(1):38–49

143. Jiang WG, Martin TA, Lewis-Russell JM, Douglas-Jones A, Ye L, Mansel RE (2008) Eplin-alpha expression in human breast cancer, the impact on cellular migration and clinical outcome. Mol Cancer 7:71

144. Kim SH, Nagalingam A, Saxena NK, Singh SV, Sharma D (2011) Benzyl isothiocyanate inhibits oncogenic actions of leptin in human breast cancer cells by suppressing activation of signal transducer and activator of transcription 3. Carcinogenesis 32(3):359–367

145. Knight BB, Oprea-Ilies GM, Nagalingam A, Yang L, Cohen C, Saxena NK, Sharma D (2011) Surviving upregulation, dependent on leptin-EGFR-Notch1 axis, is essential for leptin-induced migration of breast carcinoma cells. Endocr Relat Cancer 18(4):413–428

146. Nagalingam A, Arbiser JL, Bonner MY, Saxena NK, Sharma D (2012) Honokiol activates AMP-activated protein kinase in breast cancer cells via an LKB1-dependent pathway and inhibits breast carcinogenesis. Breast Cancer Res 14(1):R35

147. Sun PH, Ye L, Mason MD, Jiang WG (2012) Protein tyrosine phosphatase micro (PTP micro or PTPRM), a negative regulator of proliferation and invasion of breast cancer cells, is associated with disease prognosis. PLoS One 7(11):e50183

148. Yan T, Skaftnesmo KO, Leiss L, Sleire L, Wang J, Li X, Enger PO (2011) Neuronal markers are expressed in human gliomas and NSE knockdown sensitizes glioblastoma cells to radio-therapy and temozolomide. BMC Cancer 11:524

149. Zhang N, Sanders AJ, Ye L, Kynaston HG, Jiang WG (2010) Expression of vascular endothelial growth inhibitor (VEGI) in human urothelial cancer of the bladder and its effects on the adhesion and migration of bladder cancer cells in vitro. Anticancer Res 30(1):87–95

150. Allen-Gipson DS, Zimmerman MC, Zhang H, Castellanos G, O'Malley JK, Alvarez-Ramirez-H, Kharbanda K, Sisson JH, Wyatt TA (2013) Smoke extract impairs adenosine wound healing: implications of smoke-generated reactive oxygen species. Am J Respir Cell Mol Biol 48(5):665–673

151. Bosanquet DC, Harding KG, Ruge F, Sanders AJ, Jiang WG (2012) Expression of IL-24 and IL-24 receptors in human wound tissues and the biological implications of IL-24 on keratinocytes. Wound Repair Regen 20(6):896–903

152. Bosanquet DC, Ye L, Harding KG, Jiang WG (2012) Role of HuR in keratinocyte migration and wound healing. Mol Med Rep 5(2):529–534

153. Charrier L, Yan Y, Driss A, Laboisse CL, Sitaraman SV, Merlin D (2005) ADAM-15 inhibits wound healing in human intestinal epithelial cell monolayers. Am J Physiol Gastrointest Liver Physiol 288(2):G346–G353

154. Heijink IH, Brandenburg SM, Postma DS, van Oosterhout AJ (2012) Cigarette smoke impairs airway epithelial barrier function and cell-cell contact recovery. Eur Respir J 39(2):419–428

155. Hsu CC, Tsai WC, Chen CP, Lu YM, Wang JS (2010) Effects of negative pressures on epithelial tight junctions and migration in wound healing. Am J Physiol Cell Physiol 299(2): C528–C534
156. Itokazu Y, Pagano RE, Schroeder AS, O'Grady SM, Limper AH, Marks DL (2014) Reduced GM1 ganglioside in CFTR-deficient human airway cells results in decreased beta1-integrin signaling and delayed wound repair. Am J Physiol Cell Physiol 306(9):C819–C830
157. Jiang WG, Sanders AJ, Ruge F, Harding KG (2012) Influence of interleukin-8 (IL-8) and IL-8 receptors on the migration of human keratinocytes, the role of PLC-γ and potential clinical implications. Exp Ther Med 3(2):231–236
158. Jiang WG, Ye L, Patel G, Harding KG (2010) Expression of WAVEs, the WASP (Wiskott-Aldrich syndrome protein) family of verprolin homologous proteins in human wound tissues and the biological influence on human keratinocytes. Wound Repair Regen 18(6):594–604
159. Oudhoff MJ, Kroeże KL, Nazmi K, van den Keijbus PA, van't Hof W, Fernandez-Borja M, Hordijk PL, Gibbs S, Bolscher JG, Veerman EC (2009) Structure-activity analysis of histatin, a potent wound healing peptide from human saliva: cyclization of histatin potentiates molar activity 1,000-fold. FASEB J 23(11):3928–3935
160. Sanders AJ, Jiang DG, Jiang WG, Harding KG, Patel GK (2011) Activated leukocyte cell adhesion molecule impacts on clinical wound healing and inhibits HaCaT migration. Int Wound J 8(5):500–507
161. Mehta RR, Yamada T, Taylor BN, Christov K, King ML, Majumdar D, Lekmine F, Tiruppathi C, Shilkaitis A, Bratescu L, Green A, Beattie CW, Das Gupta TK (2011) A cell penetrating peptide derived from azurin inhibits angiogenesis and tumor growth by inhibiting phosphorylation of VEGFR-2, FAK and Akt. Angiogenesis 14(3):355–369
162. Ungvari Z, Tucsek Z, Sosnowska D, Toth P, Gautam T, Podlutsky A, Csiszar A, Losonczy G, Valcarcel-Ares MN, Sonntag WE, Csiszar A (2013) Aging-induced dysregulation of dicer1-dependent microRNA expression impairs angiogenic capacity of rat cerebromicrovascular endothelial cells. J Gerontol A Biol Sci Med Sci 68(8):877–891
163. Chanakira A, Kir D, Barke RA, Santilli SM, Ramakrishnan S, Roy S (2015) Hypoxia differentially regulates arterial and venous smooth muscle cell migration. PLoS One 10(9): e0138587
164. Tsapara A, Luthert P, Greenwood J, Hill CS, Matter K, Balda MS (2010) The RhoA activator GEF-H1/Lfc is a transforming growth factor-beta target gene and effector that regulates alpha-smooth muscle actin expression and cell migration. Mol Biol Cell 21(6):860–870
165. Hughes CS, Postovit LM, Lajoie GA (2010) Matrigel: a complex protein mixture required for optimal growth of cell culture. Proteomics 10(9):1886–1890
166. Chung H, Suh EK, Han IO, Oh ES (2011) Keratinocyte-derived laminin-332 promotes adhesion and migration in melanocytes and melanoma. J Biol Chem 286(15):13438–13447
167. Coutts AS, Pires IM, Weston L, Buffa FM, Milani M, Li JL, Harris AL, Hammond EM, La Thangue NB (2011) Hypoxia-driven cell motility reflects the interplay between JMY and HIF-1alpha. Oncogene 30(48):4835–4842
168. Daouti S, Li WH, Qian H, Huang KS, Holmgren J, Levin W, Reik L, McGady DL, Gillespie P, Perrotta A, Bian H, Reidhaar-Olson JF, Bliss SA, Olivier AR, Sergi JA, Fry D, Danho W, Ritland S, Fotouhi N, Heimbrook D, Niu H (2008) A selective phosphatase of regenerating liver phosphatase inhibitor suppresses tumor cell anchorage-independent growth by a novel mechanism involving p130Cas cleavage. Cancer Res 68(4):1162–1169
169. Arabzadeh A, Dupaul-Chicoine J, Breton V, Haftchenary S, Yumeen S, Turbide C, Saleh M, McGregor K, Greenwood CM, Akavia UD, Blumberg RS, Gunning PT, Beauchemin N (2016) Carcinoembryonic Antigen Cell Adhesion Molecule 1 long isoform modulates malignancy of poorly differentiated colon cancer cells. Gut 65(5):821–829
170. Grassilli S, Brugnoli F, Lattanzio R, Rossi C, Perracchio L, Mottolese M, Marchisio M, Palomba M, Nika E, Natali PG, Piantelli M, Capitani S, Bertagnolo V (2014) High nuclear level of Vav1 is a positive prognostic factor in early invasive breast tumors: a role in modulating genes related to the efficiency of metastatic process. Oncotarget 5(12):4320–4336

171. Burlacu A, Grigorescu G, Rosca A-M, Preda MB, Simionescu M (2012) Factors secreted by mesenchymal stem cells and endothelial progenitor cells have complementary effects on angiogenesis in vitro. Stem Cells Dev 22(4):643–653

172. Rutten MJ, Laraway B, Gregory CR, Xie H, Renken C, Keese C, Gregory KW (2015) Rapid assay of stem cell functionality and potency using electric cell-substrate impedance sensing. Stem Cell Res Ther 6:192

173. Neubig RR, Spedding M, Kenakin T, Christopoulos A, International Union of Pharmacology Committee on Receptor Nomenclature and Drug Classification (2003) International Union of Pharmacology Committee on Receptor Nomenclature and Drug Classification. XXXVIII. Update on terms and symbols in quantitative pharmacology. Pharmacol Rev 55(4):597–606

174. Stolwijk JA, Zhang X, Gueguinou M, Zhang W, Matrougui K, Renken C, Trebak M (2016) Calcium signaling is dispensable for receptor-regulation of endothelial barrier function. J Biol Chem 291(44):22894–22912

175. Urban C, Hamacher A, Partke HJ, Roden M, Schinner S, Christiansen E, Due-Hansen ME, Ulven T, Gohlke H, Kassack MU (2013) In vitro and mouse in vivo characterization of the potent free fatty acid 1 receptor agonist TUG-469. Naunyn Schmiedebergs Arch Pharmacol 386(12):1021–1030

176. Guo D, Mulder-Krieger T, IJzerman AP, Heitman LH (2012) Functional efficacy of adenosine A(2)A receptor agonists is positively correlated to their receptor residence time. Br J Pharmacol 166(6):1846–1859

177. Lu Q, Harrington EO, Newton J, Casserly B, Radin G, Warburton R, Zhou Y, Blackburn MR, Rounds S (2010) Adenosine protected against pulmonary edema through transporter- and receptor A2-mediated endothelial barrier enhancement. Am J Physiol Lung Cell Mol Physiol 298(6):L755–L767

178. Anthony DF, Sin YY, Vadrevu S, Advant N, Day JP, Byrne AM, Lynch MJ, Milligan G, Houslay MD, Baillie GS (2011) β-Arrestin 1 inhibits the GTPase-activating protein function of ARHGAP21, promoting activation of RhoA following angiotensin II type 1A receptor stimulation. Mol Cell Biol 31(5):1066–1075

179. Denelavas A, Weibel F, Prummer M, Imbach A, Clerc RG, Apfel CM, Hertel C (2011) Real-time cellular impedance measurements detect Ca(2+) channel-dependent oscillations of morphology in human H295R adrenoma cells. Biochim Biophys Acta 1813(5):754–762

180. Peitzman ER, Zaidman NA, Maniak PJ, O'Grady SM (2015) Agonist binding to beta-adrenergic receptors on human airway epithelial cells inhibits migration and wound repair. Am J Physiol Cell Physiol 309(12):C847–C855

181. Wegener J, Zink S, Rosen P, Galla H (1999) Use of electrochemical impedance measurements to monitor beta-adrenergic stimulation of bovine aortic endothelial cells. Pflugers Arch 437 (6):925–934

182. Watts AO, Scholten DJ, Heitman LH, Vischer HF, Leurs R (2012) Label-free impedance responses of endogenous and synthetic chemokine receptor CXCR3 agonists correlate with Gi-protein pathway activation. Biochem Biophys Res Commun 419(2):412–418

183. Adderley SP, Zhang XE, Breslin JW (2015) Involvement of the H1 histamine receptor, p38 MAP kinase, myosin light chains kinase, and Rho/ROCK in histamine-induced endothelial barrier dysfunction. Microcirculation 22(4):237–248

184. Lieb S, Littmann T, Plank N, Felixberger J, Tanaka M, Schafer T, Krief S, Elz S, Friedland K, Bernhardt G, Wegener J, Ozawa T, Buschauer A (2016) Label-free versus conventional cellular assays: functional investigations on the human histamine H1 receptor. Pharmacol Res 114:13–26

185. Stolwijk JA, Skiba M, Kade C, Bernhardt G, Buschauer A, Hubner H, Gmeiner P, Wegener J (2019) Increasing the throughput of label-free cell assays to study the activation of G-protein-coupled receptors by using a serial agonist exposure protocol. Integr Biol. https://doi.org/10. 1093/intbio/zyz010

186. Zhang X, Tan F, Brovkovych V, Zhang Y, Skidgel RA (2011) Cross-talk between carboxy-peptidase M and the kinin B1 receptor mediates a new mode of G protein-coupled receptor signaling. J Biol Chem 286(21):18547–18561

187. He D, Su Y, Usatyuk PV, Spannhake EW, Kogut P, Solway J, Natarajan V, Zhao Y (2009) Lysophosphatidic acid enhances pulmonary epithelial barrier integrity and protects endotoxin-induced epithelial barrier disruption and lung injury. J Biol Chem 284(36):24123–24132

188. Qiao J, Huang F, Naikawadi RP, Kim KS, Said T, Lum H (2006) Lysophosphatidylcholine impairs endothelial barrier function through the G protein-coupled receptor GPR4. Am J Physiol Lung Cell Mol Physiol 291(1):L91–L101

189. Kammermann M, Denelavas A, Imbach A, Grether U, Dehmlow H, Apfel CM, Hertel C (2011) Impedance measurement: a new method to detect ligand-biased receptor signaling. Biochem Biophys Res Commun 412(3):419–424

190. Meshki J, Douglas SD, Lai JP, Schwartz L, Kilpatrick LE, Tuluc F (2009) Neurokinin 1 receptor mediates membrane blebbing in HEK293 cells through a Rho/Rho-associated coiled-coil kinase-dependent mechanism. J Biol Chem 284(14):9280–9289

191. Srivastava SK, Ramaneti R, Roelse M, Tong HD, Vrouwe EX, Brinkman AG, de Smet LC, van Rijn CJ, Jongsma MA (2015) A generic microfluidic biosensor of G protein-coupled receptor activation–impedance measurements of reversible morphological changes of reverse transfected HEK293 cells on microelectrodes. RSC Adv 5(65):52563–52570

192. Reddy L, Wang HS, Keese CR, Giaever I, Smith TJ (1998) Assessment of rapid morpholog-ical changes associated with elevated cAMP levels in human orbital fibroblasts. Exp Cell Res 245(2):360–367

193. Smith TJ, Wang HS, Hogg MG, Henrikson RC, Keese CR, Giaever I (1994) Prostaglandin E2 elicits a morphological change in cultured orbital fibroblasts from patients with Graves ophthalmopathy. Proc Natl Acad Sci U S A 91(11):5094–5098

194. Wang H, Keese CR, Giaever I, Smith TJ (1995) Prostaglandin E2 alters human orbital fibroblast shape through a mechanism involving the generation of cyclic adenosine monophosphate. J Clin Endocrinol Metab 80(12):3553–3560

195. Argraves KM, Gazzolo PJ, Groh EM, Wilkerson BA, Matsuura BS, Twal WO, Hammad SM, Argraves WS (2008) High density lipoprotein-associated sphingosine 1-phosphate promotes endothelial barrier function. J Biol Chem 283(36):25074–25081

196. Garcia JG, Liu F, Verin AD, Birukova A, Dechert MA, Gerthoffer WT, Bamberg JR, English D (2001) Sphingosine 1-phosphate promotes endothelial cell barrier integrity by Edg-dependent cytoskeletal rearrangement. J Clin Invest 108(5):689–701

197. Schaphorst KL, Chiang E, Jacobs KN, Zaiman A, Natarajan V, Wigley F, Garcia JG (2003) Role of sphingosine-1 phosphate in the enhancement of endothelial barrier integrity by platelet-released products. Am J Physiol Lung Cell Mol Physiol 285(1):L258–L267

198. Wilkerson BA, Grass GD, Wing SB, Argraves WS, Argraves KM (2012) Sphingosine 1-phosphate (S1P) carrier-dependent regulation of endothelial barrier: high density lipoprotein (HDL)-S1P prolongs endothelial barrier enhancement as compared with albumin-S1P via effects on levels, trafficking, and signaling of S1P1. J Biol Chem 287(53):44645–44653

199. Becker PM, Verin AD, Booth MA, Liu F, Birukova A, Garcia JG (2001) Differential regulation of diverse physiological responses to VEGF in pulmonary endothelial cells. Am J Physiol Lung Cell Mol Physiol 281(6):L1500–L1511

200. Birukova AA, Cokic I, Moldobaeva N, Birukov KG (2009) Paxillin is involved in the differential regulation of endothelial barrier by HGF and VEGF. Am J Respir Cell Mol Biol 40(1):99–107

201. Ngok SP, Geyer R, Liu M, Kourtidis A, Agrawal S, Wu C, Seerapu HR, Lewis-Tuffin LJ, Moodie KL, Huveldt D, Marx R, Baraban JM, Storz P, Horowitz A, Anastasiadis PZ (2012) VEGF and Angiopoietin-1 exert opposing effects on cell junctions by regulating the Rho GEF Syx. J Cell Biol 199(7):1103–1115

202. Sahin O, Frohlich H, Lobke C, Korf U, Burmester S, Majety M, Mattern J, Schupp I, Chaouiya C, Thieffry D, Poustka A, Wiemann S, Beissbarth T, Arlt D (2009) Modeling ERBB receptor-regulated G1/S transition to find novel targets for de novo trastuzumab resistance. BMC Syst Biol 3:1

203. Takahashi N, Seko Y, Noiri E, Tobe K, Kadowaki T, Sabe H, Yazaki Y (1999) Vascular endothelial growth factor induces activation and subcellular translocation of focal adhesion kinase (p125FAK) in cultured rat cardiac myocytes. Circ Res 84(10):1194–1202

204. Xue M, Chow SO, Dervish S, Chan YK, Julovi SM, Jackson CJ (2011) Activated protein C enhances human keratinocyte barrier integrity via sequential activation of epidermal growth factor receptor and Tie2. J Biol Chem 286(8):6742–6750

205. Yang J, Duh EJ, Caldwell RB, Behzadian MA (2010) Antipermeability function of PEDF involves blockade of the MAP kinase/GSK/beta-catenin signaling pathway and uPAR expression. Invest Ophthalmol Vis Sci 51(6):3273–3280

206. Clark PR, Kim RK, Pober JS, Kluger MS (2015) Tumor necrosis factor disrupts claudin-5 endothelial tight junction barriers in two distinct NF-kappaB-dependent phases. PLoS One 10 (3):e0120075

207. Haines RJ, Beard Jr RS, Wu MH (2015) Protein tyrosine kinase 6 mediates TNFalpha-induced endothelial barrier dysfunction. Biochem Biophys Res Commun 456(1):190–196

208. Kakiashvili E, Dan Q, Vandermeer M, Zhang Y, Waheed F, Pham M, Szaszi K (2011) The epidermal growth factor receptor mediates tumor necrosis factor-alpha-induced activation of the ERK/GEF-H1/RhoA pathway in tubular epithelium. J Biol Chem 286(11):9268–9279

209. Pathak RR, Grover A, Malaney P, Quarni W, Pandit A, Allen-Gipson D, Dave V (2013) Loss of phosphatase and tensin homolog (PTEN) induces leptin-mediated leptin gene expression: feed-forward loop operating in the lung. J Biol Chem 288(41):29821–29835

210. Piegeler T, Votta-Velis EG, Bakhshi FR, Mao M, Carnegie G, Bonini MG, Schwartz DE, Borgeat A, Beck-Schimmer B, Minshall RD (2014) Endothelial barrier protection by local anesthetics: ropivacaine and lidocaine block tumor necrosis factor-alpha-induced endothelial cell Src activation. Anesthesiology 120(6):1414–1428

211. Rigor RR, Beard Jr RS, Litovka OP, Yuan SY (2012) Interleukin-1beta-induced barrier dysfunction is signaled through PKC-theta in human brain microvascular endothelium. Am J Physiol Cell Physiol 302(10):C1513–C1522

212. Birukova AA, Xing J, Fu P, Yakubov B, Dubrovskyi O, Fortune JA, Klibanov AM, Birukov KG (2010) Atrial natriuretic peptide attenuates LPS-induced lung vascular leak: role of PAK1. Am J Physiol Lung Cell Mol Physiol 299(5):L652–L663

213. Joshi AD, Dimitropoulou C, Thangjam G, Snead C, Feldman S, Barabutis N, Fulton D, Hou Y, Kumar S, Patel V, Gorshkov B, Verin AD, Black SM, Catravas JD (2014) Heat shock protein 90 inhibitors prevent LPS-induced endothelial barrier dysfunction by disrupting RhoA signaling. Am J Respir Cell Mol Biol 50(1):170–179

214. Clements RT, Minnear FL, Singer HA, Keller RS, Vincent PA (2005) RhoA and Rho-kinase dependent and independent signals mediate TGF-beta-induced pulmonary endothelial cytoskeletal reorganization and permeability. Am J Physiol Lung Cell Mol Physiol 288(2):L294–L306

215. Goldberg PL, MacNaughton DE, Clements RT, Minnear FL, Vincent PA (2002) p38 MAPK activation by TGF-beta1 increases MLC phosphorylation and endothelial monolayer permeability. Am J Physiol Lung Cell Mol Physiol 282(1):L146–L154

216. Sung JY, Park SY, Kim JH, Kang HG, Yoon JH, Na YS, Kim YN, Park BK (2014) Interferon consensus sequence-binding protein (ICSBP) promotes epithelial-to-mesenchymal transition (EMT)-like phenomena, cell-motility, and invasion via TGF-beta signaling in U2OS cells. Cell Death Dis 5:e1224

217. O'Donnell EF, Kopparapu PR, Koch DC, Jang HS, Phillips JL, Tanguay RL, Kerkvliet NI, Kolluri SK (2012) The aryl hydrocarbon receptor mediates leflunomide-induced growth inhibition of melanoma cells. PLoS One 7(7):e40926

218. Rotroff DM, Dix DJ, Houck KA, Kavlock RJ, Knudsen TB, Martin MT, Reif DM, Richard AM, Sipes NS, Abassi YA, Jin C, Stampfl M, Judson RS (2013) Real-time growth kinetics measuring hormone mimicry for ToxCast chemicals in T-47D human ductal carcinoma cells. Chem Res Toxicol 26(7):1097–1107

219. Tian J, Smith A, Nechtman J, Podolsky R, Aggarwal S, Snead C, Kumar S, Elgaish M, Oishi P, Goerlach A, Fratz S, Hess J, Catravas JD, Verin AD, Fineman JR, She JX, Black SM (2009) Effect of PPARgamma inhibition on pulmonary endothelial cell gene expression: gene profiling in pulmonary hypertension. Physiol Genomics 40(1):48–60
220. Wolfson RK, Chiang ET, Garcia JG (2011) HMGB1 induces human lung endothelial cell cytoskeletal rearrangement and barrier disruption. Microvasc Res 81(2):189–197
221. Yuan L, Le Bras A, Sacharidou A, Itagaki K, Zhan Y, Kondo M, Carman CV, Davis GE, Aird WC, Oettgen P (2012) ETS-related gene (ERG) controls endothelial cell permeability via transcriptional regulation of the claudin 5 (CLDN5) gene. J Biol Chem 287(9):6582–6591
222. Privratsky JR, Paddock CM, Florey O, Newman DK, Muller WA, Newman PJ (2011) Relative contribution of PECAM-1 adhesion and signaling to the maintenance of vascular integrity. J Cell Sci 124(Pt 9):1477–1485
223. Sun C, Wu MH, Guo M, Day ML, Lee ES, Yuan SY (2010) ADAM15 regulates endothelial permeability and neutrophil migration via Src/ERK1/2 signalling. Cardiovasc Res 87 (2):348–355
224. van Wetering S, van den Berk N, van Buul JD, Mul FP, Lommerse I, Mous R, ten Klooster JP, Zwaginga JJ, Hordijk PL (2003) VCAM-1-mediated Rac signaling controls endothelial cell-cell contacts and leukocyte transmigration. Am J Physiol Cell Physiol 285(2):C343–C352
225. Peters MF, Vaillancourt F, Heroux M, Valiquette M, Scott CW (2010) Comparing label-free biosensors for pharmacological screening with cell-based functional assays. Assay Drug Dev Technol 8(2):219–227
226. Gainor JP, Morton CA, Roberts JT, Vincent PA, Minnear FL (2001) Platelet-conditioned medium increases endothelial electrical resistance independently of cAMP/PKA and cGMP/PKG. Am J Physiol Heart Circ Physiol 281(5):H1992–H2001
227. Halai R, Croker DE, Suen JY, Fairlie DP, Cooper MA (2012) A comparative study of impedance versus optical label-free systems relative to labelled assays in a predominantly Gi coupled GPCR (C5aR) signalling. Biosensors 2(3):273–290
228. Konya V, Ullen A, Kampitsch N, Theiler A, Philipose S, Parzmair GP, Marsche G, Peskar BA, Schuligoi R, Sattler W, Heinemann A (2013) Endothelial E-type prostanoid 4 receptors promote barrier function and inhibit neutrophil trafficking. J Allergy Clin Immunol 131 (2):532–540.e1-2
229. Atienza JM, Yu N, Wang X, Xu X, Abassi Y (2006) Label-free and real-time cell-based kinase assay for screening selective and potent receptor tyrosine kinase inhibitors using microelectronic sensor array. J Biomol Screen 11(6):634–643
230. Xu M, Waters CL, Hu C, Wysolmerski RB, Vincent PA, Minnear FL (2007) Sphingosine 1-phosphate rapidly increases endothelial barrier function independently of VE-cadherin but requires cell spreading and Rho kinase. Am J Physiol Cell Physiol 293(4):C1309–C1318
231. Huang F, Subbaiah PV, Holian O, Zhang J, Johnson A, Gertzberg N, Lum H (2005) Lysophosphatidylcholine increases endothelial permeability: role of PKCalpha and RhoA cross talk. Am J Physiol Lung Cell Mol Physiol 289(2):L176–L185
232. McLaughlin JN, Shen L, Holinstat M, Brooks JD, Dibenedetto E, Hamm HE (2005) Functional selectivity of G protein signaling by agonist peptides and thrombin for the protease-activated receptor-1. J Biol Chem 280(26):25048–25059
233. Minshall RD, Vandenbroucke EE, Holinstat M, Place AT, Tiruppathi C, Vogel SM, van Nieuw Amerongen GP, Mehta D, Malik AB (2010) Role of protein kinase Czeta in thrombin-induced RhoA activation and inter-endothelial gap formation of human dermal microvessel endothelial cell monolayers. Microvasc Res 80(2):240–249
234. Stallaert W, Dorn JF, van der Westhuizen E, Audet M, Bouvier M (2012) Impedance responses reveal beta(2)-adrenergic receptor signaling pluridimensionality and allow classification of ligands with distinct signaling profiles. PLoS One 7(1):e29420
235. Zhang XE, Adderley SP, Breslin JW (2016) Activation of RhoA, but not Rac1, mediates early stages of S1P-induced endothelial barrier enhancement. PLoS One 11(5):e0155490

236. Ji J, Jia S, Jia Y, Ji K, Hargest R, Jiang WG (2015) WISP-2 in human gastric cancer and its potential metastatic suppressor role in gastric cancer cells mediated by JNK and PLC-gamma pathways. Br J Cancer 113(6):921–933

237. Liu S, Yu C, Yang F, Paganini-Hill A, Fisher MJ (2012) Phosphodiesterase inhibitor modulation of brain microvascular endothelial cell barrier properties. J Neurol Sci 320(1–2):45–51

238. Shinde AV, Motiani RK, Zhang X, Abdullaev IF, Adam AP, Gonzalez-Cobos JC, Zhang W, Matrougui K, Vincent PA, Trebak M (2013) STIM1 controls endothelial barrier function independently of Orai1 and Ca2+ entry. Sci Signal 6(267):ra18

239. Lorenowicz MJ, Fernandez-Borja M, Kooistra MR, Bos JL, Hordijk PL (2008) PKA and Epac1 regulate endothelial integrity and migration through parallel and independent pathways. Eur J Cell Biol 87(10):779–792

240. Moldobaeva A, Welsh-Servinsky LE, Shimoda LA, Stephens RS, Verin AD, Tuder RM, Pearse DB (2006) Role of protein kinase G in barrier-protective effects of cGMP in human pulmonary artery endothelial cells. Am J Physiol Lung Cell Mol Physiol 290(5):L919–L930

241. Adyshev DM, Dudek SM, Moldobaeva N, Kim KM, Ma SF, Kasa A, Garcia JG, Verin AD (2013) Ezrin/radixin/moesin proteins differentially regulate endothelial hyperpermeability after thrombin. Am J Physiol Lung Cell Mol Physiol 305(3):L240–L255

242. Birukova AA, Birukov KG, Smurova K, Adyshev D, Kaibuchi K, Alieva I, Garcia JG, Verin AD (2004) Novel role of microtubules in thrombin-induced endothelial barrier dysfunction. FASEB J 18(15):1879–1890

243. Rentsendorj O, Mirzapoiazova T, Adyshev D, Servinsky LE, Renne T, Verin AD, Pearse DB (2008) Role of vasodilator-stimulated phosphoprotein in cGMP-mediated protection of human pulmonary artery endothelial barrier function. Am J Physiol Lung Cell Mol Physiol 294(4): L686–L697

244. Escudero-Esparza A, Jiang WG, Martin TA (2012) Claudin-5 participates in the regulation of endothelial cell motility. Mol Cell Biochem 362(1–2):71–85

245. Finigan JH, Dudek SM, Singleton PA, Chiang ET, Jacobson JR, Camp SM, Ye SQ, Garcia JG (2005) Activated protein C mediates novel lung endothelial barrier enhancement: role of sphingosine 1-phosphate receptor transactivation. J Biol Chem 280(17):17286–17293

246. Grinnell KL, Chichger H, Braza J, Duong H, Harrington EO (2012) Protection against LPS-induced pulmonary edema through the attenuation of protein tyrosine phosphatase-1B oxidation. Am J Respir Cell Mol Biol 46(5):623–632

247. Sun C, Wu MH, Yuan SY (2011) Nonmuscle myosin light-chain kinase deficiency attenuates atherosclerosis in apolipoprotein E–deficient mice via reduced endothelial barrier dysfunction and monocyte migration. Circulation 124(1):48–57

248. Herron CR, Lowery AM, Hollister PR, Reynolds AB, Vincent PA (2011) p120 regulates endothelial permeability independently of its NH2 terminus and Rho binding. Am J Physiol Heart Circ Physiol 300(1):H36–H48

249. Sawhney RS, Liu W, Brattain MG (2009) A novel role of ERK5 in integrin-mediated cell adhesion and motility in cancer cells via Fak signaling. J Cell Physiol 219(1):152–161

250. Bilir B, Kucuk O, Moreno CS (2013) Wnt signaling blockage inhibits cell proliferation and migration, and induces apoptosis in triple-negative breast cancer cells. J Transl Med 11:280

251. Wilson JL, Taylor L, Polgar P (2012) Endothelin-1 activation of ETB receptors leads to a reduced cellular proliferative rate and an increased cellular footprint. Exp Cell Res 318 (10):1125–1133

252. Yuan L, Zhang H, Liu J, Rubin JB, Cho YJ, Shu HK, Schniederjan M, MacDonald TJ (2013) Growth factor receptor-Src-mediated suppression of GRK6 dysregulates CXCR4 signaling and promotes medulloblastoma migration. Mol Cancer 12:18

253. Berdyshev EV, Gorshkova IA, Usatyuk P, Zhao Y, Saatian B, Hubbard W, Natarajan V (2006) De novo biosynthesis of dihydrosphingosine-1-phosphate by sphingosine kinase 1 in mammalian cells. Cell Signal 18(10):1779–1792

254. Rask-Andersen M, Almen MS, Schioth HB (2011) Trends in the exploitation of novel drug targets. Nat Rev Drug Discov 10(8):579–590

255. Hoffman BB, Lefkowitz RJ (1980) Radioligand binding studies of adrenergic receptors: new insights into molecular and physiological regulation. Annu Rev Pharmacol Toxicol 20:581–608
256. Maguire JJ, Kuc RE, Davenport AP (2012) Radioligand binding assays and their analysis. Methods Mol Biol 897:31–77
257. Kenakin TP (2009) Cellular assays as portals to seven-transmembrane receptor-based drug discovery. Nat Rev Drug Discov 8(8):617–626
258. Thomsen W, Frazer J, Unett D (2005) Functional assays for screening GPCR targets. Curr Opin Biotechnol 16(6):655–665
259. Zhang R, Xie X (2012) Tools for GPCR drug discovery. Acta Pharmacol Sin 33(3):372–384
260. Cooper MA (2003) Label-free screening of bio-molecular interactions. Anal Bioanal Chem 377(5):834–842
261. Fang Y (2014) Label-free drug discovery. Front Pharmacol 5:52
262. McGuinness R (2007) Impedance-based cellular assay technologies: recent advances, future promise. Curr Opin Pharmacol 7(5):535–540
263. Miyano K, Sudo Y, Yokoyama A, Hisaoka-Nakashima K, Morioka N, Takebayashi M, Nakata Y, Higami Y, Uezono Y (2014) History of the G protein-coupled receptor (GPCR) assays from traditional to a state-of-the-art biosensor assay. J Pharmacol Sci 126(4):302–309
264. Peters MF, Scott CW (2009) Evaluating cellular impedance assays for detection of GPCR pleiotropic signaling and functional selectivity. J Biomol Screen 14(3):246–255
265. Schroder R, Schmidt J, Blattermann S, Peters L, Janssen N, Grundmann M, Seemann W, Kaufel D, Merten N, Drewke C, Gomeza J, Milligan G, Mohr K, Kostenis E (2011) Applying label-free dynamic mass redistribution technology to frame signaling of G protein-coupled receptors noninvasively in living cells. Nat Protoc 6(11):1748–1760
266. Scott CW, Peters MF (2010) Label-free whole-cell assays: expanding the scope of GPCR screening. Drug Discov Today 15(17–18):704–716
267. Xi B, Yu N, Wang X, Xu X, Abassi YA (2008) The application of cell-based label-free technology in drug discovery. Biotechnol J 3(4):484–495
268. Verdonk E, Johnson K, McGuinness R, Leung G, Chen YW, Tang HR, Michelotti JM, Liu VF (2006) Cellular dielectric spectroscopy: a label-free comprehensive platform for functional evaluation of endogenous receptors. Assay Drug Dev Technol 4(5):609–619
269. Peters MF, Knappenberger KS, Wilkins D, Sygowski LA, Lazor LA, Liu J, Scott CW (2007) Evaluation of cellular dielectric spectroscopy, a whole-cell, label-free technology for drug discovery on Gi-coupled GPCRs. J Biomol Screen 12(3):312–319
270. Flynn AN, Hoffman J, Tillu DV, Sherwood CL, Zhang Z, Patek R, Asiedu MN, Vagner J, Price TJ, Boitano S (2013) Development of highly potent protease-activated receptor 2 agonists via synthetic lipid tethering. FASEB J 27(4):1498–1510
271. Yu N, Atienza JM, Bernard J, Blanc S, Zhu J, Wang X, Xu X, Abassi YA (2006) Real-time monitoring of morphological changes in living cells by electronic cell sensor arrays: an approach to study G protein-coupled receptors. Anal Chem 78(1):35–43
272. Zweemer AJ, Nederpelt I, Vrieling H, Hafith S, Doornbos ML, de Vries H, Abt J, Gross R, Stamos D, Saunders J, Smit MJ, Ijzerman AP, Heitman LH (2013) Multiple binding sites for small-molecule antagonists at the CC chemokine receptor 2. Mol Pharmacol 84(4):551–561
273. Ciambrone GJ, Liu VF, Lin DC, McGuinness RP, Leung GK, Pitchford S (2004) Cellular dielectric spectroscopy: a powerful new approach to label-free cellular analysis. J Biomol Screen 9(6):467–480
274. Chen AN, Malone DT, Pabreja K, Sexton PM, Christopoulos A, Canals M (2015) Detection and quantification of allosteric modulation of endogenous m4 muscarinic acetylcholine receptor using impedance-based label-free technology in a neuronal cell line. J Biomol Screen 20 (5):646–654
275. Lv S, Wu L, Cheng P, Yu J, Zhang A, Zha J, Liu J, Wang L, Di W, Hu M, Qi H, Li Y, Ding G (2010) Correlation of obesity and osteoporosis: effect of free fatty acids on bone marrow-derived mesenchymal stem cell differentiation. Exp Ther Med 1(4):603–610

276. Park HE, Kim D, Koh HS, Cho S, Sung JS, Kim JY (2011) Real-time monitoring of neural differentiation of human mesenchymal stem cells by electric cell-substrate impedance sensing. J Biomed Biotechnol 2011:485173

277. Tai YY, Chen RS, Lin Y, Ling TY, Chen MH (2012) FGF-9 accelerates epithelial invagination for ectodermal organogenesis in real time bioengineered organ manipulation. Cell Commun Signal 10(1):34

278. Cohen EN, Gao H, Anfossi S, Mego M, Reddy NG, Debeb B, Giordano A, Tin S, Wu Q, Garza RJ, Cristofanilli M, Mani SA, Croix DA, Ueno NT, Woodward WA, Luthra R, Krishnamurthy S, Reuben JM (2015) Inflammation mediated metastasis: immune induced epithelial-to-mesenchymal transition in inflammatory breast cancer cells. PLoS One 10(7): e0132710

279. Stolwijk JA, Hartmann C, Balani P, Albermann S, Keese CR, Giaever I, Wegener J (2011) Impedance analysis of adherent cells after in situ electroporation: non-invasive monitoring during intracellular manipulations. Biosens Bioelectron 26(12):4720–4727

280. Chang YC, Stins MF, McCaffery MJ, Miller GF, Pare DR, Dam T, Paul-Satyasee M, Kim KS, Kwon-Chung KJ (2004) Cryptococcal yeast cells invade the central nervous system via transcellular penetration of the blood-brain barrier. Infect Immun 72(9):4985–4995

281. Ebrahimi CM, Sheen TR, Renken CW, Gottlieb RA, Doran KS (2011) Contribution of lethal toxin and edema toxin to the pathogenesis of anthrax meningitis. Infect Immun 79 (7):2510–2518

282. Grab D, Nyarko E, Nikolskaia O, Kim Y, Dumler J (2009) Human brain microvascular endothelial cell traversal by Borrelia burgdorferi requires calcium signaling. Clin Microbiol Infect 15(5):422–426

283. Lembo A, Gurney MA, Burnside K, Banerjee A, De Los Reyes M, Connelly JE, Lin WJ, Jewell KA, Vo A, Renken CW (2010) Regulation of CovR expression in Group B streptococcus impacts blood–brain barrier penetration. Mol Microbiol 77(2):431–443

284. Treeratanapiboon L, Psathaki K, Wegener J, Looareesuwan S, Galla H-J, Udomsangpetch R (2005) In vitro study of malaria parasite induced disruption of blood–brain barrier. Biochem Biophys Res Commun 335(3):810–818

285. Goc A, Al-Azayzih A, Abdalla M, Al-Husein B, Kavuri S, Lee J, Moses K, Somanath PR (2013) P21 activated kinase-1 (Pak1) promotes prostate tumor growth and microinvasion via inhibition of transforming growth factor β expression and enhanced matrix metalloproteinase 9 secretion. J Biol Chem 288(5):3025–3035

286. Goc A, Al-Husein B, Katsanevas K, Steinbach A, Lou U, Sabbineni H, DeRemer DL, Somanath PR (2014) Targeting Src-mediated Tyr216 phosphorylation and activation of GSK-3 in prostate cancer cells inhibit prostate cancer progression in vitro and in vivo. Oncotarget 5(3):775–787

287. Jiang WG, Ablin RJ, Kynaston HG, Mason MD (2009) The prostate transglutaminase (TGase-4, TGaseP) regulates the interaction of prostate cancer and vascular endothelial cells, a potential role for the ROCK pathway. Microvasc Res 77(2):150–157

288. Keese CR, Bhawe K, Wegener J, Giaever I (2002) Real-time impedance assay to follow the invasive activities of metastatic cells in culture. Biotechniques 33(4):842–844, 846, 848-850

289. Melnikova VO, Balasubramanian K, Villares GJ, Dobroff AS, Zigler M, Wang H, Petersson F, Price JE, Schroit A, Prieto VG (2009) Crosstalk between protease-activated receptor 1 and platelet-activating factor receptor regulates melanoma cell adhesion molecule (MCAM/MUC18) expression and melanoma metastasis. J Biol Chem 284(42):28845–28855

290. Ren J, Xiao Y-j, Singh LS, Zhao X, Zhao Z, Feng L, Rose TM, Prestwich GD, Xu Y (2006) Lysophosphatidic acid is constitutively produced by human peritoneal mesothelial cells and enhances adhesion, migration, and invasion of ovarian cancer cells. Cancer Res 66 (6):3006–3014

291. Saxena NK, Sharma D, Ding X, Lin S, Marra F, Merlin D, Anania FA (2007) Concomitant activation of the JAK/STAT, PI3K/AKT, and ERK signaling is involved in leptin-mediated

promotion of invasion and migration of hepatocellular carcinoma cells. Cancer Res 67 (6):2497–2507

292. Wang H-S, Hung Y, Su C-H, Peng S-T, Guo Y-J, Lai M-C, Liu C-Y, Hsu J-W (2005) CD44 cross-linking induces integrin-mediated adhesion and transendothelial migration in breast cancer cell line by up-regulation of LFA-1 (αLβ2) and VLA-4 (α4β1). Exp Cell Res 304 (1):116–126

293. Chen Y-W, Chen J-K, Wang J-S (2009) Exercise affects platelet-promoted tumor cell adhesion and invasion to endothelium. Eur J Appl Physiol 105(3):393–401

294. Tsikitis VL, Morin NA, Harrington EO, Albina JE, Reichner JS (2004) The lectin-like domain of complement receptor 3 protects endothelial barrier function from activated neutrophils. J Immunol 173(2):1284–1291

295. van Rijssel J, Kroon J, Hoogenboezem M, van Alphen FP, de Jong RJ, Kostadinova E, Geerts D, Hordijk PL, van Buul JD (2012) The Rho-guanine nucleotide exchange factor Trio controls leukocyte transendothelial migration by promoting docking structure formation. Mol Biol Cell 23(15):2831–2844

296. Zhu J, Wang X, Xu X, Abassi YA (2006) Dynamic and label-free monitoring of natural killer cell cytotoxic activity using electronic cell sensor arrays. J Immunol Methods 309(1):25–33

297. Sansing HA, Renner NA, MacLean AG (2012) An inverted blood–brain barrier model that permits interactions between glia and inflammatory stimuli. J Neurosci Methods 207(1):91–96

298. Rother J, Richter C, Turco L, Knoch F, Mey I, Luther S, Janshoff A, Bodenschatz E, Tarantola M (2015) Crosstalk of cardiomyocytes and fibroblasts in co-cultures. Open Biol 5(6):150038

299. Yamamoto Y, Goda N, Nakamura T, Kusuhara T, Maruyama T, Mohri S, Kataoka N, Kajiya F (2007) Quantitative evaluation of effect for radiation exposure to cultured cells using electrical cell-substrate impedance sensing (ECIS) method. World Congress on medical physics and biomedical engineering 2006. Springer, Heidelberg, pp 1914–1917

300. Szulcek R, van Bezu J, Boonstra J, van Loon JJ, van Nieuw Amerongen GP (2015) Transient intervals of hyper-gravity enhance endothelial barrier integrity: impact of mechanical and gravitational forces measured electrically. PLoS One 10(12):e0144269

301. DePaola N, Phelps JE, Florez L, Keese CR, Minnear FL, Giaever I, Vincent P (2001) Electrical impedance of cultured endothelium under fluid flow. Ann Biomed Eng 29 (8):648–656

302. Phelps JE, DePaola N (2000) Spatial variations in endothelial barrier function in disturbed flows in vitro. Am J Physiol Heart Circ Physiol 278(2):H469–H476

303. Seebach J, Dieterich P, Luo F, Schillers H, Vestweber D, Oberleithner H, Galla H-J, Schnittler H-J (2000) Endothelial barrier function under laminar fluid shear stress. Lab Invest 80 (12):1819–1831

304. Jaalouk DE, Lammerding J (2009) Mechanotransduction gone awry. Nat Rev Mol Cell Biol 10(1):63–73

305. Tarbell JM, Simon SI, Curry F-RE (2014) Mechanosensing at the vascular interface. Annu Rev Biomed Eng 16:505

306. Seebach J, Donnert G, Kronstein R, Werth S, Wojciak-Stothard B, Falzarano D, Mrowietz C, Hell SW, Schnittler H-J (2007) Regulation of endothelial barrier function during flow-induced conversion to an arterial phenotype. Cardiovasc Res 75(3):598–607

307. Clark PR, Jensen TJ, Kluger MS, Morelock M, Hanidu A, Qi Z, Tatake RJ, Pober JS (2011) MEK5 is activated by shear stress, activates ERK5 and induces KLF4 to modulate TNF responses in human dermal microvascular endothelial cells. Microcirculation 18(2):102–117

308. Shikata Y, Rios A, Kawkitinarong K, DePaola N, Garcia JG, Birukov KG (2005) Differential effects of shear stress and cyclic stretch on focal adhesion remodeling, site-specific FAK phosphorylation, and small GTPases in human lung endothelial cells. Exp Cell Res 304 (1):40–49

309. Bevan HS, Slater SC, Clarke H, Cahill PA, Mathieson PW, Welsh GI, Satchell SC (2011) Acute laminar shear stress reversibly increases human glomerular endothelial cell

permeability via activation of endothelial nitric oxide synthase. Am J Physiol Renal Physiol 301(4):F733–F742

310. Siddharthan V, Kim YV, Liu S, Kim KS (2007) Human astrocytes/astrocyte-conditioned medium and shear stress enhance the barrier properties of human brain microvascular endothelial cells. Brain Res 1147:39–50

311. Sircar M, Bradfield PF, Aurrand-Lions M, Fish RJ, Alcaide P, Yang L, Newton G, Lamont D, Sehrawat S, Mayadas T (2007) Neutrophil transmigration under shear flow conditions in vitro is junctional adhesion molecule-C independent. J Immunol 178(9):5879–5887

312. Slater SC, Ramnath RD, Uttridge K, Saleem MA, Cahill PA, Mathieson PW, Welsh GI, Satchell SC (2012) Chronic exposure to laminar shear stress induces Kruppel-like factor 2 in glomerular endothelial cells and modulates interactions with co-cultured podocytes. Int J Biochem Cell Biol 44(9):1482–1490

313. Ueno N, Harker KS, Clarke EV, McWhorter FY, Liu WF, Tenner AJ, Lodoen MB (2014) Real-time imaging of Toxoplasma-infected human monocytes under fluidic shear stress reveals rapid translocation of intracellular parasites across endothelial barriers. Cell Microbiol 16(4):580–595

314. Matuszak J, Zaloga J, Friedrich RP, Lyer S, Nowak J, Odenbach S, Alexiou C, Cicha I (2015) Endothelial biocompatibility and accumulation of SPION under flow conditions. J Magn Magn Mater 380:20–26

315. Curtis TM, Widder MW, Brennan LM, Schwager SJ, van der Schalie WH, Fey J, Salazar N (2009) A portable cell-based impedance sensor for toxicity testing of drinking water. Lab Chip 9(15):2176–2183

316. Zhang X, Li F, Nordin AN, Tarbell J, Voiculescu I (2015) Toxicity studies using mammalian cells and impedance spectroscopy method. Sens Biosens Res 3:112–121

317. Zhivotovsky B, Orrenius S (2010) Cell death mechanisms: cross-talk and role in disease. Exp Cell Res 316(8):1374–1383

318. Taylor RC, Cullen SP, Martin SJ (2008) Apoptosis: controlled demolition at the cellular level. Nat Rev Mol Cell Biol 9(3):231–241

319. Degterev A, Yuan J (2008) Expansion and evolution of cell death programmes. Nat Rev Mol Cell Biol 9(5):378–390

320. Martin SJ, Henry CM (2013) Distinguishing between apoptosis, necrosis, necroptosis and other cell death modalities. Methods 61(2):87–89

321. Tsujimoto Y (2012) Multiple ways to die: non-apoptotic forms of cell death. Acta Oncol 51 (3):293–300

322. Krysko DV, Berghe TV, Parthoens E, D'Herde K, Vandenabeele P (2008) Methods for distinguishing apoptotic from necrotic cells and measuring their clearance. Methods Enzymol 442:307–341

323. Fauci AS (2008) Harrison's principles of internal medicine, vol 2. McGraw-Hill, London

324. Tice RR, Austin CP, Kavlock RJ, Bucher JR (2013) Improving the human hazard characterization of chemicals: a Tox21 update. Environ Health Perspect 121(7):756

325. Nel A, Xia T, Meng H, Wang X, Lin S, Ji Z, Zhang H (2012) Nanomaterial toxicity testing in the 21st century: use of a predictive toxicological approach and high-throughput screening. Acc Chem Res 46(3):607–621

326. Niles AL, Moravec RA, Riss TL (2009) In vitro viability and cytotoxicity testing and same-well multi-parametric combinations for high throughput screening. Curr Chem Genom 3:33–41

327. Giaever I, Keese C (1992) Toxic? Cells can tell. ChemTech 22(2):116–125

328. Keese C, Karra N, Dillon B, Goldberg A, Giaever I (1998) Cell-substratum interactions as a predictor of cytotoxicity. In Vitro Mol Toxicol 11(2):183–192

329. Curtis TM, Tabb J, Romeo L, Schwager SJ, Widder MW, van der Schalie WH (2009) Improved cell sensitivity and longevity in a rapid impedance-based toxicity sensor. J Appl Toxicol 29(5):374–380

330. Kubisch R, Bohrn U, Fleischer M, Stutz E (2012) Cell-based sensor system using L6 cells for broad band continuous pollutant monitoring in aquatic environments. Sensors 12 (3):3370–3393

331. Lee WK, Torchalski B, Kohistani N, Thevenod F (2011) ABCB1 protects kidney proximal tubule cells against cadmium-induced apoptosis: roles of cadmium and ceramide transport. Toxicol Sci 121(2):343–356

332. Xing JZ, Zhu L, Jackson JA, Gabos S, Sun XJ, Wang XB, Xu X (2005) Dynamic monitoring of cytotoxicity on microelectronic sensors. Chem Res Toxicol 18(2):154–161

333. Atienzar FA, Tilmant K, Gerets HH, Toussaint G, Speeckaert S, Hanon E, Depelchin O, Dhalluin S (2011) The use of real-time cell analyzer technology in drug discovery: defining optimal cell culture conditions and assay reproducibility with different adherent cellular models. J Biomol Screen 16(6):575–587

334. Pauly D, Worbs S, Kirchner S, Shatohina O, Dorner MB, Dorner BG (2012) Real-time cytotoxicity assay for rapid and sensitive detection of ricin from complex matrices. PLoS One 7(4):e35360

335. Alborzinia H, Can S, Holenya P, Scholl C, Lederer E, Kitanovic I, Wolfl S (2011) Real-time monitoring of cisplatin-induced cell death. PLoS One 6(5):e19714

336. Otero-Gonzalez L, Sierra-Alvarez R, Boitano S, Field JA (2012) Application and validation of an impedance-based real time cell analyzer to measure the toxicity of nanoparticles impacting human bronchial epithelial cells. Environ Sci Technol 46(18):10271–10278

337. Sergent JA, Paget V, Chevillard S (2012) Toxicity and genotoxicity of nano-SiO2 on human epithelial intestinal HT-29 cell line. Ann Occup Hyg 56(5):622–630

338. Campbell CE, Laane MM, Haugarvoll E, Giaever I (2007) Monitoring viral-induced cell death using electric cell-substrate impedance sensing. Biosens Bioelectron 23(4):536–542

339. McCoy MH, Wang E (2005) Use of electric cell-substrate impedance sensing as a tool for quantifying cytopathic effect in influenza A virus infected MDCK cells in real-time. J Virol Methods 130(1–2):157–161

340. Muller J, Thirion C, Pfaffl MW (2011) Electric cell-substrate impedance sensing (ECIS) based real-time measurement of titer dependent cytotoxicity induced by adenoviral vectors in an IPI-2I cell culture model. Biosens Bioelectron 26(5):2000–2005

341. Benachour H, Bastogne T, Toussaint M, Chemli Y, Seve A, Frochot C, Lux F, Tillement O, Vanderesse R, Barberi-Heyob M (2012) Real-time monitoring of photocytotoxicity in nanoparticles-based photodynamic therapy: a model-based approach. PLoS One 7(11):e48617

342. Arndt S, Seebach J, Psathaki K, Galla HJ, Wegener J (2004) Bioelectrical impedance assay to monitor changes in cell shape during apoptosis. Biosens Bioelectron 19(6):583–594

343. Huang L, Xie L, Boyd JM, Li XF (2008) Cell-electronic sensing of particle-induced cellular responses. Analyst 133(5):643–648

344. Qiu Y, Liao R, Zhang X (2009) Impedance-based monitoring of ongoing cardiomyocyte death induced by tumor necrosis factor-alpha. Biophys J 96(5):1985–1991

345. Xiao C, Luong JH (2005) Assessment of cytotoxicity by emerging impedance spectroscopy. Toxicol Appl Pharmacol 206(2):102–112

346. Janshoff A, Wegener J, Sieber M, Galla HJ (1996) Double-mode impedance analysis of epithelial cell monolayers cultured on shear wave resonators. Eur Biophys J 25(2):93–103

347. Liu F, Voiculescu I, Nordin AN, Li F (2013) Water toxicity detection using cell-based hybrid biosensors. Sensors. IEEE, Piscataway, pp 1–5

348. Steinem C, Janshoff A, Wegener J, Ulrich WP, Willenbrink W, Sieber M, Galla HJ (1997) Impedance and shear wave resonance analysis of ligand-receptor interactions at functionalized surfaces and of cell monolayers. Biosens Bioelectron 12(8):787–808

349. Michaelis S, Wegener J, Robelek R (2013) Label-free monitoring of cell-based assays: combining impedance analysis with SPR for multiparametric cell profiling. Biosens Bioelectron 49:63–70

350. Alexander FA, Price DT, Bhansali S (2013) From cellular cultures to cellular spheroids: is impedance spectroscopy a viable tool for monitoring multicellular spheroid (MCS) drug models? IEEE Rev Biomed Eng 6:63–76
351. Kloß D, Kurz R, Jahnke H-G, Fischer M, Rothermel A, Anderegg U, Simon JC, Robitzki AA (2008) Microcavity array (MCA)-based biosensor chip for functional drug screening of 3D tissue models. Biosens Bioelectron 23(10):1473–1480
352. Thielecke H, Mack A, Robitzki A (2001) A multicellular spheroid-based sensor for anti-cancer therapeutics. Biosens Bioelectron 16(4):261–269

BIOREV (2019) 2: 77–110
https://doi.org/10.1007/11663_2017_1
© Springer International Publishing AG 2018
Published online: 6 February 2018

Transistor-Based Impedimetric Monitoring of Single Cells

F. Hempel, J. K. Y. Law, and S. Ingebrandt

Contents

Abstract As the interest in personalized medicine in general and individualized therapies in particular is increasing steadily, there is a great need for biomedical sensing platforms that offer the highest possible sensitivity, reliability, stability,

F. Hempel and J. K. Y. Law
Department of Informatics and Microsystem Technology, University of Applied Sciences Kaiserslautern, Zweibrücken, Germany

S. Ingebrandt (✉)
Department of Informatics and Microsystem Technology, University of Applied Sciences Kaiserslautern, Zweibrücken, Germany

Institute of Materials in Electrical Engineering 1, RWTH Aachen University, Aachen, Germany
e-mail: sven.ingebrandt@hs-kl.de; ingebrandt@iwe1.rwth-aachen.de

versatility, and accuracy as well as the lowest limit of detection. While not all biosensors offer the same possibilities, ion sensitive field-effect transistors (ISFET) are a well-understood and well-established platform that can be applied for many needs in the field of biomedicine. With the possibilities to modify and functionalize the device surface for cell cultures and parallel measurements with many sensing points on one device, ISFET arrays offer an optimized platform for chemical sensing (pH, ion concentration, etc.) and for the sensing of biological species like cells, DNA, and proteins. In this chapter, we present impedance sensing with ISFETs as a universal tool for cellular recordings down to single cell level and give an overview of recent measurements with this versatile approach. In accordance with the classical ECIS method we termed our technique **F**ield-**E**ffect **T**ransistor **C**ell-substrate **I**mpedance **S**ensing (FETCIS). In this chapter the fundamental technique as well as the electronic model to explain the recorded FETCIS spectra is elaborated showing several applications, which are not accessible with ECIS so far.

Keywords Cellular migration · Electric Cell-substrate Impedance Sensing (ECIS) · Field-Effect Transistor Cell-substrate Impedance Sensing (FETCIS) · Ion sensitive field-effect transistor (ISFET) · Single cell impedance sensing

1   Introduction

The interest in understanding the fundamental building blocks of life – individual cells – has been growing continuously ever since they were first cultured under laboratory conditions. Besides fundamental cell biology, the main goal of this research field nowadays is to investigate cell functions and behavior in all details and to apply this knowledge for applications like personalized drug therapy and lab-on-a-chip (LoC) devices as well as for in vitro screening purposes. Field-effect transistors (FETs) are one of the tools to make such applications become reality. Biosensing with FETs is dated back to the year 1976, when Piet Bergveld and co-workers used an ion-sensitive field-effect transistor (ISFET) to measure the electronic activity of a muscle fiber [1]. After this pioneering work, almost limitless applications for ISFET concepts in the field of biosensors and bioelectronics followed and ever since then numerous biosensors have been developed and tested for different applications. Although cell sensing is only a small part of this extended field, FET recordings are nowadays considered as a powerful tool that deepened the understanding of various cell mechanisms. General examples for biosensing applications with FETs are the detection of cancer marker proteins [2] or other kinds of proteins, antibodies, or DNA [3], and many cell-related measurements, like the detection of electrically active cells [4], and impedance measurements of cells that are not electrically active [5]. By now very large integrated FET arrays have been developed in complementary metal-oxide semiconductor (CMOS) processes and some of these systems are even commercially available [6–9]. To mention all related publications would surely go beyond the scope of this contribution and many publications are added to this pool every year.

Cells are the basic parts of all organisms and cell sensing experiments with physical transducers in general and transistors in particular can provide tremendous amounts of invaluable knowledge about cell functions and interactions. The standard method to observe effects in cell cultures is optical microscopy with fluorescence microscopy being by far the most commonly used technique. Depending on their differentiation, cells have different mechanisms that can be electronically detected as well. While neurons function as the biologically active component of the nervous system and create electronic signals in the form of action potentials by themselves, many other cell types are not able to do the same. Therefore, the effects detected by FETs vary for different cell types.

In general, there are two ways of obtaining information from cells: intracellular and extracellular recordings. The intracellular tools, such as the famous patch-clamp method [10], are invasive and are not suited for long-term measurements of the same cell, due to the penetration of the cellular plasma membrane. Furthermore, those techniques require trained and skilled personnel to perform the experiments, which is not feasible for routine applications or lab-on-a-chip (LoC) applications. In contrast to this, ISFETs belong to the family of non-invasive measurement tools, which avoid the above-mentioned drawbacks. The cells are usually cultured in the desired density right on top of the device surface and the readout is done without penetrating the membrane and touching the sensitive cell interior. Just as the invasive techniques, the measurements can be performed in real time and in parallel for all the sensing points on one device. The main drawback of this method, however, is the much smaller signal amplitudes that need to be measured with these planar devices as the coupling between cell and transducer is of capacitive nature. In a vast amount of research several model descriptions were generated and our knowledge about the ionic–electronic interface between cells and transistors was deepened. This field of research has been pioneered by the group of Peter Fromherz and a complete description could be the central theme of a book chapter itself. The readers are referred to a concise review by P. Fromherz and the references therein [11].

This chapter will focus on one specific aspect of cell-transistor sensing, in which an impedance readout technique is used to detect the adhesion strength of individual cells. Figure 1 shows a schematic of such a measurement configuration, where a cell is in contact with an ISFET device. FETs are sensors that consist of three contacts: the source, the drain, and the gate. In the case of ISFETs, a gate potential is applied via an external reference electrode contacting the electrolyte solution, which is itself in direct contact with the gate dielectric material. For very stable potential measurements with ISFETs, using a liquid junction reference electrode such as silver/silver chloride is indispensable, whereas in the field of electrophysiology it is very common to use a chloridized silver wire as a pseudo-reference electrode. Source and drain contacts are passivated towards the electrolyte in which the reference electrode is emerged. When applying a voltage between drain and source V_{DS}, current flows through the thin electronic channel formed underneath the gate dielectric layer. The current flow is regulated by a second voltage, which is applied between the electrolyte gate and the source contact. This gate-source voltage V_{GS}, which is applied via the electrolyte solution, is dropping over different parts of the circuit such as the reference electrode–electrolyte interface, the electrolyte solution

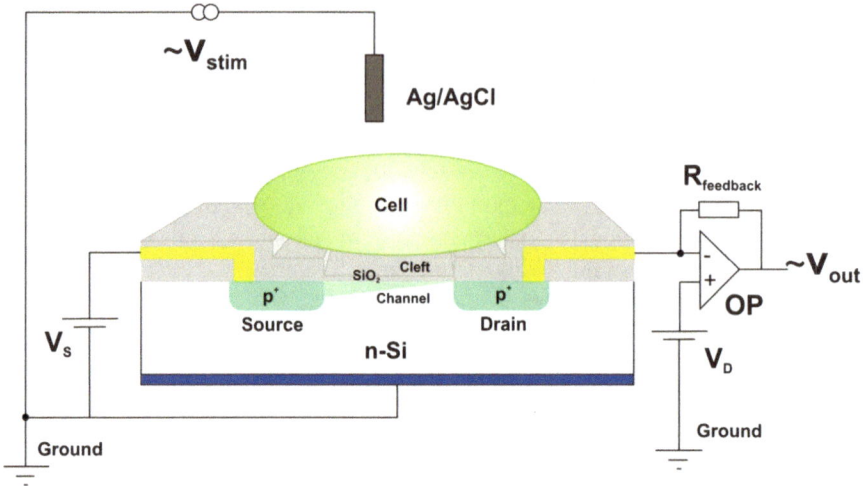

Fig. 1 Schematic of a FET device with a single adherent cell. The figure includes a schematic of the readout circuit, which is typically used in our custom-made amplifiers for single cell impedance measurements. The electrolyte solution is kept at ground contact to be compatible with patch-clamp instrumentation, which can be used in parallel experiments. A sinusoidal stimulation signal V_{stim} is fed to the reference electrode and the output voltage V_{out} after the first transimpedance amplifier stage is detected to obtain the frequency spectra. A similar layout was described earlier [13]

itself, and the electrolyte–gate dielectric interface at the ISFET input. The attachment of charged molecules such as DNA or proteins on the surface is changing the surface potential at the liquid–solid interface of the gate. Since this is part of the effectively applied gate-source voltage, the resulting differences between "empty" gates and gates covered by the charged species can be sensed with a chip with several electronically identical ISFET sensors in a transistor array [12]. Since no current flow through the gate dielectric material is allowed in FETs, this is regarded as capacitive coupling or potentiometric sensing. In this readout mode, either the differences in the transfer characteristics of the ISFETs or the changes in drain-source current I_{DS} are sensed. In the latter approach the devices are set into a working point where a constant drain-source current is flowing through the channel. When the effective gate-source voltage is changed upon biomolecule binding to the gate, this change can be sensed in the drain-source current. The relationship between changes in drain-source current caused by a distinct change in gate-source voltage is technically expressed as the *transconductance value* g_m and this value has been optimized by FET designs and choices of materials.

The second possibility of sensing with FETs is an impedimetric approach. In this case the FET is set to a working point as discussed above. Again a constant drain-source current I_{DS} is flowing through the channel. In addition to a constant DC gate-source voltage applied via the reference electrode, a small sinusoidal voltage V_{stim} with a particular frequency and amplitude is superposed. This small disturbance is transferred into a disturbance of the drain-source current in the channel. To record frequency spectra, the frequency of this stimulation voltage V_{stim} is scanned and the output voltage after the first amplifier stage of the setup V_{out} is compared to this input (Fig. 1).

A lock-in amplifier is used to sensitively record changes in amplitude and phase of the signal. This method was firstly described by Schasfoort and co-workers [14] and later on also used by others [15–17]. The applicability of the same method to sense the adhesion of individual cells on FET devices and to obtain cell responses upon exposure to chemical stimuli was first shown in 2009 [18].

This chapter introduces cell impedance sensing, as it can be performed with ISFETs on a single cell level. For cell-adhesion studies, a device with a sensing area similar to the size of a single cell is preferable and silicon-based FETs fit into these needs. Compared to the commercially available Electric Cell-substrate Impedance Sensing (ECIS) using gold microelectrodes [19, 20], the FETs are not limited to sizes of several hundred µm, but can be applied to study single cells. Typical sizes of the commercial metal microelectrodes in ECIS range from 25 to 250 µm in diameter. The reason for this is that in the case of passive metal microelectrodes, tiny currents need to be transferred through macroscopically long cables and measurements easily pick up noise from the environment. In addition, the metal–liquid interface provides considerable interface impedance, which is getting too large for standard instrumentation when scaling the sensors to individual cell sizes. In contrast to this, ISFETs offer much smaller input impedances, the tiny current changes are directly converted at the gate input and carried on top of large and strong drain-source currents so that interspersed noise is not a problem in this measurement mode.

Due to the much larger sizes of the microelectrodes in conventional ECIS, *collective* cell responses are measured instead of *individual* responses. The data obtained from such averaged cell populations might conceal the individual reactions of single cells within an ensemble of cells with heterogeneous properties [21]. Therefore, it should be of interest for specific applications to analyze individual reactions of cells. In this chapter, we describe the impedimetric sensing method for cells using ISFETs, for which we created the name Field-Effect Transistor Cell-substrate Impedance Sensing (FETCIS). We show that this technique is offering single cell resolution for ECIS.

Intensive research investigating the general characteristics of the electronic coupling between cells and ISFET transistors has been performed over many years to reach the point of knowledge of today [22]. The initially developed system consisted of 16-channel FET sensor arrays that were used in several different applications [13, 23–25]. The design of these chips was based on devices described by Offenhäusser et al. [26]. This early work, however, was not aiming for single cell impedance measurements and was focused on measuring electrically active cells [27–29]. The first investigation into the impedance sensing capabilities of these FET devices was for label-free DNA detection [30], down to the detection of single nucleotide polymorphisms [31]. These studies used a portable amplifier system for potentiometric and impedimetric measurements without the need of rack-sized amplifiers. The same setup was later on used for the first studies on cell impedance in 2009 [18]. Recently, Koppenhöfer and co-workers reported the use of the ISFETs for the impedimetric analysis of nanoparticle induced cell death of the human lung adenocarcinoma epithelial cell line H441. While H441 cells grew in colonies, a verification of the single cell resolution of the readout was not possible. In the same year, however, Susloparova and co-workers confirmed single cell impedance sensing on the same devices [24] using the Human Embryonic Kidney cell line HEK293.

The versatility of the devices with their numerous applications led to a deeper investigation of the working principles and the interpretation of the obtained data [25]. The model developed by Susloparova and co-workers allows the extraction of the cell-related parameters that are inherently included in the frequency response. Because of the inhomogeneous surface topography created by the different heights of the contact lines and transistor gates of the FET chips, a modified version of flat ISFETs was fabricated [32]. The almost completely flat topography of the chips improved the cell adhesion strength and enabled individual cell migration studies. Law and co-workers showed the capability of these improved devices for single cell adhesion and migration studies with activated human T cells [33]. In the same year the FETCIS method was used for toxicity measurements in neuronal cell cultures, which cannot form confluent tissue [34]. With these two studies beyond or at the limit of the standard ECIS capabilities, an entirely new field opened up for future works on FETCIS chips, amplifiers, and applications.

2 Materials and Methods

2.1 Field-Effect Transistors

In the initial projects for FETCIS, the ISFET devices for impedance measurements were 16-channel p-type FET microarrays in a common source layout as described in previous publications [27, 35, 36]. The arrays consisted of 16 transistor gates in a 4×4 array with a spacing of 200 μm between the individual gates, in the center of a 5×5 mm^2 silicon chip with different gate dimensions (mostly 5 μm length and 16 μm width) with a gate oxide thickness of only 8 nm. Chips were fabricated at the Institut für Mikrotechnik Mainz, Germany. The FET devices reached transconductance values between 0.2 and 0.6 mS. The devices were passivated using an oxide-nitride-oxide stack, which also ensured the re-usability of the chips for several experiments. All devices were encapsulated for biosensing purposes as described earlier [26].

Second generation devices were designed for an improved FET performance. These new devices achieved the same electrical properties as the former with the advantage of a flatter surface, which is favorable for cell growth and migration as described above. In addition, the nitride passivation layer was omitted, which reduced one lithography step and the gate oxide was thinner than before. An exact protocol can be found in the publication of Susloparova and co-workers [32]. A schematic of the fabrication principle is shown in Fig. 2. The resulting devices had several gate sizes with those used for FETCIS with 12 μm in width and 5 μm in length with a gate oxide thickness of 6 nm. The pitch between the individual gates was again 200 μm.

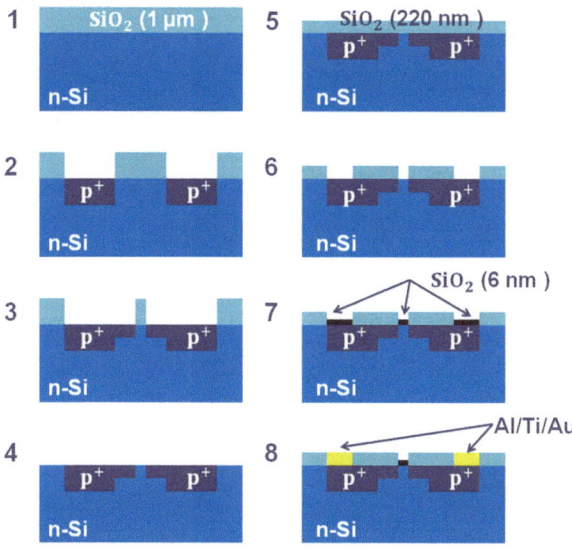

Fig. 2 Schematic of the fabrication process for the flat, open-gate ISFETs [32]. The process included two implantations, four optical lithography steps, gate oxide growth of SiO_2, and metallization of contacts. In more detail, the process steps were: (1) Growth of 1 μm silicon oxide. (2) Creation of the contact lines by first lithography including wet etching and boron implantation. (3) Definition of the source and drain contacts by a second lithography including wet etching and boron implantation. (4) Removal of the implantation oxide to achieve a quasi-planar surface topography. (5) Passivation of the surface by a SiO_2 layer of 220 nm thickness. (6) Opening of the source, drain, and gate contacts. (7) Growing of a gate oxide of 6 nm thickness by dry oxidation. (8) Deposition, structuring, and annealing of a metal-layer stack at the bond pads

The measured transfer characteristics and the transconductance of the first and second generation of FET devices are compared in Fig. 3. The top graphs in Fig. 3a, b show the transfer characteristics for a range of gate-source voltages V_{GS} from 0 to -3 V, while varying the drain-source voltage V_{DS} from 0 to -3 V in steps of 1 V. The bottom graphs (Fig. 3c, d) show the corresponding transconductance values. Compared to the first generation devices the flat FETs showed a slightly higher transconductance with an average increase of 0.4 mS. Also a saturation effect in the transfer characteristics is observed for higher gate-source voltages. This is an effect of the slightly higher resistances of the feedlines for these sensors. However, a larger bandwidth for impedance sensing was accessible with the second generation FETs as described in a former publication [32].

2.2 Measurement Setups

For FETCIS recordings with ISFET arrays several amplifier setups were developed in our group in the past few years. Some of these setups used commercial lock-in

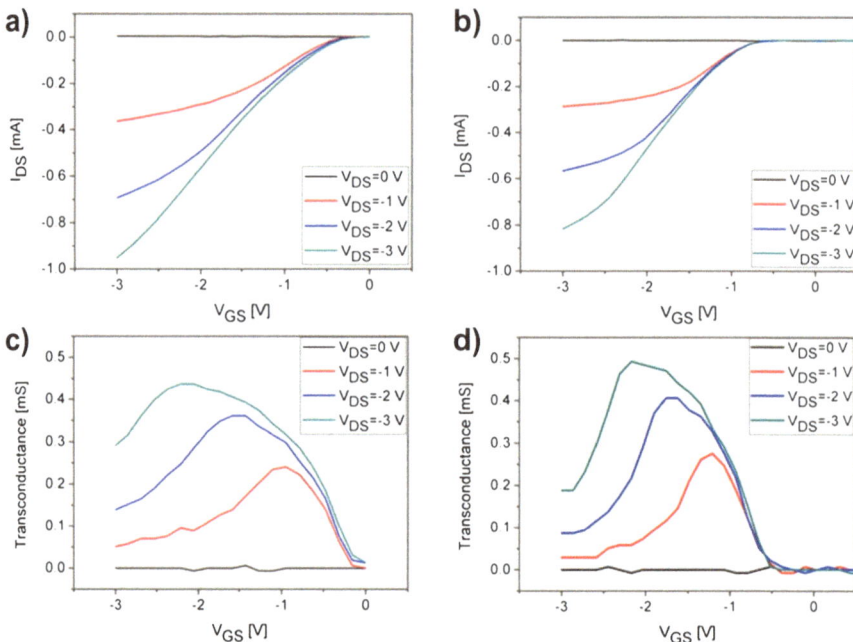

Fig. 3 (**a**) Transfer characteristics of the first generation of FET devices with their transconductance values shown in (**c**); (**b**) transfer characteristics of the second generation FET devices with flat topography and their transconductance values shown in (**d**) [32]. This generation had slightly higher transconductance values with higher resistances of the feedlines as indicated by the stronger saturation effect in the transfer characteristics

amplifiers coupled to standard transimpedance amplifier circuitry as described in Fig. 1 and some of these setups were complete, integrated, and portable solutions. The measurement setup used for the most recent studies starting from 2010 was an improved system of a previous, portable amplifier version developed for biomolecular detection [28, 29]. As the preceding portable system, it has the ability for parallel electrical characterization and time-dependent, potentiometric readout and contained 16 independent phase selective amplifiers for parallel impedimetric readout as well. The FETs are operated in saturation and the working point is set by applying constant drain-source voltages V_{DS} and gate-source voltages V_{GS}. After the characterization of the chips, the first derivative of the transfer characteristics is calculated to provide the transconductance values for the individual channels, which allows setting the working point at the point of the highest transconductance, meaning the point of the highest sensitivity of the device. A direct-digital-synthesis device is providing a sinusoidal stimulation signal V_{stim} with exact amplitude, frequency, and phase for the measurements. The frequency-selective detections are then realized by mixers and a low-pass filter as in standard lock-in instrumentations. In signal theory, the impedimetric measurement principle can be compared to the recording of a transfer function. The V_{stim} signal is scanned over the frequency range from typically 1 Hz to 1 MHz resulting in a band pass measurement of the complete system consisting of reference electrode, cell, FET, and transimpedance amplifier. Here, a very basic description for this measurement method is provided. It was named the recording of the "transistor-transfer function" (TTF) and the corresponding amplifier is referred to as the TTF amplifier box.

The very general definition of a transfer function is the ratio of the output to the input voltage of the system as stated in Eq. (1).

$$H(j\omega) = \frac{v_{out}(j\omega)}{v_{in}(j\omega)} = |H(j\omega)| \cdot e^{j\varphi(\omega)} \tag{1}$$

This can be separated into the amplitude part $|H(j\omega)|$ and the corresponding phase part $e^{j\phi(\omega)}$. For measurements with the TTF amplifier box only the real part of the impedance is recorded, providing Eq. (2).

$$H(j\omega) = \frac{v_{out}(j\omega)}{v_{in}(j\omega)} = |H(j\omega)| \cdot \cos(\varphi(\omega)) \tag{2}$$

The recorded TTF spectrum is then a combination of the bandwidth limiting effects of reference electrode, electrolyte solution, the cell covering the transistor, the transistor itself including contact lines on chip, and the first amplifier stage. By applying a sinusoidal signal with 10 mV amplitude and varying frequency from 1 Hz to 1 MHz via the reference electrode, changes in bandwidth caused by attaching different biological samples to the gate oxide of the FETs were identified [18, 31]. Responses of the biological sample to the V_{stim} voltage are frequency-dependent as well and can be extracted from the spectra. However, the lock-in based nature of the amplifier also enables a time-dependent approach, where the amplifier is set to a fixed frequency and changes over time at that particular frequency are recorded. Using a lock-in amplifier for this technique is greatly improving the signal-to-noise ratio of the recordings. This has been discussed in detail before. Based on the recorded frequency spectrum, the dielectric properties of the attached biological sample can be extracted by equivalent circuit modeling [25]. This model includes cellular properties as well as parasitic effects as it will be discussed in the following paragraphs. A later setup developed in 2010 included a fast lock-in amplifier (HF2LI Lock-in Amplifier, Zürich Instruments, Switzerland) that provided the full amplitude and phase information of the spectrum as the basis for the subsequent in-depth analysis of the TTF spectra [24, 25, 32].

2.3   Chip Encapsulation

For FETCIS experiments with cell types that require a longer culture period in a humidified incubator, some basic requirements for the ISFET devices must be fulfilled. Most importantly they need to be encapsulated appropriately with

Fig. 4 Encapsulated chip for cell measurements (here 16-channel silicon, open-gate ISFETs). The encapsulation is a standard process and applied for all our devices. The cavity inside of the glass ring holds about 400 μL of cell medium and the cells are cultured on the silicon oxide chip surface in the center. The silicone resin provides a stable electrical isolation of the bond wires from the cell culture medium

5 mm

biocompatible epoxy resins. Sylgard® resins were used for this purpose (Sygard® 184 and 96-083, Dow Corning, Germany), which have proven perfect biocompatibility even for more delicate, primary neuron cultures.

During electronic readout, the resins need to ensure that no electrical short circuit is generated between the bond wires, while the measurement medium is kept in a petri dish like liquid container formed with a glass ring on top of the chips. Figure 4 shows one of the encapsulated devices.

2.4   Preparation of Devices and Cells

Over the past years, the mechanical stability of the gate oxide and the oxide passivation layers of the contact lines were optimized as well. Nowadays the FET sensors withstand a relatively harsh cleaning procedure and can be reused many times for cell culture experiments. The ISFETs were mechanically cleaned with cotton buds and ethanol before they were ultrasonicated for 5 min in 2% Hellmanex solution (Helma GmbH & Co., Müllheim, Germany). Afterwards the devices were ultrasonicated in de-ionized (DI) water for 5 more minutes before they were blown dry using nitrogen. Sulfuric acid (H_2SO_4) with a concentration of 20% was used for activation of the chip surfaces. The device surface was covered with H_2SO_4 and kept at 80°C for 30 min. Finally the chips were rinsed with DI water, sterilized with 70% ethanol and functionalized with different surface coatings like laminin, fibronectin, poly-lysine, etc. depending on the desired cell line prior to the seeding of the cells. Except for the primary T-cell experiments discussed in paragraph 3.6, all chips were coated with 10 μL of 0.1 μg/mL fibronectin (AppliChem GmbH, Germany) and incubated at 37°C for 3 h. The fibronectin solution was then removed and chips were washed with sterile buffer solution before the cells were seeded. For the primary T-cell experiments, different surface coatings were used for the chip surface. Details are described in the corresponding paragraph. The following cell lines were used in the experiments described here:

Human Malign Melanoma SkMel28 Cells

For human malign melanoma SkMel28 cells 100,000 cells were seeded in 20 μL of medium. Dulbecco's Modified Eagle's Medium (supplemented with 10% Fetal Calf Serum, 100 U/mL Penicillin and 0.1 mg/mL Streptomycin and 2 mM L-Glutamine) was used as culture and measurement medium.

Human Lung Adenocarcinoma Epithelial H441 Cells

For human lung adenocarcinoma epithelial H441 cells 100,000 cells were seeded in 20 μL of medium. Roswell Park Memorial Institute medium (supplemented with 10% Fetal Calf Serum, 100 U/mL Penicillin and 0.1 mg/mL Streptomycin and 2 mM L-Glutamine) was used for culturing.

Human Embryonic Kidney HEK293 Cells

In case of the human embryonic kidney HEK293 cells, standard Minimal Essential Medium (MEM) supplemented with 10% Fetal Calf Serum (FCS), 1% Non-Essential Amino Acid (NEAA), 100 U/mL Penicillin, and 0.1 mg/mL Streptomycin and 2 mM L-Glutamine (all PAN Biotech GmbH, Germany) was used as culture and measurement medium. A cell suspension of 1.5×10^5 cells/mL was added to each chip and cultured for 1 day before the cultures were experimentally studied.

Neuro2A Cells

The Neuro2A murine neuroblastoma cell line, which shows neuronal and amoeboid stem cell morphology was grown in Dulbecco's Modified Eagle's Medium (DMEM), 10% FCS, 2 mM L-Glutamine, 100 U/mL Penicillin and 0.1 mg/mL Streptomycin, and 1% of non-essential amino acids (NEAA).

Primary SVZ Cells

The primary cells from the subventricular zones of 1–3 days old postnatal BALB/c mice were grown in proliferation medium consisting of DMEM/F12 containing 0.35% (w/v) bovine serum albumin (BSA) in PBS, 2 mM L-Glutamine, 0.05 mM β-mercaptoethanol, 100 U/mL Penicillin and 0.1 mg/mL Streptomycin, 2% neural cell culture supplement B27 (without retinoic acid), 20 ng/mL recombinant human fibroblast growth factor, 10 ng/mL recombinant human epidermal growth factor for 5 days. This was followed by dissociation and cultivation in differential medium consisting of DMEM/F12 containing 0.35% bovine serum albumin (BSA) in PBS, 2 mM L-Glutamine, 50 μM β-mercaptoethanol, 100 U/mL Penicillin and 0.1 mg/mL Streptomycin, 2% neural cell culture supplement B27 (with retinoic acid) for 7 days before the measurements.

Primary Human T Cells

For the experiments with primary human T-cells, the FETs were coated with different surface coatings: (1) 0.1 mg/mL fibronectin, (2) 30 μg/mL CD3 antibody, or (3) 30 μg/mL LFA-1 antibody for 3 h before removing the unbound residue by rinsing the surfaces with sterile buffer solution. The TTF spectra were recorded 15 min after the incubation with the T-cells. The effect of the different coatings on the adhesion strength of the cells was investigated in this study. In experiments using human T-cells [33], peripheral blood mononuclear cells (PBMCs) were purified from the blood of the human donors. $CD8^+$ T-cells were negatively isolated from PBMC by Dynabeads® Untouched™ Human CD8 T Cells Kit (Life technologies, USA). In brief, a mixture of mouse IgG antibodies against the non-$CD8^+$ T-cells was added to the PBMCs and allowed to bind to the cells. Dynabeads® were added and allowed to bind to the antibody labeled cells during a short incubation. All bead-bound non-$CD8^+$ T-cells were then separated from the $CD8^+$ T-cells using a magnet. Afterwards the $CD8^+$ T-cells were stimulated by Dynabeads® Human T-Activator CD3/CD28 (Life Technologies, USA) for 3 days, before the beads were removed. The remaining cells were re-suspended in AIM V® medium supplemented with 10% FCS and 100 U/mL recombinant human IL-2 and 100,000 cells were applied in medium into the FET device.

3   Results and Discussion

3.1   Low Density Cultures

Low density cell cultures were used to prove that the sensitivity of the FET devices and the amplifier system can reach down to single cell resolution. SkMel28 cells with polygonal morphology are shown as an example for this type of experiments. As it can be seen in Fig. 5, the FET surface was completely or partially covered by the cells (Fig. 5a, b). Using our FETs, significant differences in the spectra were observed between the cell-free and cell-covered transistors in both confluent and sparse cultures (Fig. 5c, d).

The results show that the FET devices and amplifier system are suitable for detecting cell adhesion from confluent down to single cell cultures. As it can be seen from Fig. 5, the changes in the TTF spectra are very similar when comparing confluent and single cell cultures. The reason for this is that the tiny FET gates used in this platform are already covered completely by a single cell. In the ECIS model derived for confluent cell layers, the cell–cell junctions also play a major role in data interpretation of the measured impedance spectra. In the model described in this chapter we focused on single cell experiments only so that cell–cell junctions are not included. Of course, by using larger FET gates, the same cell–cell contributions like in the standard ECIS approach including cell–cell junctions would be visible in the TTF spectra, but the single cell FET approach has the capability of showing differences from cell to cell in complex cell cultures that can even consist of different cell types.

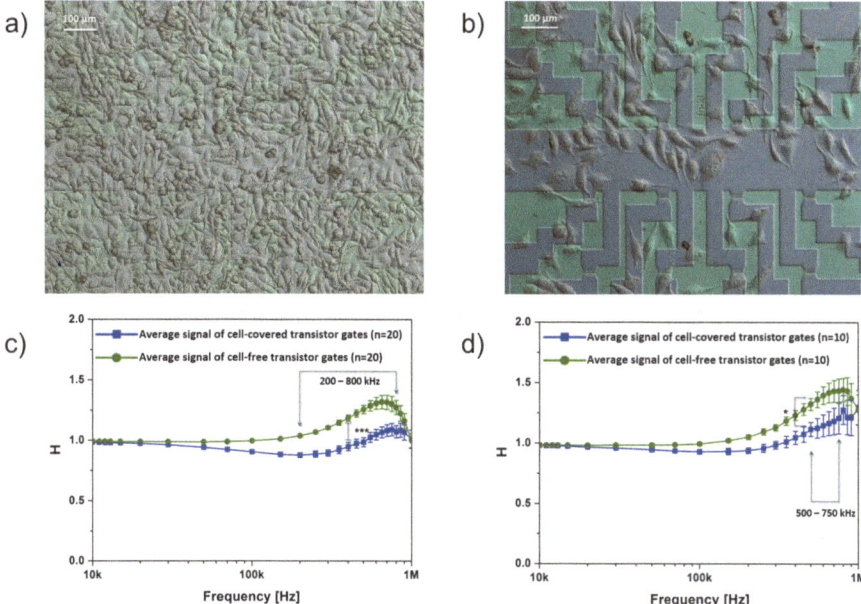

Fig. 5 FET device of the new design with a gate size of 16×5 μm^2 covered by a dense (**a**) or a sparse layer of SkMel28 cells (**b**) with the corresponding measurements of cell-free and cell-covered transistor gates (**c**, **d**). The graphs show the averaged curve of the measurements with the respective standard deviations represented in the error bars. In the indicated frequency ranges statistically significant differences were observed for cell-free and cell-covered transistor gates

3.2 Single Cell Impedance

The HEK293 cell line was used for many single cell impedance experiments with FETs. This is a very well-studied cell type also in electronic cell-transistor coupling experiments using a parallel patch-clamp configuration [23, 37, 38]. In complementary studies, the adhesion region between cells and transistor surfaces and also the morphology of the HEK293 cells was characterized using optical microscopy [39] and transmission electron microscopy [40]. For single cell adhesion experiments, the HEK293 cells were cultured under standard conditions and the experiments were performed using the chips and the portable amplifier system inside of an incubator at 37°C and 5% CO_2.

Figure 6 shows an SEM image of a single HEK293 cell on a transistor gate. The gate opening size was chosen to be approximately the same as the cell size (16×5 μm^2, green area in Fig. 6) to ensure the best electronic coupling.

Figure 7 shows the results of the TTF box measurements. In Fig. 7b the real part of the impedance for four individual channels is plotted. These TTF spectra reveal distinct differences between cell-covered and uncovered gates for frequencies above 50 kHz. At frequencies above 100 kHz the TTF spectra are rising in values, which is

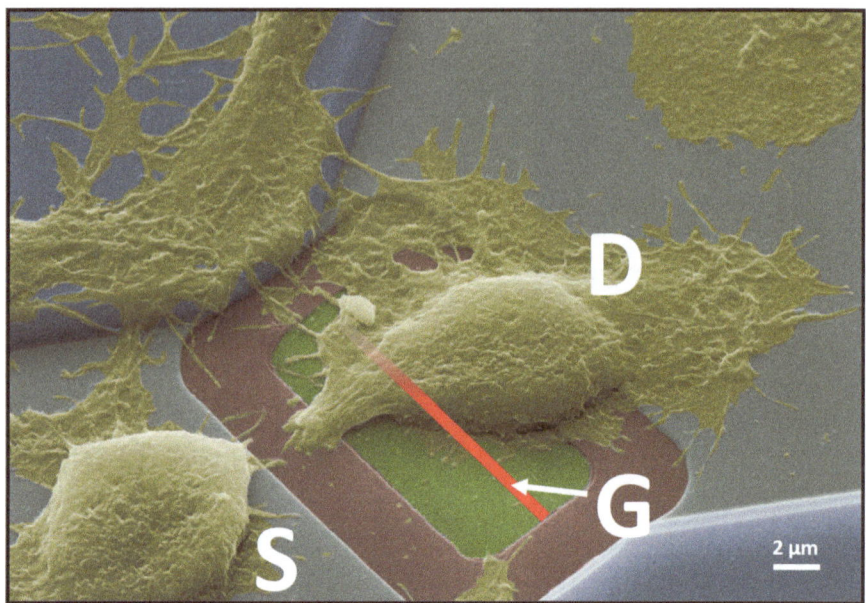

Fig. 6 Scanning electron microscopy image of a single HEK293 cell attached to a transistor gate (colored image). The letters S, D, and G represent the structures of the transistor, namely source, drain, and gate [24]. Please note, when properly positioned a single cell can cover one transistor gate

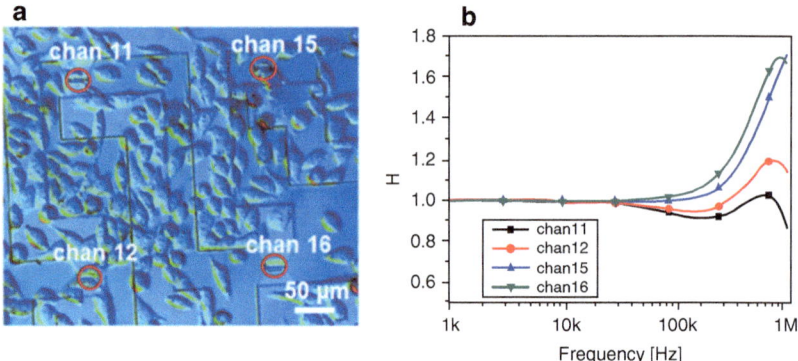

Fig. 7 (**a**) Microscopic image of transistor gates (indicated by the red circles) of the test chip covered by HEK293 cells. Gate 11 and gate 12 are cell-covered, while gate 16 is cell-free and gate 15 is only partially covered. (**b**) Normalized TTF spectra measured by the TTF amplifier box [24]

due to a swing-effect with the respective operational amplifier in the circuit. Cell-covered gates (no. 11 and 12 in Fig. 7) experience a stronger damping compared to cell-free gates (no. 15 and 16 in Fig. 7). Therefore this graph demonstrates the influence of cellular adhesion on TTF readings, which leads to a low-pass effect in the TTF spectra. A direct comparison for the impact of one single cell on the impedance spectra is provided in Fig. 8. The TTF spectrum was measured with the adherent cell and after this cell was removed mechanically by a patch-clamp pipette. The spectra show a clear difference for the cell-covered and the cell-free gate starting from 50 kHz and peaking at around 500 kHz.

3.3   TTF Spectra and Modeling

To fully explain the recorded TTF spectra, the FET device in contact with an adherent cell is described by an electrical-equivalent circuit (EEC) model (Fig. 9), which is derived from the most simple Point-Contact Model, first used for the explanation of signal shapes of extracellularly recorded action potentials of neurons [41]. This model was used in the early days of this direction of research as the easiest approach to explain extracellularly recorded signal shapes from neurons [29, 42] or cardiac myocytes [27]. For a first approximation of the recorded effects in FETCIS, this model is sufficient as well.

In brief, the capacitance of the adherent cell membrane C_M, the resistance of the cell membrane R_M, and the resistance of the thin, electrolyte-filled cleft between the cell and transistor R_{seal} represent the passive elements of the EEC model. It is to be noted that in this FETCIS model the C_M value of the membrane does not represent the complete capacitance of the cell membrane as typically measured from whole-cell patch-clamp experiments, but rather a "projected" capacitance formed by the ratio of the surface-attached part and the free part of the cellular membrane. In the EEC this can be modeled by dividing the cellular membrane into free (capacity of the free membrane C_{FM}, resistance of the free membrane R_{FM}) and attached parts (capacitance of the junction membrane C_{JM}, resistance of the junction membrane R_{JM}) including the inner resistance of the cell (R_i) and the seal resistance (R_{seal}) as well.

The contact line capacitances and resistances of source and drain are included in the EEC with C_{source}, C_{drain}, and R_{source}, R_{drain}, respectively. A combined resistance R_{el} is representing the reference electrode impedance (ideally only a resistor) and the electrolyte resistance in series. The transistor is set into its working point by adjusting the voltages V_{source} and V_{drain} with respect to the electrolyte solution at ground potential. The TTF spectrum is then measured between the input voltage V_{in} and the output voltage V_{out} of the transimpedance amplifier to the right (Fig. 9). In this EEC, the FET can be modeled by the drain-source resistance R_{DS} in parallel to a current source formed by V_{GS} and the transconductance in the working point g_m. In order to optimize the FETCIS sensor for highest resolution in TTF recordings, different simulations were performed using the EEC model implemented in a PSpice circuit simulation (PSpice® 9.1 student version, OrCAD® now by Cadence Design Systems, Inc., USA) [32]. Sensitive FETCIS sensors show large TTF differences between the cell-covered and the cell-free case and in addition these effects are

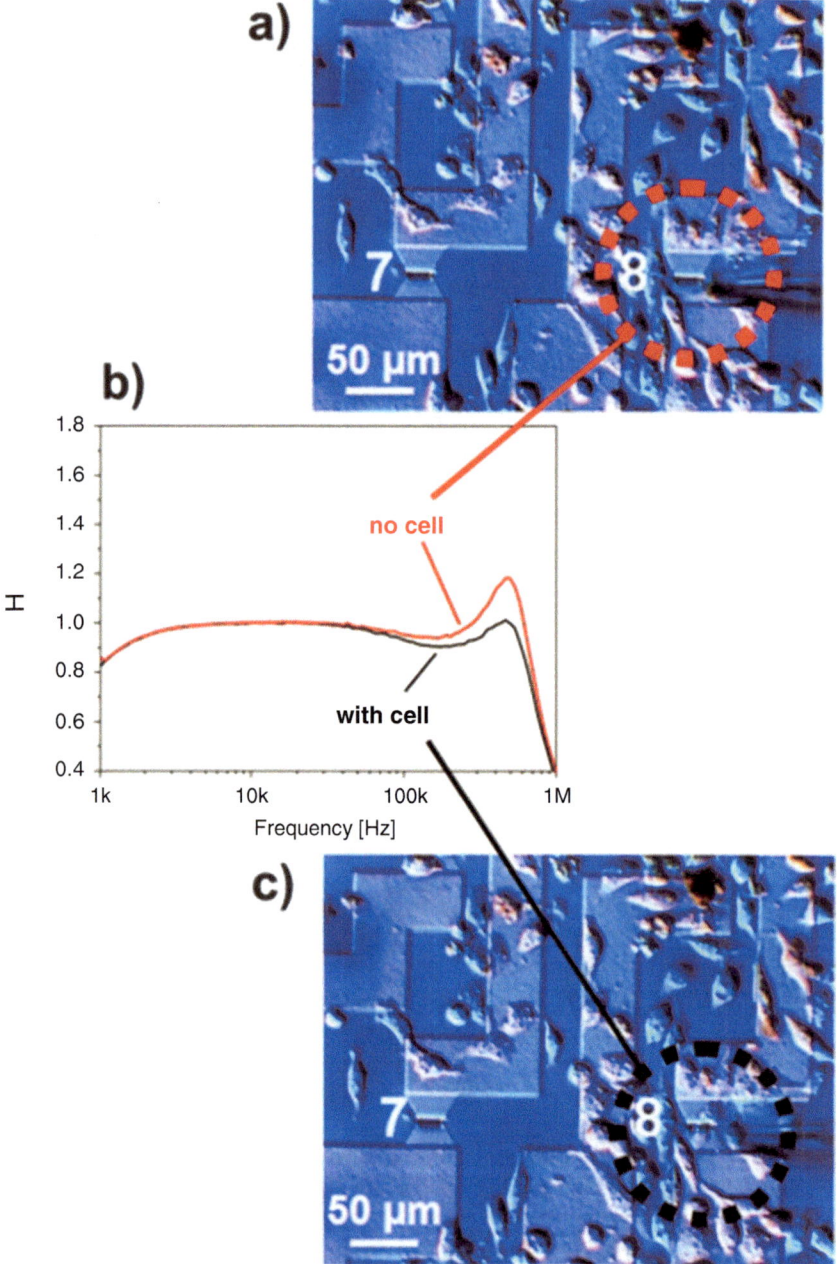

Fig. 8 Images of the mechanical removal of a cell from a single transistor gate (no. 8) with the corresponding TTF spectra of this transistor for the cell-covered (black) and the cell-free (red) case [24]

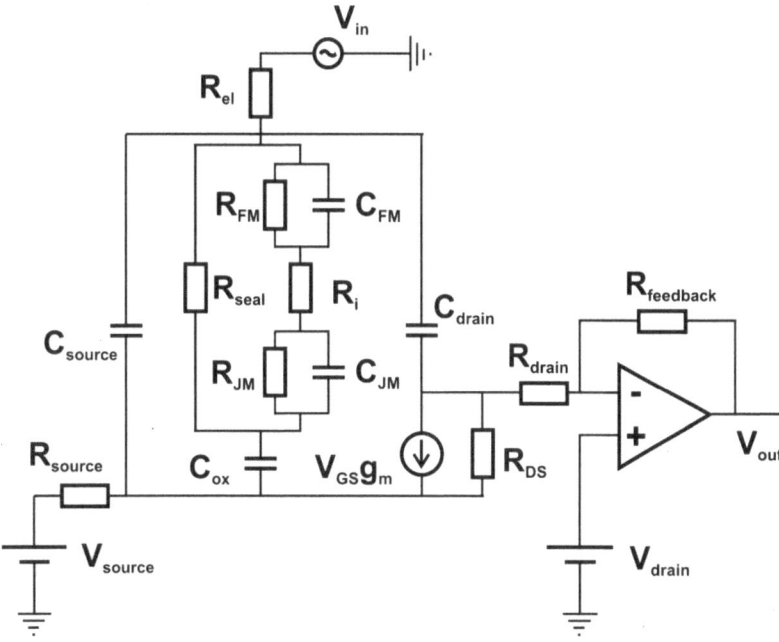

Fig. 9 EEC of an ISFET with adherent cell on the transistor gate [25]: The FET can be modeled in its simple form by the source-drain resistance in the channel R_{DS} in parallel to a current source formed by V_{GS} and g_m of the working point (V_{GS}, g_m) The gate oxide is represented by C_{ox}. The FET is set into its working point by applying voltages V_{source} and V_{drain} with respect to the electrolyte solution, which is grounded by the reference electrode. To fully model the transistor response, capacitances and resistances of the contact lines (R_{source}, C_{source} and R_{drain}, C_{drain}) need to be added as well. The transistor currents are converted and amplified by the transimpedance amplifier formed by an operational amplifier and a feedback resistance $R_{feedback}$ (right side). The cell is modeled by a seal resistance R_{seal} and two passive RC elements for the junctional membrane (R_{JM} and C_{JM}) and the free membrane (R_{JM} and C_{JM}) connected by an inner resistance R_i. A stimulation signal V_{in} is fed to the reference electrode, which forms a resistor together with the electrolyte solution R_{el}. FET and circuit amplify and attenuate this input signal. The cell-related parameters R_{seal} and C_M (as a combination of C_{JM} and C_{FM}) are extracted from the obtained TTF spectra

favorably observed at lower frequencies, which can be realized by cheaper instrumentations.

In order to fully understand the TTF spectra, an analytical equation representing the transfer function of the EEC was derived [25]. In this analytical equation, several parameters were considered, which influence the typical shape of the measured TTF spectra in individual ways. These parameters include seven chip-related parameters (g_m, R_{source}, C_{source}, R_{drain}, C_{drain}, R_{DS}, C_{ox}), two transimpedance circuit parameters (R_{el}, $R_{feedback}$), and six cell-related parameters (R_{seal}, R_{JM}, C_{JM}, R_{JM}, C_{JM}, R_i). To derive an analytical model, the resistive path through the cell was neglected (R_i, R_{JM} and R_{FM}) and only the capacitive components of the free and attached membrane were combined to a cell membrane capacitance C_M. By fitting the resulting equation to the TTF spectra (procedure and equation is discussed below), the cell-related parameters (R_{seal} and C_M) are obtained, which indicate the status of the cell. The R_{seal} value is a measure how tightly the cell adheres to the surface of the transistor, while the C_M value indicates the shape of the cell and the dielectric properties of the membrane. These two cell parameters are changing during cell attachment and detachment processes. In Fig. 10 an idealized cell detachment process from the transistor surface is modeled for further analysis with the cell not changing its shape during detachment. It is only lifted up in a stepwise manner. The figure shows the simulated TTF spectra upon varying R_{seal} in a range of 1 MΩ to 200 kΩ.

By increasing R_{seal}, a regular increase of the signal amplitude between 20 and 200 kHz is observed, while the complete removal of the cell can be clearly distinguished from the cell-attached case.

On the other hand, the projected membrane capacitance C_M has different values when the morphology of the cells is changing during detachment or as a response to external stimuli. In Fig. 11, an adherent cell is gradually deformed during the detachment process indicated by the cell height h_{cell}. Simulations were performed by modeling an adherent cell as a hemisphere and a gradually detaching cell from the transistor surface with different spherical caps keeping the total surface area of the cell membrane constant, but changing the ratios of junctional and free membrane. In this model, the C_M value is then decreased stepwise. The change in the ratio of attached membrane area to free membrane area during detachment changes the shape of the TTF spectra, accordingly. The simulated TTF spectra show that decreasing C_M values lead to smaller TTF values between 200 and 500 kHz (Fig. 11).

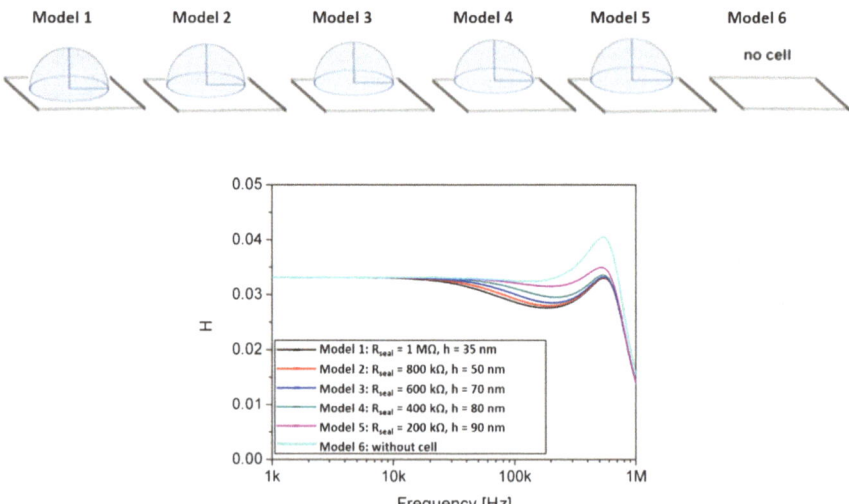

Fig. 10 Changes in the calculated frequency-dependent TTF for different values of R_{seal}. The data simulates the gradual detachment of the cell from the gate. In the graph the different values for the seal resistance are shown for the respective cell-substrate distances h

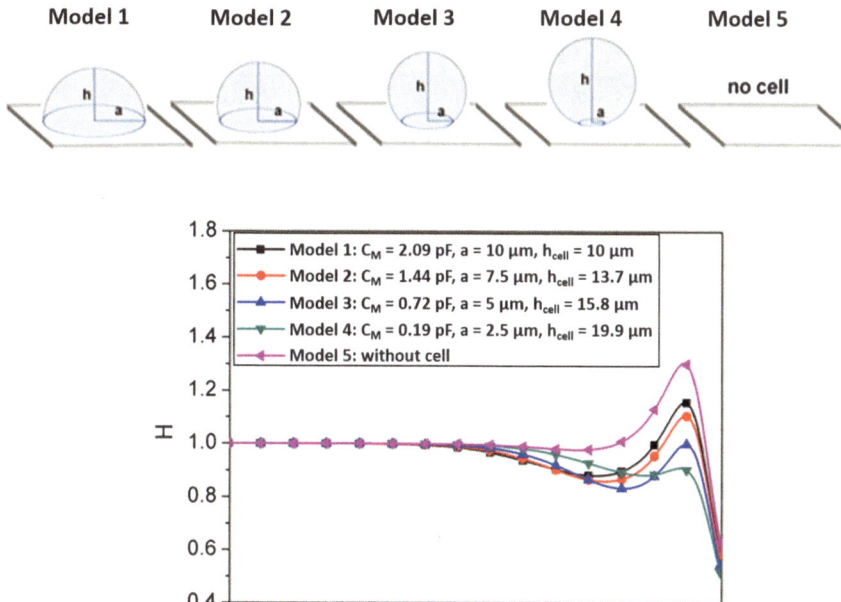

Fig. 11 Changes in TTF spectra during systematically changing C_M from values of 2.09 to 0.19 pF mimicking cell release from the gate surface. The respective C_M values corresponding to different cell heights h_{cell} are indicated in the figure. The response of the TTF spectra to capacitance changes is not as regular as to the changes in R_{seal} shown in Fig. 10

According to these simulations, the TTF spectral changes during cell attachment and detachment are assigned to changes in R_{seal} and C_M. For a complete picture of the changes on the gate, however, recordings at a fixed frequency are not enough. For time-dependent processes, recordings of the full spectra and fitting the model to the data should be done in order to extract the time evolution of the cell-related parameters from the model [25, 43], like it was earlier described for ECIS. For further studies of this topic and for a more detailed description of these effects the readers are referred to publication [25].

3.4 Optimization of Devices for Single Cell FETCIS

To be able to improve the devices with respect to sensitivity for single cell impedance analysis, the device-related parameters influencing the TTF spectra were evaluated using simulations [32].

$$R_{\text{seal}} = \frac{k_B T}{4\pi \cdot e_0^2 \cdot n_{\text{tot}}^B D_{\text{K}^+} h} \left[1 + 4 \frac{L_D^2}{r^2} \left(\frac{1}{I_0(r/L_D)} - 1 \right) \right] \tag{3}$$

$$H(j\omega) = R_{\text{FB}} g_m \frac{1}{\sqrt{1 - \left(\frac{f}{f_g} \right)^2}} \cdot$$

$$\frac{1 + j\omega \left(C_M R_{\text{seal}} - \frac{C_L}{g_m} \right) - \omega^2 \frac{C_L R_{\text{seal}}}{g_m} (C_M + C_{\text{ox}})}{1 + j\omega (R_{\text{seal}}(C_{\text{ox}} + C_M) + R_{\text{el}}(C_L + C_{\text{ox}})) - \omega^2 R_{\text{seal}} R_{\text{el}}(C_L(C_{\text{ox}} + C_M) + C_{\text{ox}} C_M)} \tag{4}$$

Equation (3) was derived by Pabst et al. [37] to calculate R_{seal} for cell-transistor coupling experiments. The parameters in the equation are the Boltzmann constant k_B, the temperature T, the elementary charge e_0, the ion density of the surrounding bath solution n_{tot}^B, given in numbers per cubic meter, and the diffusion coefficient of potassium D_{K^+}. This approach only considers potassium currents from patch-clamp data like, for instance, measured in Wrobel et al. [23]. The main two parameters, which are influencing the value of R_{seal}, are the cleft height h and the total ion density in the bath solution n_{tot}^B. In the right part of Eq. (3) the correction term including I_0 (modified Bessel function) has to be considered only if the cell radius r is of the same size as the Debye length L_D. Since in our measurements we are working in physiological buffer solutions, L_D will be in the nm range, while the cell size is in the μm range. Therefore, the expression in brackets is equal to one, which is largely simplifying this equation. Equation (4) was described by Susloparova et al. [25]. A detailed derivation of Eq. (4) is given in the supplementary material of that study. In this publication, the equation was used to fit the actual impedimetric measurements and to extract the corresponding R_{seal} and C_M values out of the TTF spectra. This equation shows the complex relation of chip-related parameters, transimpedance-circuit parameters, and cell-related parameters and was derived by applying Kirchhoff's laws and the basic rules impedance circuitry. It should be noted that in this case the underlying model is a modification of the Point-Contact Model including parasitic parameters of the chip. Our theory so far cannot account for area-related effects as already pointed out for the interpretation of electronic recordings by others [11]. However, in contrast to all these earlier studies where cells were recorded by patch-clamp pipettes in whole-cell mode we had to include parasitic effects of the contact lines in our model, since the V_{stim} signal is provided by the reference electrode to the electrolyte bath globally.

The transistor transconductance g_m is usually calculated as shown in Eq. (5). It is depending on the gate oxide capacitance (C_{ox}), the channel width (W) and length (L), the mobility of the charge carriers in the channel (μ_p), and the drain-source voltage (V_{DS}).

$$g_m = \mu_p C_{ox} \frac{W}{L} V_{DS} \tag{5}$$

By increasing the width to length ratio of the gate (from 3.2 to 5) the transistor transconductance increases (from 0.22 to 0.33 mS). This results in a shift of the spectra to lower frequencies and increases the difference that can be measured between cell-free and cell-covered gates. From systematic simulations, it was concluded that for optimum impedance performance of the flat ISFETs the thickness of the contact line passivation has to be decreased. To simultaneously increase the transconductance, the implantation dose of the source and drain contacts was increased. This then leads to devices with a decreased passivation layer thickness to 220 nm silicon oxide and 6 nm gate oxide thickness with an increased W/L ratio of the gate design [32]. With these device modifications the chip-related and amplifier-related parameters were perfectly matching and the cell-related parameters were mainly influencing the shape of the TTF spectra.

3.5 Recordings from Primary Neuronal Cultures

Usually the ECIS method is applied to cell types, which are forming confluent cell layers such as the epithelial cell line Madin-Darby canine kidney (MDCK). The relatively large metal electrodes in ECIS are covered by confluent tissue so that its barrier function and the influences of toxins on the cell ensemble are regularly considered. Since the FETCIS technique can reach down to the single cell level, also studies with sparse cultures such as neurons are possible without loss of sensitivity. In a study with the Neuro2A murine neuroblastoma cell line and with primary neurons from the subventricular zones (SVZ) of 1–3 days old postnatal-BALB/c mice, we combined two FET chips having eight readout channels each (Fig. 12) to monitor drug effects on a one cell population and use another one as the control under exactly the same conditions [34].

In this proof-of-concept study, the neurons were treated with different concentrations of hydrogen peroxide (H_2O_2) in one culture dish whereas the control was maintained in standard culture medium. The results from the FETCIS recordings compared favorably with standard MTT assays (biochemical viability tests) with respect to the observed toxicity. Figure 13 shows the cytotoxic effects of a H_2O_2 exposure of Neuro2A and SVZ neuronal cells. This approach may of course be transferred to other chemical or pharmaceutical compounds to evaluate their direct effects to neurons. The main advantage of the FETCIS method in comparison to the standard MTT assays is the option to perform time-resolved recordings. As demonstrated for the toxic effect of silica nanoparticles on human lung adenocarcinoma epithelial H441 cells, fast effects happening in a few minutes can be monitored by the FETCIS method. These studies demonstrate a powerful new approach for personalized medicine in the field of toxicity [13] or drug tolerance in the area of cancer therapy [24].

Fig. 12 Two 8-channel
FETs were encapsulated in a
combined configuration for
neuronal toxicity studies in
parallel to corresponding
control measurements under
exactly the same conditions

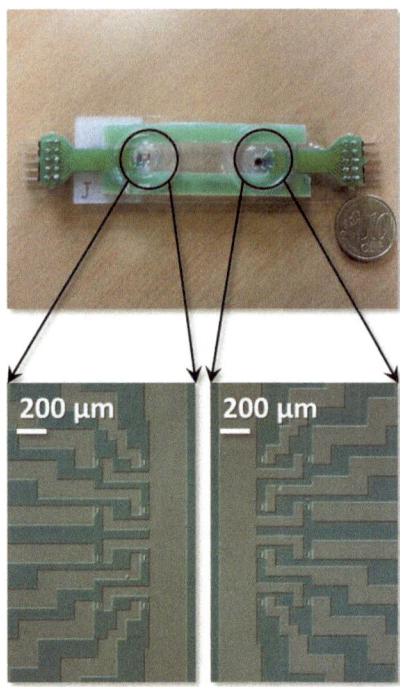

3.6 Human T-Cell Experiments

The capability of the FETCIS method opens up new opportunities to study the
migration of individual cells. In a recent study human cytotoxic T-cells were used as
an example demonstrating the capabilities of the FETCIS technique [33]. Cytotoxic
T-cells play an important role in the immune system by recognizing and eliminating
pathogens. In order to fulfill their physiological role, T-cells have to migrate
throughout the whole body and identify their respective targets. Once the target is
identified, an immunological synapse (IS) is formed in between the two cell types,
the cytotoxic T-cell and the antigen-presenting cell (APC). Then the cytotoxic action
of the T-cell is triggered and a "killing" mechanism is pursued. In the present study,
FETs were used for measuring the adhesion of CD8$^+$ T-cells when different surface
coatings (fibronectin, CD3 antibodies, and LFA-1 antibodies) were applied on the
FET surfaces and the single cell adhesion strengths were evaluated by their TTF
spectra. The cell-related parameters were extracted by fitting Eq. (4) to the experimental data [25, 33].

Fibronectin serves as the extracellular matrix for T-cell migration and signaling.
By using fibronectin-coated FET devices, significant differences between the cell-
covered transistor gates and the cell-free transistor gates were observed (Fig. 14a).
The frequency range with the most significant changes was between 100 and
500 kHz (Fig. 14a). A comparison of TTF spectra for differently coated chips was

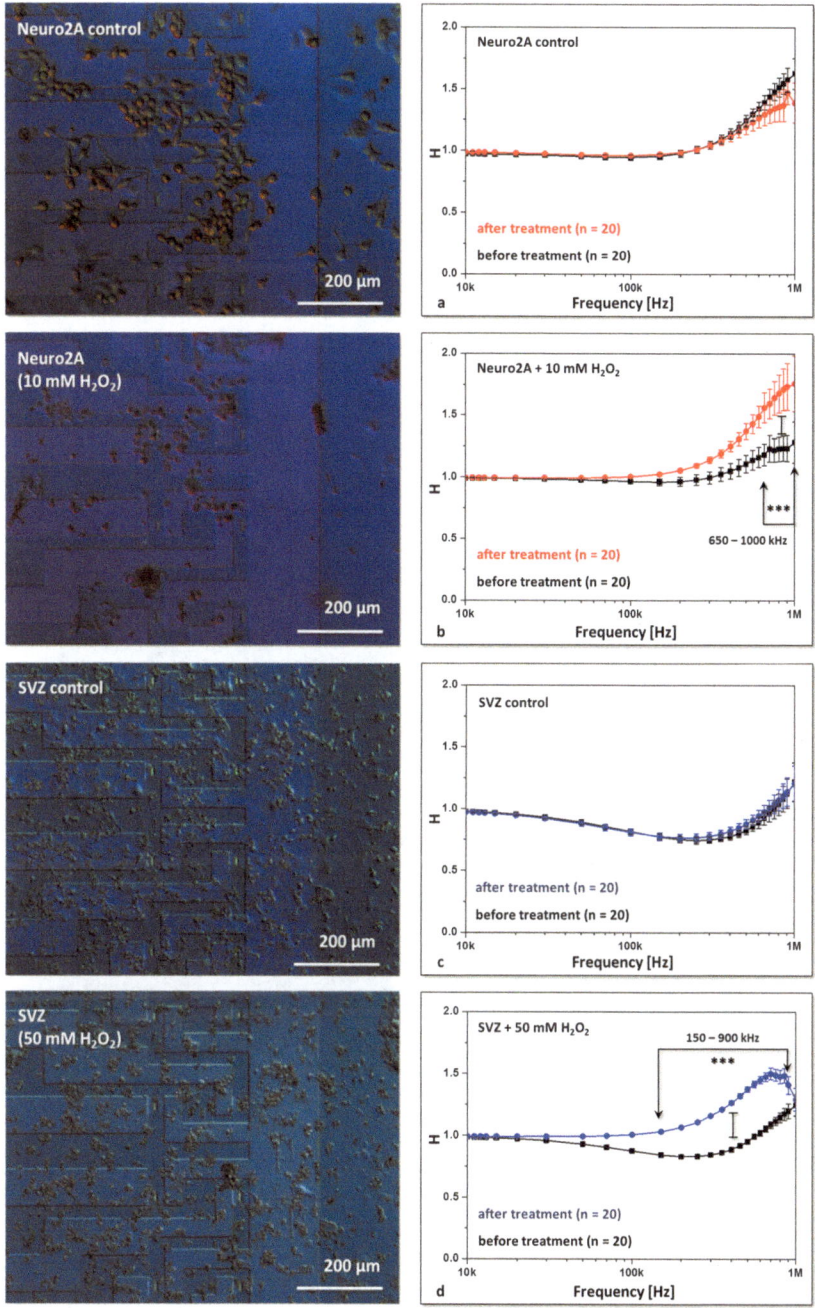

Fig. 13 FETCIS measurements reporting on the cytotoxic effect of H_2O_2 on Neuro2A (top) and SVZ neuronal cultures (bottom). The graphs show the average signal with the corresponding standard deviation for $n = 20$ channels of five independent chips. The internal controls showed no effect whereas the treated cultures showed significant effects after 24 h exposure to H_2O_2. Ranges for significance test (*** for $p < 0.001$) are indicated in the TTF spectra

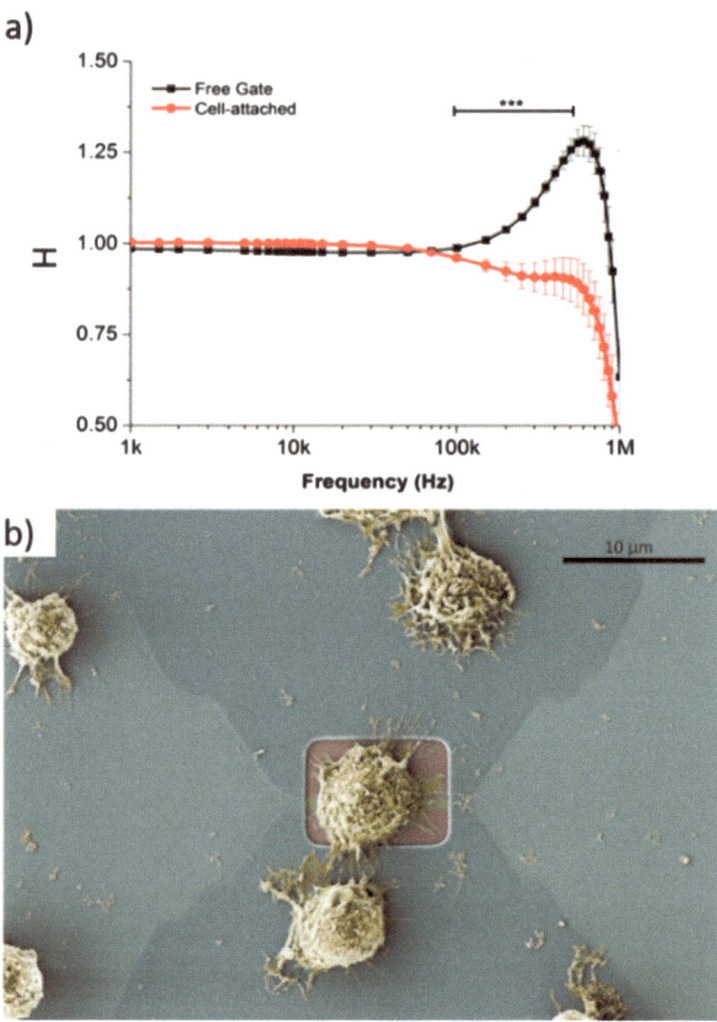

Fig. 14 (**a**) Significant differences between cell-free transistor gates (black) and cell-covered transistor gates (red) detected at frequencies ranging from 100 to 500 kHz (***$p < 0.001$). (**b**) A cytotoxic T cell adheres on top of a transistor with a dimension of $12 \times 5\ \mu m^2$. The size of the transistor gate is compatible with a single T cell [33]

performed. More significant differences were observed for the CD3 antibody-coated transistors compared to the LFA-1 antibody-coated transistors (Fig. 15). Obviously, T-cells adhered stronger to the CD3 antibody-coated FET devices than to the LFA-1 antibody-coated FET devices.

By fitting the EEC model to the TTF spectra, R_{seal} and C_M of the T cells attached on different coatings were derived (Table 1). R_{seal} can be used as an indicator on how tightly the T cells are adhering on the transistor gates. Compared to the

Fig. 15 CD8$^+$ T-cells form a tighter contact with the CD3 antibody-coated transistors than to the LFA-1 antibody-coated transistors. Significant changes in the TTF spectra are observed when the T-cells adhere on CD3 antibody-coated surfaces compared to the LFA-1 antibody-coated surfaces (***$p < 0.001$)

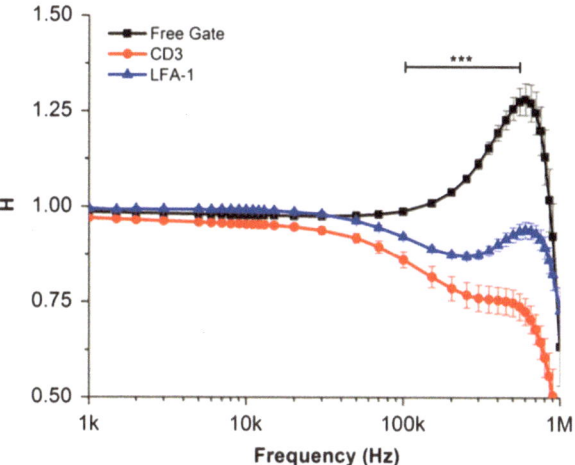

Table 1 Cell-related parameters, R_{seal} and C_M, derived by fitting the EEC model to the recorded TTF spectra

	R_{seal} (MΩ)	C_M (pF)
Fibronectin	1.23 ± 0.17	1.52 ± 0.14
Anti-CD3	4.71 ± 0.24	0.50 ± 0.06
Anti-LFA-1	2.95 ± 0.90	2.16 ± 0.09

fibronectin-coated FET devices, the R_{seal} between CD3 antibody-coated gate and the T-cells has the highest value of 4.71 MΩ, while the R_{seal} between LFA-1 antibody-coated gate and the T-cells provides a value of 2.95 MΩ. The C_M, which reflects the shape of the adhered cell, was also derived by fitting the EEC model to the TTF spectra. The average C_M values for CD3 antibody- and LFA-1 antibody-coated FETs were 0.49 pF and 2.16 pF, respectively. From discussions and simulations as shown for the two extreme models in Figs. 10 and 11, it can be derived that larger C_M values represent a larger fraction of the total cell membrane being attached membrane close to the gate relative to the free membrane. Our proof-of-concept study showed that the FETCIS method has the potential to serve as valuable tool in single cell research using cytotoxic T-cells and is capable of extracting the adhesion strength of these cells in dependence of specific surface coatings. Here, the FET devices act as an artificial APC, with CD3 antibody- and LFA-1 antibody-coated transistor surfaces. By measuring the TTF changes resulting from the cell presence on the differently coated gates and by extracting the cell-related parameters from the measurements, the adhesion strength for these two important receptor types for synapse formation was compared.

This quantitative EEC-based analysis provides a far greater insight in biological interactions than a simple comparison of the TTF spectra in a qualitative way. The R_{seal} value presented here can be used as an adhesion strength index for adhesion studies and the projected membrane capacitance C_M can be used to reflect T-cell morphologies under different conditions. The higher the C_M value, the flatter is the adhered cell on the transistor. According to the quantitative analysis, we found that T-cells adhered to the CD3 antibody coatings had a much smaller C_M value

$(C_M = 0.50 \text{ pF})$ than those adhered to fibronectin $(C_M = 1.52 \text{ pF})$ or LFA-1 antibody coatings $(C_M = 2.16 \text{ pF})$. Additional studies indicated that the T cells maintained a slightly elongated cell shape, when they adhered on the CD3 antibody surface. In contrast to this, the T-cells were highly spread in morphology when they adhered on the LFA-1 antibody-coated surfaces. When the crosslinking and the formation of an immunological synapse happens, the T-cells will change from a round morphology to a slightly elongated cell body due to a polarization effect within the T-cells [44]. That explains why the C_M value of the T-cells on CD3 antibody-coated surface is rather small. For T-cells adhering on the surfaces with LFA-1 antibody-coatings, they exhibited widely spread and flat morphologies. Several studies have shown that T-cells plated on LFA-1 ligands generate a broad contact area [45, 46]. Upon adhesion to the APC, actin rearrangement occurs in the T-cell. The rearrangement of actin results in changes in the T-cell morphology and in the start of their activation process [47, 48]. The C_M value in the present FETCIS study provides information about the T-cell morphologies under different conditions [33].

In order to capture dynamic motions of single T-cells, TTF amplitudes of a transistor exposed to a migrating T-cell were measured over time at one fixed frequency. The TTF amplitude was changing accordingly when the single T-cell was migrating towards the transistor gate and covering it partially or completely (Fig. 16). The TTF amplitude remained unchanged when there was no cell on the

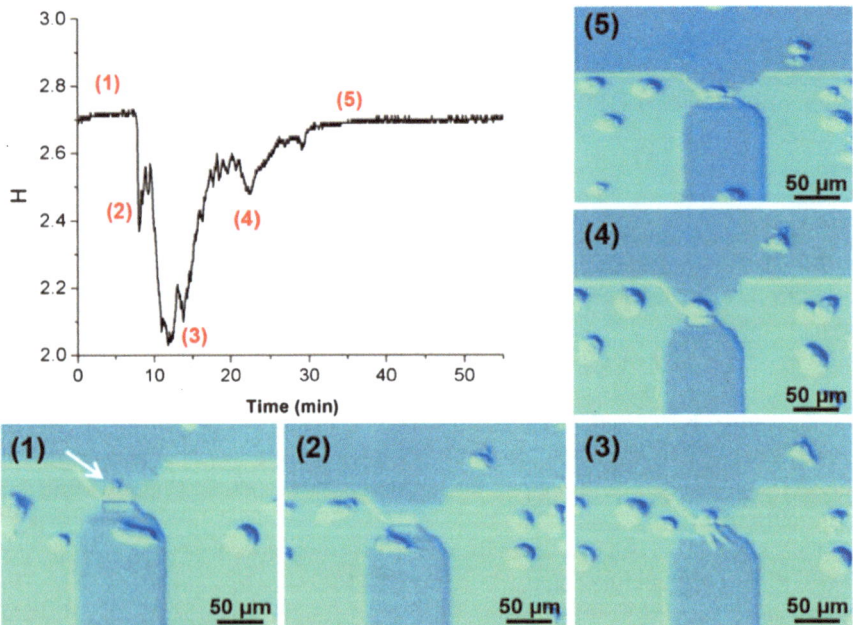

Fig. 16 Migration study of a single CD8[+] T-cell along one transistor gate during approximately 60 min. The changes in TTF amplitude were measured during cell migration and adhesion at a fixed frequency of 300 kHz. Fluctuations in the signals are caused by the cell's micromotions including undulations of the cell membrane on top of the transistor during migration

transistor gate ($T = 420$ s). It was gradually decreased when the transistor gate was covered by the T-cell ($T = 420$–$1,000$ s). The T-cell was moving on top of the transistor and its micromotion were detected and reflected by the fluctuations in TTF amplitude (T about $1,200$ s). The TTF amplitude returned to its baseline when the T-cell was not adhering tightly upon the transistor gate any more (T around $1,600$ s). Almost all T-cells, including the one on the recording transistor gate, became round in shape and lost migration mobility after about 60 min in this particular experiment. Even though the transistor gate was partially covered by the T-cell until the end of the recording, the TTF method recorded it as non-covered indicating a very loosely bound, dying cell ($T = 4,000$ s). During the adherent period on the transistor gate, however, slow fluctuations in TTF amplitude were recorded which can be attributed to membrane fluctuations during migration. The data suggest that those fluctuations are mainly due to the micromotion of the cells on top of the transistor during the migration. This was to our knowledge the first time that such micromotions have been detected on a single cell level. Eventually these micromotions could be used as a measure for cell viability as described for the ECIS technique for confluent cell cultures many years ago [49].

3.7 Adjustment of Medium Conductivity for Improved FETCIS

With the better understanding of the TTF spectra and the ability to calculate not only the chip-related parameters but also the cell-related parameters, it was possible to improve the FETCIS method even further. While the evaluation and optimization of the chip-related parameters made a significant improvement of the devices possible, the next step was to further investigate cell-related parameters. For this investigation HEK293 cells were used as test system. In total 20,000 cells were seeded in 100 µL on each chip and cultured for 1 day before measurements were performed. To investigate the influence of the medium conductivity, two measurement solutions of different conductivity were prepared and tested: (1) A sodium-containing buffer (NaCl buffer) with high conductivity (135 mM NaCl, 1.8 mM $CaCl_2$, 1 mM $MgCl_2$) and (2) a sodium-free buffer (KCl buffer) with low conductivity (5 mM KCl, 1.8 mM $CaCl_2$, 1 mM $MgCl_2$). The conductivity of both solutions was measured by a conductivity meter and they were adjusted to a pH value of 7.4. In addition, the osmolarity was set to 340 mOsm/kg. As shown previously, the seal resistance value R_{seal} is primarily dependent on the distance between cell and transistor gate h and on the total concentration of ions in the electrolyte solution. According to Eq. (3) R_{seal} is inversely proportional to the ion concentration. This seal resistance has a major influence on the TTF spectra of a single cell. The tighter the binding of the cell to the transistor gate, the bigger will be the difference in the measured spectra with and without cells. Looking at the equations for data fitting, it is obvious that R_{seal} should scale inversely proportional with the conductivity of the measurement medium. With the correct assumption that the cell diameter is much larger than the Debye length L_D of the solution, which is true for both solutions, Eq. (3) can be simplified to:

$$R_{seal} = \frac{k_B T}{4\pi \cdot e_0^2 \cdot n_{tot}^B D_{K^+} h} \quad (6)$$

It is well known that the Debye length L_D in a solution is given by:

$$L_D = \sqrt{\frac{\varepsilon_0 \varepsilon_r k_B T}{e_0^2 n_{tot}^B}} \quad (7)$$

Re-sorting the parameters of Eq. (3) provides Eq. (8):

$$R_{seal} = \frac{k_B T}{e_0^2 \cdot n_{tot}^B} \cdot \frac{1}{4\pi \cdot D_{K^+} h} \quad (8)$$

Combining Eqs. (7) and (8) allows expressing the seal resistance in a simple equation including the Debye length L_D:

$$R_{seal} = \frac{L_D^2}{4\pi \varepsilon_0 \varepsilon_r \cdot D_{K^+} h} \quad (9)$$

Another important value is the ion density of the surrounding bath solution n_{tot}^B, which can be calculated from:

$$n_{tot}^B = 2N_A \cdot I \quad (10)$$

with Avogadro's number N_A and ionic strength I:

$$I = \frac{1}{2} \cdot \sum c_i \cdot z_i^2 \quad (11)$$

Here the concentration c has to be given in mol/m^3 and z is the charge magnitude of the ions. Since all values to solve this equation are known, the ionic strength and Debye length values for both solutions can be calculated:

Sodium-free: $I = 13.4$ mol/m^3
$L_D = 2.67$ nm
Sodium-containing: $I = 143.4$ mol/m^3
$L_D = 0.82$ nm.

When comparing these results, the ionic strength values reflect a tenfold difference as confirmed in the conductivity tests, while the Debye length values differ about a factor of three due to the square root dependence between both parameters. The TTF measurements with the same cells in both solutions clearly showed a significant filtering effect in impedance spectra for cells attached on the transistor gates. The sodium-free buffer shifts the cut off frequency about one order of magnitude to lower frequencies, thus increasing the measurement range of the experimental setup (Fig. 17). Most of the parameters involved are constant values. Therefore this effect has to be caused by the cell-related parameters R_{seal} and C_M. Following Eq. (3), R_{seal} is inversely proportional to the ionic strength of the measurement buffer. This is mirrored in the results of the data analysis as presented in Table 2. R_{seal} increased significantly for the low conductivity KCl buffer.

The shift of the TTF spectra to lower frequencies is a direct cause of the different buffer compositions, since the resistance of the solution and the resistance within the cleft between the cell and the transistor surface are directly correlated. The theoretically calculated value of R_{seal} does agree with the experimental result, if the values for the distance between cell and chip surface h are adjusted accordingly. With a distance of $h = 35$ nm the value for the KCl buffer is matching almost exactly. The

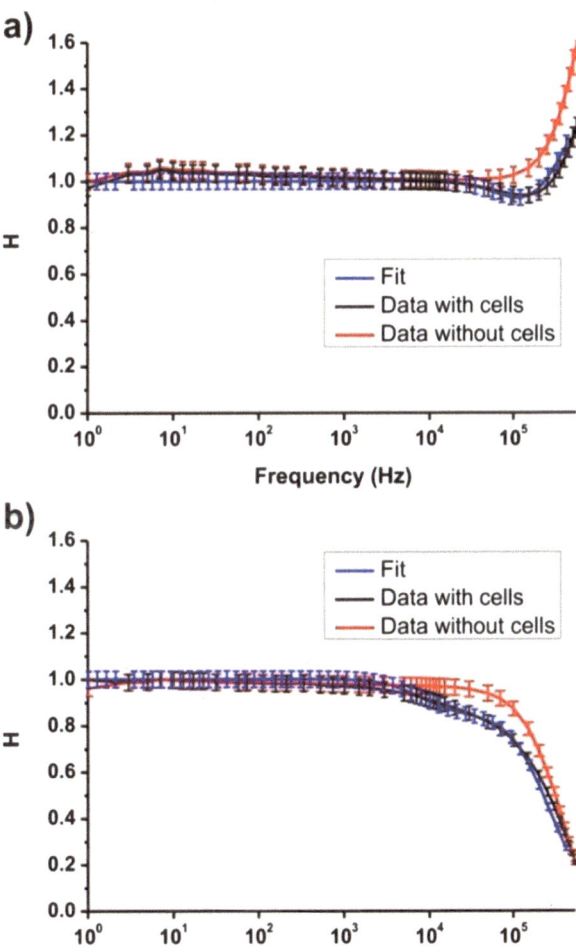

Fig. 17 TTF spectra of experiments conducted in two different buffer solutions, (**a**) Sodium-containing buffer and (**b**) Sodium-free buffer. The TTF spectra are compared between cell-covered and cell-free transistor gates [50]. The graphs show the average signals with the corresponding standard deviation for $n = 16$ channels from four independent chips

Table 2 Comparison of fitted and calculated cell-related parameters for the measurements with buffer solutions of different conductivity

	NaCl buffer	KCl buffer
Fitted C_M (F)	1.77×10^{-12}	1.57×10^{-12}
Fitted R_{seal} (Ω)	9.75×10^5	1.16×10^7
Calculated R_{seal} (Ω)	9.60×10^5 ($h = 60$ nm)	1.20×10^7 ($h = 35$ nm)

difference to the value of the NaCl buffer can be compensated by adjusting this parameter to $h = 60$ nm, which results in $R_{seal,Na+} = 9.60 \times 10^5$ Ω. However, as shown in previous publications, the distance of the HEK293 cells to the transistor varies with the cell diameter as well [40]. In another report a different approach was taken to investigate the seal resistance in the contact area between HEK293 cells and a substrate. Two optical measurements with a voltage-sensitive dye and fluorescence interference contrast microscopy were combined. Cells were electronically stimulated and with the two independent optical techniques a local mapping of the potential distribution and the distance of the cell membrane from the surface was realized. By this technique a *bulk resistivity in cell adhesion* (BRICA) model was suggested, where the seal resistance is directly coupled to the geometry of the contact and the resistivity of the bath electrolyte [51]. The inaccuracy of the resulting distance values in our FETCIS study is most likely a direct result of our simplified Point-Contact Model approach to interpret the recorded TTF spectra. Since the diameter of the cell covering the transistor gate cannot be determined exactly and usually the HEK 293 cells show elongated morphologies [52], our model should be regarded as an approximation. Typical distance values of 50 nm up to 200 nm were earlier reported and obtained by different methods [40]. The differences in our experiments for the TTF spectra measured with buffer solutions of varying conductivity leading to different cleft heights after fitting might be due to changes in distance h of the cells or due to a failure of the BRICA assumption in this particular frequency domain or experimental condition. Additional experiments are needed to deepen our understanding of the recorded TTF spectra.

Our FETCIS study showed that for an exact description of the damping effect in the TTF spectra upon cell adhesion two main directions should be followed in the future. In order to further use the simple Point-Contact Model, the transistor needs to be much smaller than the cell. This could be done by using nanowire transistors as described in different studies [53]. To fully explain the TTF spectra with the micro sized ISFETs, an extension of this model towards a planar contact model is required. This could be similar to the models, which were earlier developed to explain extracellular signal shapes from neurons recorded by FET devices [11], but would also need an extension for chip-related and amplifier-related parameters similar to our analytical model [25].

4 Conclusion and Outlook

Label-free biosensors are important in many different areas of biosensing and bioelectronics. Our present FETCIS platform offers a simple and low-cost tool for cell adhesion and cell migration studies down to single cell resolution, which is a promising feature complementing the traditional ECIS method. FETCIS allows for studies of individual cell responses rather than collective responses from a population of cells. Due to the flat topography of the FET devices, cells can freely migrate over the sensor surface without topographic obstacles. This is also a great improvement over classical ECIS devices. While not only the migration of the cells is un-hindered by the device, the cellular outgrowth is less restricted as well. Especially when studying the behavior of neurons forming a dense network in culture. The flat topography can be expected to be beneficial for this particular application. So far our EEC model is a strong tool to calculate cell-related parameters that can be used as an index for comparison of different conditions. By optimizing the parameters in this model several improvements for FETCIS devices and amplifiers were achieved. Only with this EEC it was possible to adjust the relevant chip-related parameters towards better device performance, which resulted in a new design of ISFETs that already have proven their value. It was also by applying the EEC that the correlation between medium conductivity and the TTF spectra was demonstrated. Basically all parameters involved can be studied exactly and therefore help to deepen the understanding of the measured spectra. For the future, transistor arrays with more measurement points are planned. These sensors would greatly enhance the capability to study cell migration and outgrowth on the ISFET devices. However, the parallel readout, which is one of the key features of the presented setup, is not easy to scale to very high numbers of sensor spots. Nevertheless, the FETCIS method holds great promises for further improvements and adaptations to various cellular studies which have been inaccessible till today.

Acknowledgements Over the past 10 years, many people were involved in the development of the FETCIS technique. The first, 16-channel FET sensor devices were fabricated between 1999 and 2000 in the Institute for Microtechnology (IMM), Mainz, during a project at the Max-Planck Institute for Polymer Research (MPIP), Mainz. This was done in the research group of W. Knoll. A. Offenhäusser, M. Krause, S. Ingebrandt (all MPIP), T. Zetterer, and W. Staab (both IMM) were involved in FET design and fabrication. The first devices were optimized for low noise recordings from electrogenic cells in the audio frequency range. First portable amplifiers for TTF recordings were later realized in the Institute for Bio- and Nanosystems, Institute 2: Bioelectronics of A. Offenhäusser at the Research Center Jülich (FZJ). The new generations of FET devices for TTF recordings were fabricated by R. Stockmann and S. Schäfer and amplifier circuits were realized by R. Otto and optimized by N. Wolters, T. Dufaux, and S. Ingebrandt (all FZJ). After transfer of the project to the University of Applied Sciences Kaiserslautern (UASK) new devices were fabricated by X. T. Vu with the help of D. Cassel (both UASK). To maintain and expand the FETCIS subject funding came between the years 2010 and 2014 from the UASK and from three different BMBF projects (Fkz.: 17N2110, 17008X10, and 17042X11). At the UASK besides the three authors of this contribution the following persons were involved: X. T. Vu, M. Weil, V. Pachauri, T. Oberbillig, A. Susloparova, D. Koppenhöfer, T. C. Nguyen, M. Schwartz, L. Delle, R. Lanche, D. Rani, X. Lu, W. M. Munief, A. Müller, E. Lajko (guest researcher) as scientists, D. Cassel, P. Molter and

R. Lilischkis as technical staff and several students in their bachelor and master studies (all UASK). The FETCIS technique is currently maintained and expanded in scientific cooperation with colleagues from the University of Saarland in Homburg, Germany, and at the Semmelweis University, Budapest, Hungary.

References

1. Bergveld P, Wiersma J, Meertens H (1976) Extracellular potential recordings by means of a field-effect transistor without gate metal, called osfet. IEEE Trans Biomed Eng 23(2):136–144
2. Bergveld P (1991) A critical evaluation of direct electrical protein detection methods. Biosens Bioelectron 6(1):55–72
3. Souteyrand E, Cloarec J, Martin J, Wilson C, Lawrence I, Mikkelsen S, Lawrence M (1997) Direct detection of the hybridization of synthetic homo-oligomer DNA sequences by field effect. J Phys Chem B 101(15):2980–2985
4. Offenhäusser A, Ingebrandt S, Pabst M, Wrobel G (2009) Interfacing neurons and silicon-based devices. Nanobioelectronics – for electronics, biology, and medicine. Springer, New York, pp 287–301
5. Martinoia S, Rosso N, Grattarola M, Lorenzelli L, Margesin B, Zen M (2001) Development of ISFET array-based microsystems for bioelectrochemical measurements of cell populations. Biosens Bioelectron 16(9–12):1043–1050
6. Eversmann B, Jenkner M, Hofmann F, Paulus C, Brederlow R, Holzapfl B, Fromherz P, Merz M, Brenner M, Schreiter M, Gabl R, Plehnert K, Steinhauser M, Eckstein G, Schmitt-Landsiedel D, Thewes R (2003) A 128x128 CMOS biosensor array for extracellular recording of neural activity. IEEE J Solid State Circuits 38(12):2306–2317
7. Frey U, Sanchez-Bustamante C, Ugniwenko T, Heer F, Sedivy J, Hafizovic S, Roscic B, Fussenegger M, Blau A, Egert U, Hierlemann A (2007) Cell recordings with a CMOS high-density microelectrode array. Conf Proc IEEE Eng Med Biol Soc 2007:167–170
8. Heer F, Franks W, Blau A, Taschini S, Ziegler C, Hierlemann A, Baltes H (2004) CMOS microelectrode array for the monitoring of electrogenic cells. Biosens Bioelectron 20(2):358–366
9. Müller J, Ballini M, Livi P, Chen Y, Radivojevic M, Shadmani A, Viswam V, Jones I, Fiscella M, Diggelmann R, Stettler A, Frey U, Bakkum D, Hierlemann A (2015) High-resolution CMOS MEA platform to study neurons at subcellular, cellular, and network levels. Lab Chip 15(13):2767–2780
10. Neher E, Sakmann B (1992) The patch clamp technique. Sci Am 266(3):44–51
11. Fromherz P (2016) Current-induced transistor sensorics with electrogenic cells. Biosensors (Basel) 6(2):18
12. Uslu F, Ingebrandt S, Mayer D, Böcker-Meffert S, Odenthal M, Offenhäusser A (2004) Labelfree fully electronic nucleic acid detection system based on a field-effect transistor device. Biosens Bioelectron 19(12):1723–1731
13. Koppenhöfer D, Susloparova A, Docter D, Stauber R, Ingebrandt S (2013) Monitoring nanoparticle induced cell death in H441 cells using field-effect transistors. Biosens Bioelectron 40(1):89–95
14. Schasfoort R, Streekstra G, Bergveld P, Kooyman R, Greve J (1989) Influence of an immunological precipitate on Dc and Ac behavior of an isfet. Sensors Actuators 18(2):119–129
15. Antonisse M, Snellink-Ruël B, Lugtenberg R, Engbersen J, van den Berg A, Reinhoudt D (2000) Membrane characterization of anion-selective CHEMFETs by impedance spectroscopy. Anal Chem 72(2):343–348

16. Kharitonov A, Wasserman J, Katz E, Willner I (2001) The use of impedance spectroscopy for the characterization of protein-modified ISFET devices: application of the method for the analysis of biorecognition processes. J Phys Chem B 105(19):4205–4213

17. Kruise J, Rispens J, Bergveld P, Kremer F, Starmans D, Haak J, Feijen J, Reinhoudt D (1992) Detection of charged proteins by means of impedance measurements. Sens Actuators B Chem 6:101–105

18. Schäfer S, Eick S, Hofmann B, Dufaux T, Stockmann R, Wrobel G, Offenhäusser A, Ingebrandt S (2009) Time-dependent observation of individual cellular binding events to field-effect transistors. Biosens Bioelectron 24(5):1201–1208

19. Giaever I, Keese C (1993) A morphological biosensor for mammalian cells. Nature 366(6455):591–592

20. Wegener J, Keese C, Giaever I (2000) Electric cell–substrate impedance sensing (ECIS) as a noninvasive means to monitor the kinetics of cell spreading to artificial surfaces. Exp Cell Res 259(1):158–166

21. Levsky J, Singer R (2003) Gene expression and the myth of the average cell. Trends Cell Biol 13(1):4–6

22. Ingebrandt S (2001) Charakterisierung der Zell-Transistor Kopplung. Doctoral thesis, Johannes Gutenberg University Mainz, Mainz

23. Wrobel G, Seifert R, Ingebrandt S, Enderlein J, Ecken H, Baumann A, Kaupp U, Offenhäusser A (2005) Cell-transistor coupling: investigation of potassium currents recorded with p-and n-channel FETs. Biophys J 89(5):3628–3638

24. Susloparova A, Koppenhöfer D, Vu X, Weil M, Ingebrandt S (2013) Impedance spectroscopy with field-effect transistor arrays for the analysis of anti-cancer drug action on individual cells. Biosens Bioelectron 40(1):50–56

25. Susloparova A, Koppenhöfer D, Law J, Vu X, Ingebrandt S (2015) Electrical cell-substrate impedance sensing with field-effect transistors is able to unravel cellular adhesion and detachment processes on a single cell level. Lab Chip 15(3):668–679

26. Offenhäusser A, Sprössler C, Matsuzawa M, Knoll W (1997) Field-effect transistor array for monitoring electrical activity from mammalian neurons in culture. Biosens Bioelectron 12(8):819–826

27. Ingebrandt S, Yeung C, Krause M, Offenhäusser A (2001) Cardiomyocyte-transistor-hybrids for sensor application. Biosens Bioelectron 16(7-8):565–570

28. Ingebrandt S, Yeung C, Staab W, Zetterer T, Offenhäusser A (2003) Backside contacted field effect transistor array for extracellular signal recording. Biosens Bioelectron 18(4):429–435

29. Ingebrandt S, Yeung C, Krause M, Offenhäusser A (2005) Neuron–transistor coupling: interpretation of individual extracellular recorded signals. Eur Biophys J 34(2):144–154

30. Ingebrandt S, Offenhäusser A (2006) Label-free detection of DNA using field-effect transistors. Phys Status Solidi A 203(14):3399–3411

31. Ingebrandt S, Han Y, Nakamura F, Poghossian A, Schöning M, Offenhäusser A (2007) Label-free detection of single nucleotide polymorphisms utilizing the differential transfer function of field-effect transistors. Biosens Bioelectron 22(12):2834–2840

32. Susloparova A, Vu X, Koppenhöfer D, Law J, Ingebrandt S (2014) Investigation of ISFET device parameters to optimize for impedimetric sensing of cellular adhesion. Phys Status Solidi A 211(6):1395–1403

33. Law J, Susloparova A, Vu X, Zhou X, Hempel F, Qu B, Hoth M, Ingebrandt S (2015) Human T cells monitored by impedance spectrometry using field-effect transistor arrays: a novel tool for single-cell adhesion and migration studies. Biosens Bioelectron 67:170–176

34. Koppenhöfer D, Kettenbaum F, Susloparova A, Law J, Vu X, Schwab T, Schafer K, Ingebrandt S (2015) Neurodegeneration through oxidative stress: monitoring hydrogen peroxide induced apoptosis in primary cells from the subventricular zone of BALB/c mice using field-effect transistors. Biosens Bioelectron 67:490–496

35. Sprössler C, Richter D, Denyer M, Offenhäusser A (1998) Long-term recording system based on field-effect transistor arrays for monitoring electrogenic cells in culture. Biosens Bioelectron 13(6):613–618

36. Sprössler C, Denyer M, Britland S, Knoll W, Offenhäusser A (1999) Electrical recordings from rat cardiac muscle cells using field-effect transistors. Phys Rev E 60(2):2171–2176

37. Pabst M, Wrobel G, Ingebrandt S, Sommerhage F, Offenhäusser A (2007) Solution of the Poisson-Nernst-Planck equations in the cell-substrate interface. Eur Phys J E Soft Matter 24(1):1–8

38. Sommerhage F, Baumann A, Wrobel G, Ingebrandt S, Offenhäusser A (2010) Extracellular recording of glycine receptor chloride channel activity as a prototype for biohybrid sensors. Biosens Bioelectron 26(1):155–161

39. Sommerhage F, Helpenstein R, Rauf A, Wrobel G, Offenhäusser A, Ingebrandt S (2008) Membrane allocation profiling: a method to characterize three-dimensional cell shape and attachment based on surface reconstruction. Biomaterials 29:3927–3935

40. Wrobel G, Höller M, Ingebrandt S, Dieluweit S, Sommerhage F, Bochem H, Offenhäusser A (2008) Transmission electron microscopy study of the cell–sensor interface. J R Soc Interface 5(19):213–222

41. Regehr W, Pine J, Cohan C, Mischke M, Tank D (1989) Sealing cultured invertebrate neurons to embedded dish electrodes facilitates long-term stimulation and recording. J Neurosci Methods 30(2):91–106

42. Schätzthauer R, Fromherz P (1998) Neuron–silicon junction with voltage-gated ionic currents. Eur J Neurosci 10(6):1956–1962

43. Arndt S, Seebach J, Psathaki K, Galla H-J, Wegener J (2004) Bioelectrical impedance assay to monitor changes in cell shape during apoptosis. Biosens Bioelectron 19(6):583–594

44. Huse M (2012) Microtubule-organizing center polarity and the immunological synapse: protein kinase C and beyond. Front Immunol 235(3):37–47

45. Dustin M, Springer T (1988) Lymphocyte function-associated antigen-1 (LFA-1) interaction with intercellular adhesion molecule-1 (ICAM-1) is one of at least three mechanisms for lymphocyte adhesion to cultured endothelial cells. J Cell Biol 107(1):321–331

46. Shaw A, Dustin M (1997) Making the T cell receptor go the distance: a topological view of T cell activation. Immunity 6(4):361–369

47. Beemiller P, Krummel M (2010) Mediation of T-cell activation by actin meshworks. Cold Spring Harb Perspect Biol 2(9):a002444

48. Yu Y, Smoligovets A, Groves J (2013) Modulation of T cell signaling by the actin cytoskeleton. J Cell Sci 126(5):1049–1058

49. Giaever I, Keese C (1991) Micromotion of mammalian-cells measured electrically. Proc Natl Acad Sci U S A 88(17):7896–7900

50. Hempel F, Nguyen T, Law J, Ingebrandt S (2015) The influence of medium conductivity on ECIS measurements with field-effect transistor arrays. Phys Status Solidi A 212(6):1260–1265

51. Gleixner R, Fromherz P (2006) The extracellular electrical resistivity in cell adhesion. Biophys J 90(7):2600–2611

52. Sommerhage F (2011) Chloride versus protons: ion currents in the cell transistor junction. Doctoral thesis, Fachgruppe Physik, RWTH Aachen, Aachen

53. Eschermann J, Stockmann R, Hueske M, Vu X, Ingebrandt S, Offenhäusser A (2009) Action potentials of HL-1 cells recorded with silicon nanowire transistors. Appl Phys Lett 95(8):083703

BIOREV (2019) 2: 111–134
https://doi.org/10.1007/11663_2018_5
© Springer Nature Switzerland AG 2019
Published online: 13 April 2019

Label-Free Monitoring of 3D Tissue Models via Electrical Impedance Spectroscopy

Frank Alexander Jr., Sebastian Eggert, and Dorielle Price

Contents

Abstract There is a strong tendency in in vitro testing of drugs or toxins to use 3D tissue models as biological test objects rather than 2D cell monolayers. The latter have been used with increasing success throughout the last decades even though their limitations were well-known. Two-dimensional cell layers cannot fully mimic

F. Alexander Jr. (✉)
The Geneva Foundation, Rockville, MD, USA
e-mail: falexander@genevausa.org

S. Eggert
Technische Universität München, München, Germany
e-mail: sebastian.eggert@mytum.de

D. Price
University of South Florida, Tampa, FL, USA
e-mail: dorielle@mail.usf.edu

the complex architecture of real tissue mainly because cell-cell interactions, cell communication, cell signaling, extracellular matrix composition, and all the physicochemical properties of the microenvironment are quite different. Motivated by these shortcoming, 3D tissue models have been developed over the years capable of overcoming some of the limitations of 2D tissue. However, for successful screening campaigns and toxicology assessment, it is not sufficient to only have an appropriate tissue model, and it also takes analytical techniques to read the biological response to a given exposure to drugs or toxins. It has been notoriously difficult to find experimental approaches that are capable of reporting from the inside of a 3D tissue model without destroying it. Because of its noninvasive nature and the ability of AC currents to permeate through tissue, impedance analysis has emerged as one potential technique to fill this gap. The current chapter will summarize the state of the art in impedance-based monitoring of 3D tissue models with particular focus on electrode design and the constraints that are associated with it.

Keywords Drug screening · Impedance analysis · Label-free · Multi-electrode arrays · Spheroids · Tissue engineering · Tissue models

1   Introduction

To translate in vitro toxicity testing to in vivo applications, it is crucial to conduct in vitro studies using model systems which emulate the in vivo situation as close as possible with respect to molecular composition and architecture. In conventional two-dimensional (2D) models, cells have been cultured in monolayers on rigid 2D substrates, such as tissue culture-grade polystyrene, for several decades [1]. Two-dimensional cell culture had a pioneering role fostering our understanding of biology and led the way to outstanding findings like in disease mechanisms or drug discovery. Although some 2D model cultures are sufficient for toxicity assessments, many fail to emulate tissue complexity in vivo sufficiently well due to the removal of influential proteins contained in the three-dimensional (3D) extracellular matrix (ECM) [2] or the absence of cell-cell contacts in three dimensions, to mention just a few factors. It is known that 3D tissue models develop a more in vivo-like environment that supports a more realistic exchange of oxygen, nutrients, and waste compared to 2D cell culture [3]. Furthermore, 3D models exhibit many structural features that mimic the complexity of the in vivo situation more closely and thereby provide a new dimension for in vitro research [4].

To this end, the current drive to convert existing toxicity testing protocols into the more in vivo-like 3D milieu has resulted in a number of researchers investigating ways to integrate these cultures into label-free monitoring systems based on electrical impedance spectroscopy (EIS). Over the past two decades, impedance-based techniques have been gaining ground as an important tool for toxicity assessments using a multitude of different biological tissues. The emergence of commercial systems such as electric cell-substrate impedance sensing (ECIS;

Applied BioPhysics Inc.), xCELLigence (ACEA Biosciences Inc.), and EVOM (World Precision Instruments Inc.) offers a complement and advantage over traditional endpoint assays, capable of reporting on the time course of candidate drug action within many planar cellular models. In EIS small constant AC current is typically applied at multiple frequencies, and the resultant voltage is measured to obtain the frequency-dependent impedance of a sample under test. It has been initially used to investigate tissue or planar cultures in assays addressing cytotoxicity, cell proliferation, cell migration, and cell shape dynamics [5]. EIS applicable to 3D tissue models have emerged more recently monitoring tissue responses to chemotherapeutics [6], differentiation of osteogenic cells [7], and Alzheimer's disease-like neurodegeneration of neuroblastoma spheroids [8]. An inherent flaw in these systems is their lack of spatial resolution, inhibiting spatially resolved analysis as well as investigation of drug penetration and diffusion. Furthermore, the mechanisms leading to cell death and morphology changes are difficult to identify based on impedance measurements alone. Within this context, the current chapter addresses (a) the influence of electrode design on the measured impedance and (b) recent EIS investigations in two classes of 3D tissue models.

1.1   Microelectrode Design and Spatial Resolution

EIS measurements have been used to characterize suspended biological cells over a wide range of frequencies extensively since the 1950s. Analysis of adherent cells grown directly on the electrode surface began in the 1980s. Information about cell morphology, adhesion state, and cell movement is obtained at low- to mid-range frequencies (100 Hz to 10 kHz) since most of the current flows around the dielectric cell bodies under these conditions. At higher frequencies, the current penetrates the membrane and provides information about the membrane themselves or the cell interior. Impedance characterization of biological cells, using planar microelectrodes, has been used as a diagnostic tool to monitor and analyze electrophysiological and biophysical changes of the cell bodies due to viral infections [9], malignant transformation [10], or drug exposure [11]. Microelectrodes offer many advantages over their conventional counterparts including (1) manufacturing is economic due to reduced material cost and batch fabrication [12]; (2) higher sensitivity in monitoring cell responses to any kind of stressor [13]; and the ability to integrate electrodes in microfluidic or portable measurement systems [14]. The small radii of circular microelectrodes enable the study of low resistivity samples that are otherwise masked by the solution resistance [15].

There are also disadvantages associated with microelectrodes, commonly resulting in measurement error. At low frequencies, microelectrodes are challenged with interfacial polarization impedance in two-electrode setups. Interfacial or double-layer capacitance is indirectly proportional to interfacial impedance and arises from the inability of charge carriers to move across the solid-liquid interface [16]. The result of this barrier is the accumulation of charges in response to an applied potential to the electrode, thus giving rise to a capacitive effect. Since capacitance is directly proportional to surface area, in the case of microelectrodes, this effect can lead to very large impedances, particularly at low frequencies.

In two-electrode measurement setups, current is passed, and voltage is measured between the same set of electrodes. One electrode is called the working electrode and the other, usually larger, electrode is named the counter electrode. If the area of the counter electrode is a t least 300 times larger than the area of the working electrode, the impedance of the system is dominated by the impedance of the working electrode [17], thus simplifying the analysis of the system [18]. Alternatively, four-electrode measurement setups pass current through one pair of electrodes, and voltage is measured between the other pair. Normally, the voltage sensing electrodes are placed linearly between the current-carrying electrodes.

Two-electrode measurements are notoriously sensitive to changes in electrode polarization impedance, caused by the adsorption of ions or molecules on the electrode surface, which is geometry dependent. As electrode area decreases, the effect of electrode polarization impedance becomes greater, causing relevant impedance contributions of the system under test to be masked at lower frequencies. On the other hand, electrode polarization impedance has negligible effects in four-electrode measurement setups since the current passing and voltage sensing electrodes are separate. However, with this configuration, multi-electrode array configurations are not easily achieved for increased spatial resolution. Thus, it is important to carefully design electrodes to help alleviate the known measurement parasitics.

1.2 Optimization of Measurement Parameters

Previous work [7–21] focusing on optimizing electrode designs, measurement setups, has laid a solid foundation for 3D studies, as spatial resolution, statistically relevant technical impacts, and reduced measurement parasitics are all essential for electrically characterizing the effects of toxins and chemotherapeutic drugs on 3D cultures. Researchers have performed systematic studies in different directions to optimize the measurement performance for various applications. Fosdick and Anderson [19] optimized the geometry of a microelectrode array flow detector with respect to amperometric response. Min and Baeumner [20] investigated geometric parameters (i.e., electrode height, material, gap size, and electrode width) of interdigitated ultramicroelectrode arrays (IDUAs) to optimize oxidation and reduction reactions of ferro/ferrihexacyanide. Sandison and coworkers [21] studied electrode array geometry (center-to-center spacing and diameter) and porosity of electrodes made of Si_3N_4-coated silicon substrates. Lempka and coworkers [22] optimized silicon-based microelectrodes for extracellular field potential recordings in neuron cultures. Wang and colleagues investigated the sensitivity and frequency characteristics of interdigitated microelectrode arrays for electric cell-substrate impedance sensing (ECIS) [23]. Other studies have optimized numerical recipes and algorithms for cell data analysis [24] or identified the most robust and sensitive cell lines for field-portable toxicology studies [25].

1.3 Literature Review on the Optimization of Electrode Design

Design optimization of microelectrodes is critical to the efficient employment of impedance-based monitoring of 2D or 3D tissue models in drug discovery, cancer research, and toxicology studies. Pejcic and De Marco reiterate that electrode optimization is one of the most crucial steps in the realization of an electroanalytical device [26].

In reference [18], a set of design rules was derived for optimization of micro-electrodes used to study cell monolayers in ECIS up to 10 MHz. The effect of electrode layout (electrode area, widths of contact leads, and passivation coating thickness) on the contribution of the passivation coating impedance was experimentally evaluated using impedance measurements. The parasitic coating impedance was successfully minimized by designing electrodes with either a thicker coating layer or contact leads with small surface area.

The effect of microelectrode sensor diameters was studied in [27], using human umbilical vein endothelial cell (HUVEC) cultures and excised human skin tissue. Four diameters of circular electrodes were used: 500, 250, 100, and 50 μm. It was concluded that below 100–50 μm, nonuniform current distribution effects are noticeable and skin tissue is not well-characterized with electrodes smaller than 250 μm, due to its considerable heterogeneity.

Rahman et al. designed a two-electrode-based impedance device that allowed for continuous, quantitative, and label-free monitoring of 2D cell cultures with spatio-temporal resolution (impedance mapping). Cell adhesion/detachment, proliferation, and spreading were monitored and visualized with this system [28]. More recently, Canali et al. developed a bioimpedance system with two-, three-, and four-terminal electrode configurations to monitor large 3D cell cultures with spatial resolution [29]. They stated that electrode design to improve spatial resolution has become more significant within the past decade due to the increasing interest in studying 3D tissue models in all kinds of biomedical experiments. These studies take advantage of the knowledge and experience gathered for electrodes suitable to study 2D cell cultures.

2 Real-Time Monitoring of 3D Tissue Models

In order to overcome the shortcomings of 2D cell culture models with respect to their limits as in vivo model, various methods were developed for culturing cells in vitro in a 3D environment aiming to achieve an in vivo-like behavior and bridge the gap between 2D models and animal studies. The various methods that have been developed for producing such 3D models can be categorized into two approaches that will be discussed individually in the next paragraphs: scaffold-based and scaffold-free culture techniques. The ultimate goal of all 3D tissue models is inclusion into preclinical studies for drug development such as in vitro drug or toxicity screening campaigns. The farthest developed 3D tissue models are promising since they mimic biological processes like metabolism, signal transduction, and cell-cell interactions, as known from in vivo tissue, already quite realistically. To be suitable for screening, such engineered tissue models must be easily fabricated in high amounts at low cost and high reproducibility. But it is not enough to have appropriate tissue models, and it also takes readout systems capable of monitoring the response of such 3D models to drug or toxin exposure.

Due to their noninvasive nature and the ability of high-frequency AC currents to penetrate tissue, multi-frequency impedance measurements are well suited to monitor 3D models during drug and toxicity testing. As highlighted in the above section, label-free impedance monitoring was first derived primarily for planar 2D cell cultures with relatively few investigations into small-scale 3D cultures. With the recent advances in preparing 3D models, several groups have sought to interrogate these models with impedance monitoring techniques.

2.1 Scaffold-Based 3D Cultures

A *scaffold* is a 3D, sponge-like construct prepared from man-made materials providing mechanical support for cells to proliferate and develop their specific, distinct functions in a 3D functional network [30]. The available network provides a structural framework for cell adhesion, growth, and extracellular matrix production [31]. Mimicking the natural ECM architecture with its fibers and networks, the scaffold provides the mechanical support and the 3D environment for cell proliferation, differentiation, and secretion of cell-type-specific ECM proteins. Different scaffold features, e.g., pore size, channel branching, and shape, are adjusted and tailored in order to control medium perfusion or cell proliferation. These tissue engineering scaffolds are used for a diversity of biomedical applications, such as guiding the tissue to adopt a certain external shape or internal structure [32] or to provide a mechanical support similar to the ECM in toxicology screening [33].

Scaffold materials range from ceramics, which are widely used in medical tissue engineering and focus on applications in orthopedics, to polymers, which are often applied in conventional in vitro settings due to easy processing and control over biodegradability [34]. Polymers include biologically derived polymers such as collagen, chitosan, agarose, and alginate, as well as synthetic polymers like Poly (α-hydroxyacids), such as poly(glycolic acid) (PGA) and poly (lactic acid) (PLA), or Poly(ε-caprolactone) (PCL) [30]. These polymers provide easy and flexible use leading to validated materials. Several commercial products are available, such as BD Matrigel (BD Biosciences), AlgiMatrix® 3D Cell Culture System (Invitrogen), 3D-Insert PCL (3D Biotek), or PureCol® (Advanced BioMatrix) [35]. Commonly, cells are either seeded atop an acellular scaffold or mixed into the liquid matrix which is then allowed to solidify.

Scaffold materials are either nondegradable or degradable. In the context of biomedical applications, biodegradability of polymer-based scaffolds means the "break down due to macromolecular degradation with dispersions in vivo" [36]. Degradable materials provide function for a limited time period and then degrade due to enzymatic degradation or hydrolysis in a controlled process with well-defined loss of mechanical properties, release of byproducts during degradation, or overall degradation kinetics [37]. Therefore, it is crucial to understand the degradation process in detail and the effects of the degradation on cell proliferation, metabolism, or other phenotypes, especially when toxicity tests are performed with degradable scaffolds to ensure that there is no impact of the scaffold itself.

Another attractive approach for creating 3D tissue models in scaffold-based tissue engineering is the use of hydrogels, as they have mechanical and structural properties similar to the ECM of many tissues [38]. Hydrogels are hydrophilic polymer networks arranged in a physical or/and chemical mode and are typically used for cell encapsulation [39]. Due to the ease of their fabrication, hydrogels may be produced by rapid prototyping techniques such as laser-based or printer-based systems which allow for an easy control of the external and internal geometry of such constructs [40].

Drawback of using any kind of exogenous material for creating a 3D cellular model is the risk of toxicity and immunogenicity induced by the scaffold material during scaffold preparation with the cells present, upon seeding or during degradation. Furthermore, unknown interactions between cells and scaffold material have to be considered when using scaffold-based approaches produce results that are otherwise hard to interpret. More detailed reviews regarding scaffold design, degradation mechanism, and applications for drug screening are available [41–43].

2.2   EIS in Scaffold-Based 3D Cultures

As bio-derived polymer scaffolds (e.g., Agarose and Matrigel) are the primary type of scaffold used in conjunction with impedance measurements, most experimental attempts to apply impedance sensing to cell-laden scaffolds make use of dual-electrode approaches that sandwich the entire hydrogel scaffold between the electrodes and monitor changes in impedance over time. Prior to any attempt to use such systems for monitoring tissue responses, it is critical to ensure the stability of the scaffold. Scaffold degradation can cause impedance changes that obscure and mask those that may be imposed by cellular activities and cell responses to external stimuli. Such an overlay of signal contributions makes interpretations difficult and reduces the measurement sensitivity. Lin et al. performed a study using NIH3T3 fibroblasts and cortical neurons that were suspended in polyethylene glycol (PEG) hydrogels. The latter were placed on planar microelectrode arrays (MEAs) [44]. Prior to performing impedance measurements, studies were conducted to show that the degradation rate was initially rapid over the first 14 days and then slowed down. Cells suspended in the hydrogels caused increases in resistance, reactance, and total impedance when measured at 1 kHz over 8 days of monitoring. Interestingly, comparing measurements between 3T3 fibroblasts and cortical neurons revealed that the presence of fibroblasts induced a faster increase in reactance and total impedance notable already after day 3. This is possibly due to a higher growth rate of fibroblast cells in comparison to the stagnant growth of neural cells. Cells were killed at the end of the experiment by abruptly changing the pH of the medium; the latter resulted in a significant reduction in all impedance components. Lin et al. assume the effects of matrix degradation were negligible based on prior unreported characterization studies. However, this assumption can prove unreliable when attempting to use impedance data to quantify cellular growth rates over extended time periods. By measuring the basal impedance of the scaffold alone over the total duration of the experiment, one can normalize the impedance values of the cell-laden scaffold to this and attempt to infer growth rates of the cells. It is of note that the impedance changes observed by these authors are the largest relative increases in cellular impedance in scaffold-based culture to date and no cellular counting was presented as an attempt to correlate increased cell number with the impedance growth curves.

Other groups have made attempts to quantify and correlate the measured impedance magnitude with the number of cells embedded within scaffold materials. Lei et al. have reported several impedance devices designed for quantifying cancer cells within hydrogel scaffolds. Their earliest attempt utilized copper electrodes on printed circuit boards. A polydimethylsiloxane (PDMS) culture chamber and fluidic channels allowed perfusion above a layer of cells embedded in an agarose gel which was placed on the bottom electrode of the chamber. The cells were then sandwiched between this bottom electrode and a top electrode cover, and impedance was recorded at discrete time points. Despite varying the concentration of cells from 10,000 to 10 million cells per milliliter, the resulting change in impedance remained less than 30 Ω for all measured frequencies [45].

Another study reported the development of a microfluidic perfusion chip with an array of "vertical" copper electrodes for making measurements of 3D hydrogel cultures of oral cancer cells (OEC-M1). Gel cultures positioned between the electrodes were perfused by a path defined in a channel above the electrodes. Several linear relationships were established to show the correlation between cell density, measurement frequency, and impedance magnitude. As expected, the dielectric nature of the cellular membrane contributes heavily to the overall impedance. The experiments revealed impedance measurements at lower frequencies (500 Hz) to be more sensitive to the concentration of cells. The authors normalized their data to a reference time of 6 h prior to perfusion. A proliferation assay performed over 120 h showed the impedance doubling after ~60 h. However, exposing the cells to cisplatin – well-known from chemotherapy – counterintuitively showed *increases* of the impedance in response to cellular death. This data contrasts previous studies that have identified impedance increases as an indicator for cellular proliferation. An explanation for this unexpected observation may be the mechanism of death induced by the specific concentration of cisplatin or unknown interactions with the matrix [46].

A major disadvantage in these studies is the use of copper as an electrode material. Not only does it corrode during extended electrochemical measurements it is also particularly susceptible to biofouling [47]. Lei et al.'s latest study overcomes this limitation by utilizing a combination of indium tin oxide or gold electrodes deposited on glass [48]. Hepatoma cells (Huh-7) encapsulated within methyl cellulose (MC) hydrogels were suspended in an agar gel, and static measurements of the gel-bound cells were taken over time without perfusion. Stacked ITO electrodes separated by a PDMS/hydrogel chamber were used to make measurements over various frequencies, and changes to magnitude and phase at 100 Hz were termed *colony index* (Δ magnitude) and *colony size index* (Δ phase), respectively. Over time, an increase in the *colony index* was observed indicating an increased number of cells in the gel. Toxicity tests where cells were exposed to doxorubicin reduced the growth rate of the *colony index* with a new stable value at 100 μg/mL.

As mentioned earlier, work has also been done utilizing more sophisticated macroelectrode designs in an attempt to spatially resolve the measured impedance in specific regions of 3D cell cultures. Canali et al. presented both simulations and experimental work in order to build a platform more amenable for use in tissue engineering applications [29]. Comparison studies of two-electrode (2T) and three-electrode (3T) measurements using rectangular gold electrodes were performed to measure conductivity standards (10 × 25 mm and a thickness of 200 μm). Three-electrode measurements eliminate the polarization impedance, which develops near the surface of polarizable electrodes in two-electrode measurements. Impedance measurements showing the individual conductivity values of different gelatin concentrations were presented together with the observed change of impedance that occurs during the cross-linking process. Finally, the change of impedance from a blank gelatin scaffold and those with 1.5×10^5 cells/mL and 1.5×10^6 cells/mL were compared. Two-electrode measurements did a better job of discriminating between cell-laden hydrogels and blank hydrogels. Discrimination was found to be best at 4 kHz in a time-lapse experiment meant to monitor cell growth. The two-electrode setup returned an impedance increase of ~10 Ω over 48 h, a time that corresponds to a population doubling of the cells while in culture.

Another study by Canali utilized an eight-electrode array and several tetrapolar electrode configurations to measure the impedance of 5% gelatin constructs with and without embedded HepG2 cells in different regions of a chamber [49]. When gelatin constructs were placed in the top left corner of the measurement chamber, readings from that region showed higher impedance magnitudes than when cell constructs were placed in the center of the chamber, demonstrating the possibility to discriminate areas with denser cell populations.

Lee et al. presented a multi-electrode capacitance sensor to monitor cells encapsulated in alginate hydrogels. Only capacitance was measured between four pairs of gold electrodes (1 × 2 mm). The authors were able to show that the capacitance increased over time due to an increased numbers of cells. In addition, cell deaths induced decreases in the measured capacitance when cells were exposed to doxorubicin. Finally, migration of human Mesenchymal stem cells through hydrogels was shown by exposing hydrogels to SDF-1-alpha. Cells unexposed to this growth factor show slower migration rates. This study showed the benefits of having a multi-electrode array and the accompanying spatial resolution of electrical measurements of such cell-laden scaffolds. Moreover, it illustrates the use of a simplified capacitance sensor that only returns the overall capacitance of the electrode/hydrogel system [50].

Cell growth is not the only phenotype that has been correlated with changes in impedance of cell-loaded scaffolds. Valero et al. proposed a microbioreactor to monitor the impedance of neuroblastoma (N2a) embedded in collagen-laminin (CLG) scaffolds in order to detect cell differentiation by neurite outgrowth. N2a cells were found to differentiate and to form neurites when cultured in CLG scaffolds, while simple collagen (CG) did not induce differentiation. Comparing the impedance after 12 h of monitoring in CG and CLG revealed significantly different impedance behaviors. Researchers observed a decrease in impedance over the first 2 h (most likely due to thermal equilibration) followed by steady increases over the next 10 h. These changes are thought to correlate with a combination of axon formation and changes within the collagen-laminin scaffolds [51] seen in separate 96-well plate studies. In CG, the impedance magnitude decreased by 7 Ω in the first 2 h and then stabilized. The initial decrease was deemed to be due to thermal equilibration. Unlike Lin et al., no prior studies were conducted on hydrogel degradation in this study. For this reason, it is difficult to rule out the influence that hydrogel degradation may impose on the measured impedance changes in the 12 h time span. In addition, no comparisons between the impedance behavior of composite CLG scaffolds and CG scaffolds were reported. Differences in initial stabilization periods may have a major impact on the measured impedance, as ions released from the gel overtime will contribute to the basal impedance of the gel. Furthermore, the medium perfusion on the apical side of the gel may further accelerate its chemical breakdown. For this reason, it is critical to perform side-by-side impedance analysis of both seeded and unseeded scaffolds in order to normalize data and avoid artifacts originating from the degradation of the scaffold. Overall, all attempts to monitor the impedance of cells in scaffold-based systems have had only marginal success till today due to their rather low sensitivity. Electrode design has played a dominant role in scientific efforts to improve the sensitivity of impedance measurements in the past and may still be the key to more powerful impedance devices for scaffold-based 3D cultures.

2.3 Scaffold-Free 3D Cultures

Not all 3D culture models require a supporting noncellular structure for developing their 3D environment, and technologies for creating a scaffold-free 3D tissue models have emerged. The scaffold-free approach, also described as *scaffold-less tissue engineering*, refers "to any platform that does not require cell seeding or adherence within an exogenous, three-dimensional material" [52]. Scaffold-free culture avoids the exposure of the cells to potentially harmful processes known from scaffold-based culture, such as the use of bioreactors or polymerization of chemicals, and it is believed to increase cell viability. Scaffold-free culture allows a high degree of cell-cell communications which may significantly affect the cell phenotype. Scaffold-free approaches are grouped in *self-organization* techniques and *self-assembling* processes. The term *self-organization* technique refers to "a process in which order appears when external energy or forces are input in the system," whereas in contrast "no external forces are required to promote order" in *self-assembling* processes [52]. A widespread example for *self-organization* techniques is cell sheet engineering, i.e., cultured cells are harvested in sheets with their functional ECM followed by directly transplantation to tissue beds or stacking of several cell sheets [53]. The formation of articular cartilage and fibrocartilage is a good example of *self-assembling* processes. More details on the topic are discussed by Athanasiou et al. [52].

A widely used scaffold-free 3D tissue model is the multicellular spheroid, which provides a powerful tool for in vitro testing, and is assembled by the compaction of loose cell aggregates into compressed spheroidal structures. The diffusion limit of app. 200 μm can lead to inefficient nutrient support and waste accumulation which may result in a necrotic core [54]. Commonly, spheroids are produced by seeding cells on nonadhesive surfaces like agarose or inside hanging drops [55]. The absence of sites suitable for cell attachment and spreading leads to aggregation and spheroid formation. However, these traditional spheroid culture methods have some limitations like low-throughput or producing tissue models of unrealistic spherical geometry, so that researchers are developing new technologies for producing higher numbers of spheroids in shorter times and with increased geometrical control [57–59]. In pellet culture the cells are forced to concentrate at the bottom of a test tube using centrifugal forces. The downside of this easy to conduct approach is the fact that it introduces shear stress that may lead to cell damage [56].

2.4 EIS in Scaffold-Free 3D Cultures

In contrast to scaffold-based cultures, the emergence of scaffold-free 3D cultures has resulted in the rapid development of powerful experimental devices to study these tissue models by means of impedance techniques that show higher sensitivity and are more amenable to be used in high-throughput applications. As early as 2000, researchers began leveraging the in vivo similarities of spheroids in impedance studies. Molckovsky et al. devised a holder that measured the impedance of liver tumor spheroids with Teflon-coated stainless steel electrodes. Following protocols that mimic photodynamic therapy (PDT) in vitro, impedance measured in a low-conductivity sucrose solution was found to be capable of discriminating whether cells died from apoptosis or necrosis based on the dispersion curves [60]. Hildebrandt et al. produced a capillary-based holder for making impedance measurements on spheroids comprised of mesenchymal stem cells. With this, they found that osteogenic differentiation caused significant increases in the basal impedance of spheroids. Necrotic spheroids show a decreased impedance magnitude at low frequencies and increases at higher frequencies when compared to undifferentiated spheroids [7].

The Robitzki group has performed extensive work developing a microcavity array-based (MCA) impedance sensing system that has been used in conjunction with 2D impedance systems to develop preclinical models for neurodegeneration [8], cardiotoxicity [61], immunotoxicity [62], and drug efficacy [6]. The MCA-based system is composed of electrodes prepared on the side walls of microcavities (100–400 μm) etched into silicon. In the original study, a normalized relative impedance change was tracked where the measured impedance in the presence of the spheroid is normalized to measurements of the impedance of the surrounding culture medium. Relative impedance changes produced by spheroids from several cell types were recorded as well as impedance changes induced by glass and zeolite beads for comparison [6].

More recent studies that employed an optimized version of the MCA chip for toxicity assays used basal impedance values of spheroids to normalize the impedance of spheroids exposed to xenobiotics. For instance, investigations of neuronal spheroids exposed to okadaic acid revealed a dose-dependent decrease in impedance due to degeneration processes as observed in Alzheimer's disease associated with hyperphosphorylation of intracellular target proteins [63]. A more detailed analysis was presented by Seidel et al. who investigated spheroids of three individual neuronal cell lines (wild-type, single, and quadruple point mutations in tau associated genes). Incubation with okadaic acid (OA) induced drops in impedance that correlated with the number of mutations within each cell type, i.e., more drastic impedance drops for more point mutations. Later incubation with inhibitors (e.g., hymenialdisine) allowed cells to regain cellular integrity and caused a corresponding restoration of impedance [8].

Jahnke et al. utilized clusters of human embryonic stem cell-derived cardiomyocytes (hCMCs) to study cardiotoxicity of doxorubicin. A small change to the design of the MCA device included vacuum holes at the bottom of the MCA chamber to allow hydrodynamic positioning of the hCMC prior to impedance analysis. In addition to impedance monitoring, the MCAs were utilized to monitor action potentials generated by the cardiospheres over the course of 35 days. Ninety percent of cardiospheres in MCAs maintained stable beating rates over this time course. In comparison, when the hCMCs were allowed to adhere upon standard planar multi-electrode arrays (MEAs), only 10% of the cardiospheres showed stable beating for a time period of 14 days. Comparison with 2D monolayers of cardiomyocytes showed even less sustained electrical activity. Finally, doxorubicin was tested as a modulator of cardiospheres' overall impedance as well as changes in beating rate. After 3 h of exposure, the beating rates decreased in a dose- and time-dependent manner, whereas only the highest concentration of 100 μM induced an immediate effect within the first hour. Significant relative impedance changes were detected only for the highest concentration of doxorubicin (100 μM) down to 75% of the starting values within 3 h, indicating a lower sensitivity of the overall impedance to detect cell responses compared to the beating rate of hCMCs [61].

Recently, Poenick et al. performed comparison studies with 2D and 3D MCF-7 mouse mammary tumor and human fibroblast (HDFa) models to determine the efficacy of a novel immunotoxin (B3-PE38) engineered to be specific to the Lewis Y antigen. Responses to B3-PE38, paclitaxel, and a nonspecific PE38 immunotoxin were analyzed. Impedance measurements reported cell death trends that indicated specificity of B3-PE38 for cancer cell lines which express the Lewis Y antigen in large amounts compared to the tested fibroblasts. Moreover, MCF7 cultures were more susceptible to toxins when cultured in a 2D format than in spheroidal format resulting in lower IC50 values. In addition, the efficacy of the B3-PE38 immunotoxin was shown to be significantly higher in 3D cultures of MCF7 cells compared to paclitaxel and PE38 [62].

Interestingly, the advances in label-free impedance monitoring techniques have not been exclusive to singular scaffold-based or scaffold-free 3D cultures. Combinations of the two techniques have been presented as a way to perform invasion assays. The commercial RTCA system (ACEA biosciences) was used to monitor cellular invasion of ovarian cancer spheroids generated from several different cell lines using methylcellulose media into a monolayer of mesothelial LP9 cells cultured on a layer of Matrigel separated from chemoattractant medium by a porous membrane. Special plates were used that separated a top chamber containing serum-free medium and a bottom chamber containing medium enriched with fetal calf serum as a chemoattractant. Invasion of the spheroids into the LP9 monolayer was detected by changes in the *cell index*, a measure of the relative change in measured cell impedance as an indicator of the cell status. All cell lines seem to invade when chemoattractants are placed in the bottom chamber over a time course of 24 h [64]. Seemingly, the incorporation of multiple tissue models into a co-culture device was able to form a novel invasion assay that can be used to monitor real-time movement of cancer cells introduced into the system as spheroids.

3 A Linear Electrode Array Integrated in a Microfluidic Channel for Spheroid Measurements

Spatial resolution is an essential improvement for impedance-based assays on 3D tissue models that can help significantly to interpret the results of toxicology assessments with 3D tissues. Utilizing linear arrays of microelectrodes can effectively introduce a spatial resolution of impedance sensors. Linderholm et al. created a simple linear array of electrodes in an effort to image an adherent 3D multilayer cell culture. An equation was derived to indicate the "penetration depth" of a measurement defined by the point above the electrodes where 50% of the electrical power was dissipated for four-electrode configurations [65, 66]. This technique can enable a single set of electrodes to probe multiple layers of a single tissue without the need for moving the electrodes to different positions. To accurately define the area being interrogated, proper electrode design must be

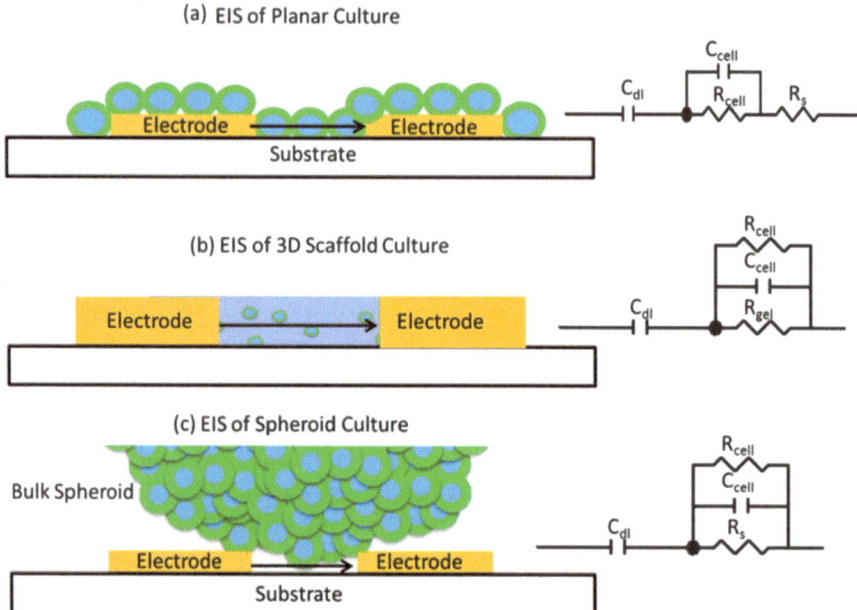

Fig. 1 Measuring the impedance of 2D (**a**) and 3D tissue models (**b, c**) by two-electrode setups. Since the cells do no longer adhere to the electrode surface in 3D models, a shunt pathway for the current is created that reduces the sensitivity for the cells under study and may dominate the overall readout

thoroughly considered and validated. Looking in spheroidal cultures, it is equally important to consider the change in current flow caused by the cell-cell contacts, which are very different from 2D cultures, and the removal of an adherent surface. Adherent cell lines that attach directly to the electrode surface in ECIS-based systems create a dielectric barrier to current flow in addition to the resistivity of the bulk R_s as they reach confluence (cf. Fig. 1a). This forces the current to flow between tiny intercellular spaces or through the cellular membrane prior to escaping into the culture medium. In contrast, when 3D tissue models are considered, cells attach to each other or to the scaffold and do not directly adhere to the measuring electrodes (Fig. 1b, c). This allows the culture medium to come into direct contact with the measuring electrodes, causing a shunt current that bypasses the cell bodies. This shunt current is presented by the shunt resistance R_s in Fig. 1. As discussed above, normalization can be used to factor out the influence of the scaffold. However, in scaffold-free cultures, it is more feasible to mitigate this shunt resistance by using electrode designs that maximize electrode contact with the cellular sample and minimize contact with the culture medium.

Microfluidic techniques offer a useful solution to overcome this disadvantage. Small microfluidic channels can increase the contact area between planar electrodes and the sample, reducing the shunt resistance and acting as a pseudo-passivation layer. In addition, microfluidics provide a useful mechanism for controlling parameters such

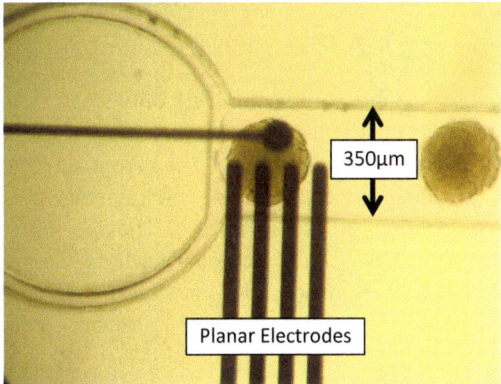

Fig. 2 Microfluidic EIS sensor for long-term monitoring of tumor spheroids. Designing the microfluidic channel to approximate the size of the spheroid allows the sample to be hydrodynamically positioned over the planar electrodes. In theory, the limited height of the channel causes the sample to be pressed onto the electrodes effectively minimizing the effect of the shunt resistance

as nutrient perfusion rate, sheer force, and chemical gradients and are a great platform for long-term observations of the tissue models. However, continuous medium perfusion is necessary to maintain the viability of the cells in the 3D model. One such microfluidic channel with integrated electrode array for live EIS monitoring is presented in Fig. 2. Variations in electrode structure and composition controlled by the individual fabrication techniques provide different electrode performances that can be tailored to the needs of the experiment. For this reason, validation measurements using conductivity standards such as potassium chloride solutions and DMEM culture medium should be performed to characterize electrode sensitivity before the tissue models are placed into the microfluidic structure.

For an electrolyte solution in contact with a planar metal electrode, a serial R_s-C_{dl} circuit model can accurately describe the system. The double-layer capacitance, C_{dl}, defines the high impedance at lower frequencies. In a log-log plot of $|Z|$ (f), the impedance caused by the interface capacitance falls with a slope of (-1) with increasing frequencies. This capacitance is well-characterized for gold electrodes and is known to range between 10 and 20 $\mu F/cm^2$ (14.25 $\mu F/cm^2$ according to [67]). By measuring the magnitude of the impedance, $|Z|$, at a known frequency, f, within the linear frequency regime mentioned above, the capacitance is found by solving the following equation:

$$|Z| = \frac{1}{2\pi f c} \rightarrow c = \frac{1}{2\pi f |Z|}$$

The measured capacitance has to be normalized by the surface area of the electrodes to estimate the value of the area-specific double-layer capacitance, C_{dl} as follows:

$$C_{dl} = \frac{c}{\text{Area}}.$$

When two conductive solutions are compared, only small changes in the impedance at lower frequencies occur because the conductivity of the solution is the only parameter changed and the system capacitance should remain largely unchanged. At higher frequencies, the double-layer capacitance no longer dominates the impedance of the system but the solution resistance instead. The specific conductivity of the solution allows calculating the solution resistance within the system. Because the semicircular electrode immersed in solution is similar to the rotating disk electrode, the solution resistance is approximated using the following equation:

$$R_s = 2\frac{\rho}{4r} = \frac{\rho}{2r},$$

which is taken from Newman's analysis on the spinning disk electrode [68]. Because there are two electrodes required for the measurement the total resistance doubles.

Due to adsorption and desorption of biomolecules to the electrode surfaces, chemical changes in the microenvironment of the tissue models and the most significant biological variations, it is troublesome to compare and interpret absolute impedance values of 3D cultures. For this reason the majority of groups have reported relative changes in impedance by normalizing time course data to impedance values recorded at the beginning of the experiment. Sensitivity of the measurements is also commonly reported by normalizing measurements with the 3D tissue model introduced into the system relative to baseline measurements of the electrodes in pure culture medium without any biological structure. Multiple measurements using redundant electrodes can improve statistical significance and spatial resolution of the measurements [69]. An example of this is shown in Fig. 3a, in which the impedance of Hs578T spheroids recorded by a linear array of planar electrodes is reported. From this it can be seen that increasing the space between electrodes causes the measured impedance of the spheroid to increase. When the relative change of impedance magnitude |Z| is calculated with respect to the individual medium baselines (Fig. 3b), the maximum change was observed for a 250 μm interelectrode distance. For these electrodes a 60% increase in impedance is seen in comparison to medium baselines at frequencies higher than 1 MHz.

The greatest increase in impedance is seen at frequencies above 100 kHz and reaches a plateau around 1 MHz, revealing that the mechanism causing the impedance change is the tumor spheroid physically displacing the conductive medium in the microfluidic chamber. Although the channel size approximates the size of the tumor spheroid, the medium has still access to the uncovered portions of the electrodes. Because the electrodes are not fully insulated from the medium, current flows through both the spheroid and medium (shunt pathway). Despite this, impedance changes begin to increase at 30 kHz, indicating that the cause of the impedance change from 30 kHz to 1 MHz is in response to the presence of the tumor spheroid.

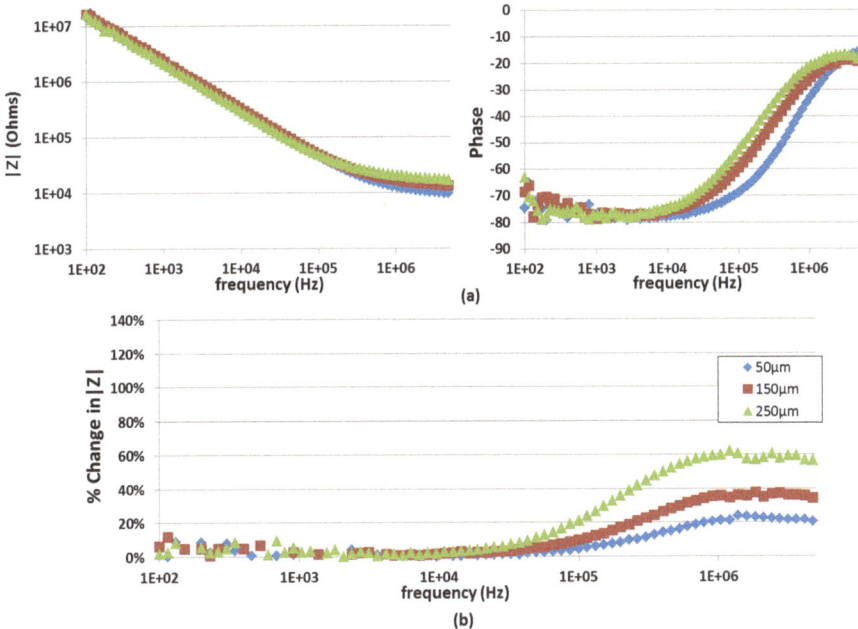

Fig. 3 (**a**) Bode plot of impedance and phase when a spheroid is measured in a microfluidic channel in a two-electrode setup. The interelectrode distance increases from 50 to 250 µm. (**b**) Relative impedance change caused by the presence of a spheroid in the microfluidic channel above the impedance sensing electrodes

This falls into the region of the β-dispersion where the impedance of biological tissue is dominated by the dielectric properties of the cellular plasma membrane.

4 Modeling of 3D Microelectrodes

Finite element simulation has proven to be helpful in effectively designing electrodes for analyzing various 3D tissue models. Canali et al. used such simulation studies to enhance spatial resolution of measurements of larger hydrogel cultures to be used in tissue engineering applications [29, 49]. Unfortunately, due to size limitations, these devices suffered from low sensitivity. Furthermore, the size of the macro-electrodes used in these studies will not allow accurate spatially resolved analysis of scaffold-free spheroid models. Microelectrodes are as a more suitable solution for measuring spheroidal models as demonstrated by the Robitzki group [6, 8, 61–63, 70, 71]. But microelectrode performance is significantly affected by shunt pathways for the current that may reduce the sensitivity drastically.

Fig. 4 Microfabricated
silicon microneedles can be
metallized for use as
microelectrodes to probe 3D
cultures. Linear arrays of
these electrodes can be used
to increase spatial resolution
of measurements

Three-dimensional microneedle electrodes are a particularly efficient electrode design to reduce or even eliminate shunting of the solution resistance. Sharpened silicon microneedles (cf. Fig. 4) are fabricated with a combination of anisotropic and isotropic deep reactive ion etching (DRIE) techniques [72].

By metallizing these pillars, electrodes are formed that can pierce the surface of larger spheroidal models. This eliminates the shunt resistance by bringing the electrode in direct and intimate contact with the sample. One clear and obvious disadvantage of this technique is the invasiveness of puncturing prefabricated tissue models with the electrodes. This may be mitigated by utilizing sharpened needle structures that can minimize the damage inflicted on the spheroid [73]. Simulation results are shown in Fig. 5 for the tips of two microneedle electrodes (approximated as 2D triangles) with progressive spacing. The calculation returns the penetration depth of such electrodes into the tissue indicating that spatial resolution is possible by using different electrode distances.

The calculation shows that midway between the two terminals, the magnitude of the E-field reaches a minimum. At this point in space, a sample will experience a minimum exposure to the applied electric field. For this reason this midpoint is used to find the effective penetration depth of the electric field into the sample. Above the midpoints of the electrodes, the field decreases as the distance above the electrodes increases. This indicates that the majority of the current will be passed directly between the electrodes but maintains a high intensity for as far above the electrodes as the electrodes are spaced apart. Increasing the electrode spacing causes a significant decrease in the maximum magnitude of the electric field as the field becomes more spread.

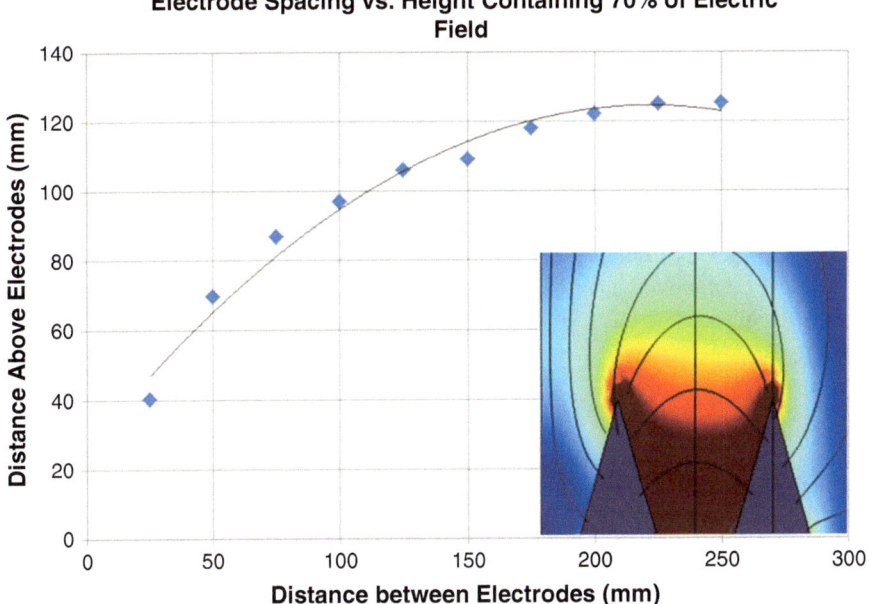

Fig. 5 A plot of the intensity of the electric field emanating from the tip of microneedle electrodes simplified as 2D triangles. As the distance between microneedle tips increases, so does the extension of the electric field above the base of the needle tips

5 The Future of EIS-Based Monitoring of 3D Tissue Models

Future toxicology assessments will likely continue to leverage the predictive advantage of scaffold-based and scaffold-free 3D tissue models as a method for preclinical testing. In particular, scaffold-free spheroid cultures are likely to grow in prominence due to expanding capabilities of commercial systems for mid- to high-throughput generation of spheroids. This will allow continued translation of many techniques used in planar cell cultures into 3D modalities, including the many gains made in the past decade in EIS monitoring of planar cultures.

Despite improvements in real-time impedance analysis for cellular cultures, to date, EIS measurements alone may be inadequate for assessing the status of such 3D tissue models. In order to more accurately assess the underlying mechanisms causing cellular stressors/changes, a micro-physiometry-based approach employing parallel sensors for direct or indirect metabolic parameters may be a suitable extension for monitoring 3D cultures. Combining impedance-based sensors with optical or semiconductor pH detectors has been successfully explored [74, 75] and has further strengthened the case for using label-free techniques by exposing changes in cellular metabolism caused by some colorimetric labels [76]. The incorporation of 3D tissue models into these systems may provide a potentially powerful tool for preclinical toxicology assessments.

Moving forward, analyzing the complete impedance spectrum will be crucial to correlating underlying mechanistic changes on the cellular level with changes in the measured impedance. Many groups have chosen a simplified approach of tracking maximum changes in impedance magnitude at a specific frequency as opposed to applying equivalent circuit modeling. While sufficient for simple live/dead assays, this treatment can be troublesome when trying to determine the cause of cellular death or even more details. To get to more informative assessments of the toxic effect new compounds have on 3D cultures, we will have to make use of spatially resolved real and imaginary impedance and pinpoint the observed changes to cellular phenotypes.

References

1. Freshney R (2011) Culture of animal cells: a manual of basic technique and specialized applications. Wiley, Hoboken
2. Fitzgerald KA, Malhotra M, Curtin CM, O'Brien FJ, O'Driscoll CM (2015) Life in 3D is never flat: 3D models to optimise drug delivery. J Control Release 215:39–54. https://doi.org/10.1016/j.jconrel.2015.07.020
3. Edmondson R, Broglie JJ, Adcock AF, Yang L (2014) Three-dimensional cell culture systems and their applications in drug discovery and cell-based biosensors. Assay Drug Dev Technol 12:207–218. https://doi.org/10.1089/adt.2014.573
4. Hartung T (2014) 3D – a new dimension of in vitro research. Adv Drug Deliv Rev 69–70:vi. https://doi.org/10.1016/j.addr.2014.04.003
5. Alexander FA, Price DT, Bhansali S (2013) From cellular cultures to cellular spheroids: is impedance spectroscopy a viable tool for monitoring multicellular spheroid (MCS) drug models? IEEE Rev Biomed Eng 6:63–76. https://doi.org/10.1109/RBME.2012.2222023
6. Kloss D, Fischer M, Rothermel A, Simon JC, Robitzki AA (2008) Drug testing on 3D in vitro tissues trapped on a microcavity chip. Lab Chip 8(6):879–884
7. Hildebrandt C, Büth H, Cho S, Impidjati, Thielecke H (2010) Detection of the osteogenic differentiation of mesenchymal stem cells in 2D and 3D cultures by electrochemical impedance spectroscopy. J Biotechnol 148(1):83–90. https://doi.org/10.1016/j.jbiotec.2010.01.007
8. Seidel D, Krinke D, Jahnke H-G, Hirche A, Kloß D, Mack TG, Striggow F, Robitzki A (2012) Induced tauopathy in a novel 3D-culture model mediates neurodegenerative processes: a real-time study on biochips. PLoS One 7(11):e49150
9. McCoy MH, Wang E (2005) Use of electric cell-substrate impedance sensing as a tool for quantifying cytopathic effect in influenza A virus infected MDCK cells in real-time. J Virol Methods 130(1):157–161
10. Åberg P, Nicander I, Hansson J, Geladi P, Holmgren U, Ollmar S (2004) Skin cancer identification using multifrequency electrical impedance-a potential screening tool. IEEE Trans Biomed Eng 51(12):2097–2102
11. Huang X, Greve D, Nguyen D, Domach M (2003) Impedance based biosensor array for monitoring mammalian cell behavior. Sensors, 2003. Proceedings of IEEE, vol 1. IEEE, Piscataway

12. Judy JW (2001) Microelectromechanical systems (MEMS): fabrication, design and applications. Smart Mater Struct 10(6):1115
13. Rahman ARA, Justin G, Guiseppi-Elie A (2009) Towards an implantable biochip for glucose and lactate monitoring using microdisc electrode arrays (MDEAs). Biomed Microdevices 11 (1):75–85
14. Park TH, Shuler ML (2003) Integration of cell culture and microfabrication technology. Biotechnol Prog 19(2):243–253
15. Matysik F-M, Meister A, Werner G (1995) Electrochemical detection with microelectrodes in capillary flow systems. Anal Chim Acta 305(1):114–120
16. Bard AJ, Faulkner LR, Leddy J, Zoski CG (1980) Electrochemical methods: fundamentals and applications, vol 2. Wiley, New York
17. Wegener J, Keese CR, Giaever I (2000) Electric cell–substrate impedance sensing (ECIS) as a noninvasive means to monitor the kinetics of cell spreading to artificial surfaces. Exp Cell Res 259(1):158–166
18. Price DT, Rahman ARA, Bhansali S (2009) Design rule for optimization of microelectrodes used in electric cell-substrate impedance sensing (ECIS). Biosens Bioelectron 24(7):2071–2076
19. Fosdick LE, Anderson JL (1986) Optimization of microelectrode array geometry in a rectangular flow channel detector. Anal Chem 58(12):2481–2485
20. Min J, Baeumner AJ (2004) Characterization and optimization of interdigitated ultramicroelectrode arrays as electrochemical biosensor transducers. Electroanalysis 16 (9):724–729
21. Sandison ME, Anicet N, Glidle A, Cooper JM (2002) Optimization of the geometry and porosity of microelectrode arrays for sensor design. Anal Chem 74(22):5717–5725
22. Lempka SF, Johnson MD, Barnett DW, Moffitt MA, Otto KJ, Kipke DR, McIntyre CC (2006) Optimization of microelectrode design for cortical recording based on thermal noise considerations. Engineering in medicine and biology society, 2006. EMBS'06. 28th annual international conference of the IEEE. IEEE, Piscataway
23. Wang L, Wang H, Mitchelson K, Yu Z, Cheng J (2008) Analysis of the sensitivity and frequency characteristics of coplanar electrical cell–substrate impedance sensors. Biosens Bioelectron 24(1):14–21
24. English AE, Moy AB, Kruse KL, Ward RC, Kirkpatrick SS, Goldman MH (2009) Instrumental noise estimates stabilize and quantify endothelial cell micro-impedance barrier function parameter estimates. Biomed Signal Proc Control 4(2):86–93
25. Curtis TM, Tabb J, Romeo L, Schwager SJ, Widder MW, Van der Schalie WH (2009) Improved cell sensitivity and longevity in a rapid impedance-based toxicity sensor. DTIC document. J Appl Toxicol 29:374–380
26. Pejcic B, De Marco R (2006) Impedance spectroscopy: over 35 years of electrochemical sensor optimization. Electrochim Acta 51(28):6217–6229
27. Rahman ARA, Price DT, Bhansali S (2007) Effect of electrode geometry on the impedance evaluation of tissue and cell culture. Sensors Actuators B Chem 127(1):89–96
28. Rahman A, Register J, Vuppala G, Bhansali S (2008) Cell culture monitoring by impedance mapping using a multielectrode scanning impedance spectroscopy system (CellMap). Physiol Meas 29(6):S227
29. Canali C, Heiskanen A, Muhammad HB, Høyum P, Pettersen F-J, Hemmingsen M, Wolff A, Dufva M, Martinsen ØG, Emnéus J (2015) Bioimpedance monitoring of 3D cell culturing – complementary electrode configurations for enhanced spatial sensitivity. Biosens Bioelectron 63:72–79
30. Hutmacher DW (2000) Scaffolds in tissue engineering bone and cartilage. Biomaterials 21:2529–2543. https://doi.org/10.1016/S0142-9612(00)00121-6
31. Carletti E, Motta A, Migliaresi C (2011) Scaffolds for tissue engineering and 3D cell culture. Methods Mol Biol 695:17–39. https://doi.org/10.1007/978-1-60761-984-0_2

32. Lanza R, Langer R, Vacanti J (2013) Principles of tissue engineering. Academic Press, San Diego
33. Dutta RC, Dutta AK (2009) Cell-interactive 3D-scaffold; advances and applications. Biotechnol Adv 27:334–339. https://doi.org/10.1016/j.biotechadv.2009.02.002
34. O'Brien FJ (2011) Biomaterials & scaffolds for tissue engineering. Mater Today 14:88–95. https://doi.org/10.1016/S1369-7021(11)70058-X
35. Rimann M, Graf-Hausner U (2012) Synthetic 3D multicellular systems for drug development. Curr Opin Biotechnol 23:803–809. https://doi.org/10.1016/j.copbio.2012.01.011
36. Hutmacher DW (2001) Scaffold design and fabrication technologies for engineering tissues – state of the art and future perspectives. J Biomater Sci Polym Ed 12:107–124. https://doi.org/10.1163/156856201744489
37. Nair LS, Laurencin CT (2007) Biodegradable polymers as biomaterials. Prog Polym Sci 32:762–798. https://doi.org/10.1016/j.progpolymsci.2007.05.017
38. Drury JL, Mooney DJ (2003) Hydrogels for tissue engineering: scaffold design variables and applications. Biomaterials 24:4337–4351
39. Tibbitt MW, Anseth KS (2009) Hydrogels as extracellular matrix mimics for 3D cell culture. Biotechnol Bioeng 103:655–663. https://doi.org/10.1002/bit.22361
40. Billiet T, Vandenhaute M, Schelfhout J, Van Vlierberghe S, Dubruel P (2012) A review of trends and limitations in hydrogel-rapid prototyping for tissue engineering. Biomaterials 33:6020–6041. https://doi.org/10.1016/j.biomaterials.2012.04.050
41. Fisher RJ, Peattie RA (2007) Controlling tissue microenvironments: biomimetics, transport phenomena, and reacting systems. Adv Biochem Eng Biotechnol 103:1–73
42. Nair LS, Laurencin CT (2006) Polymers as biomaterials for tissue engineering and controlled drug delivery. Adv Biochem Eng Biotechnol 102:47–90
43. Velema J, Kaplan D (2006) Biopolymer-based biomaterials as scaffolds for tissue engineering. Adv Biochem Eng Biotechnol 102:187–238
44. Lin S-P, Kyriakides TR, Chen J-JJ (2009) On-line observation of cell growth in a three-dimensional matrix on surface-modified microelectrode arrays. Biomaterials 30(17):3110–3117
45. Lei KF, Wu M-H, Liao P-Y, Chen Y-M, Pan T-M (2011) Development of a micro-scale perfusion 3D cell culture biochip with an incorporated electrical impedance measurement scheme for the quantification of cell number in a 3D cell culture construct. Microfluid Nanofluid 12(1):117–125. https://doi.org/10.1007/s10404-011-0854-x
46. Lei KF, Wu M-H, Hsu C-W, Chen Y-D (2014) Real-time and non-invasive impedimetric monitoring of cell proliferation and chemosensitivity in a perfusion 3D cell culture microfluidic chip. Biosens Bioelectron 51:16–21. https://doi.org/10.1016/j.bios.2013.07.031
47. Antonijevic M, Petrovic M (2008) Copper corrosion inhibitors. A review. Int J Electrochem Sci 3(1):1–28
48. Lei KF, Wu Z-M, Huang C-H (2015) Impedimetric quantification of the formation process and the chemosensitivity of cancer cell colonies suspended in 3D environment. Biosens Bioelectron 74:878–885
49. Canali C, Mazzoni C, Larsen LB, Heiskanen A, Martinsen Ø, Wolff A, Dufva M, Emnéus J (2015) An impedance method for spatial sensing of 3D cell constructs – towards applications in tissue engineering. Analyst 140(17):6079–6088
50. Lee S-M, Han N, Lee R, Choi I-H, Park Y-B, Shin J-S, Yoo K-H (2016) Real-time monitoring of 3D cell culture using a 3D capacitance biosensor. Biosens Bioelectron 77:56–61
51. Valero T, Moschopoulou G, Kintzios S, Hauptmann P, Naumann M, Jacobs T (2010) Studies on neuronal differentiation and signalling processes with a novel impedimetric biosensor. Biosens Bioelectron 26(4):1407–1413
52. Athanasiou KA, Eswaramoorthy R, Hadidi P, Hu JC (2013) Self-organization and the self-assembling process in tissue engineering. Annu Rev Biomed Eng 15:115–136. https://doi.org/10.1146/annurev-bioeng-071812-152423
53. Chen G, Qi Y, Niu L, DI T, Zhong J, Fang T, Yan W (2015) Application of the cell sheet technique in tissue engineering. Biomed Rep 3:749–757. https://doi.org/10.3892/br.2015.522

54. Lin R-Z, Lin R-Z, Chang H-Y (2008) Recent advances in three-dimensional multicellular spheroid culture for biomedical research. Biotechnol J 3:1172–1184. https://doi.org/10.1002/biot.200700228

55. Khetani SR, Berger DR, Ballinger KR, Davidson MD, Lin C, Ware BR (2015) Microengineered liver tissues for drug testing. J Lab Autom 20:216–250. https://doi.org/10.1177/2211068214566939

56. Yoon No D, Lee K-H, Lee J, Lee S-H (2015) 3D liver models on a microplatform: well-defined culture, engineering of liver tissue and liver-on-a-chip. Lab Chip 15:3822–3837. https://doi.org/10.1039/C5LC00611B

57. Fukuda J, Nakazawa K (2011) Hepatocyte spheroid arrays inside microwells connected with microchannels. Biomicrofluidics 5:22205. https://doi.org/10.1063/1.3576905

58. Gevaert E, Dollé L, Billiet T, Dubruel P, van Grunsven L, van Apeldoorn A, Cornelissen R (2014) High throughput micro-well generation of hepatocyte micro-aggregates for tissue engineering. PLoS One 9:e105171. https://doi.org/10.1371/journal.pone.0105171

59. Napolitano AP, Dean DM, Man AJ, Youssef J, Ho DN, Rago AP, Lech MP, Morgan JR (2007) Scaffold-free three-dimensional cell culture utilizing micromolded nonadhesive hydrogels. Biotechniques 43:494–500. https://doi.org/10.2144/000112591

60. Molckovsky A, Wilson B (2001) Monitoring of cell and tissue responses to photodynamic therapy by electrical impedance spectroscopy. Phys Med Biol 46(4):983

61. Jahnke H-G, Steel D, Fleischer S, Seidel D, Kurz R, Vinz S, Dahlenborg K, Sartipy P, Robitzki AA (2013) A novel 3D label-free monitoring system of hES-derived cardiomyocyte clusters: a step forward to in vitro cardiotoxicity testing. PLoS One 8(7):e68971

62. Poenick S, Jahnke H-G, Eichler M, Frost S, Lilie H, Robitzki AA (2014) Comparative label-free monitoring of immunotoxin efficacy in 2D and 3D mamma carcinoma in vitro models by impedance spectroscopy. Biosens Bioelectron 53:370–376

63. Krinke D, Jahnke H-G, Mack TG, Hirche A, Striggow F, Robitzki AA (2010) A novel organotypic tauopathy model on a new microcavity chip for bioelectronic label-free and real time monitoring. Biosens Bioelectron 26(1):162–168

64. Bilandzic M, Stenvers KL (2014) Assessment of ovarian cancer spheroid attachment and invasion of mesothelial cells in real time. J Vis Exp 87. https://www.jove.com/video/51655/assessment-ovarian-cancer-spheroid-attachment-invasion-mesothelial

65. Linderholm P, Braschler T, Vannod J, Barrandon Y, Brouard M, Renaud P (2006) Two-dimensional impedance imaging of cell migration and epithelial stratification. Lab Chip 6(9):1155–1162

66. Linderholm P, Marescot L, Loke MH, Renaud P (2008) Cell culture imaging using microimpedance tomography. IEEE Trans Biomed Eng 55(1):138–146

67. Moulton S, Barisci J, Bath A, Stella R, Wallace G (2004) Studies of double layer capacitance and electron transfer at a gold electrode exposed to protein solutions. Electrochim Acta 49(24):4223–4230

68. Newman J (1966) Resistance for flow of current to a disk. J Electrochem Soc 113(5):501–502

69. Alexander FA Jr, Huey EG, Price DT, Bhansali S (2012) Real-time impedance analysis of silica nanowire toxicity on epithelial breast cancer cells. Analyst 137(24):5823–5828

70. Eichler M, Jahnke H-G, Krinke D, Müller A, Schmidt S, Azendorf R, Robitzki AA (2015) A novel 96-well multielectrode array based impedimetric monitoring platform for comparative drug efficacy analysis on 2D and 3D brain tumor cultures. Biosens Bioelectron 67:582–589

71. Thielecke H, Mack A, Robitzki A (2001) Biohybrid microarrays – impedimetric biosensors with 3D in vitro tissues for toxicological and biomedical screening. Fresenius J Anal Chem 369(1):23–29

72. Khanna P, Luongo K, Strom JA, Bhansali S (2010) Sharpening of hollow silicon microneedles to reduce skin penetration force. J Micromech Microeng 20(4):045011

73. Alexander F (2014) RTEMIS: real-time tumoroid and environment monitoring using impedance spectroscopy and pH sensing. Graduate Theses and Dissertations. https://scholarcommons.usf.edu/etd/5168

74. Wiest J, Stadthagen T, Schmidhuber M, Brischwein M, Ressler J, Raeder U, Grothe H, Melzer A, Wolf B (2006) Intelligent mobile lab for metabolics in environmental monitoring. Anal Lett 39(8):1759–1771
75. Otto AM, Brischwein M, Niendorf A, Henning T, Motrescu E, Wolf B (2003) Microphysiological testing for chemosensitivity of living tumor cells with multiparametric microsensor chips. Cancer Detect Prev 27(4):291–296
76. Shinawi TF, Kimmel DW, Cliffel DE (2013) Multianalyte microphysiometry reveals changes in cellular bioenergetics upon exposure to fluorescent dyes. Anal Chem 85(24):11677–11680

BIOREV (2019) 2: 135–162
https://doi.org/10.1007/11663_2018_4
© Springer Nature Switzerland AG 2018
Published online: 11 April 2019

On the Use of the Quartz Crystal Microbalance for Whole-Cell-Based Biosensing

D. Johannsmann

Contents

Abstract The quartz crystal microbalance (QCM) can be used and has often been used to study the interactions of cells with man-made surfaces. The instrument as such is simple. For screening purposes, one can easily run numerous resonators in parallel. The main problem is the interpretation of experimental data. Living cells by their very nature are enormously complicated and the limited amount of information obtained from a QCM experiment therefore is not easily turned into a meaningful

D. Johannsmann (✉)
Institute of Physical Chemistry, Clausthal Universtity of Technology, Clausthal-Zellerfeld,
Germany
e-mail: johannsmann@pc.tu-clausthal.de

diagnostic statement. In the first part, the text elaborates on the technical background with special emphasis on quantitative modeling. While thorough quantitative modeling is difficult, simplified models (which have a limited scope and which provide limited answers) can be applied. The text provides checks on consistency and applicability. These simple models are the Sauerbrey film (only applicable to biofilms on torsional resonators), the semi-infinite viscoelastic medium, and the coupled resonance. In search for more depth of information, one may explore novel sensing dimensions, which include the variation of amplitude, exploitation of piezoelectric stiffening, the analysis of temporal variations, and temperature sweeps. Given the instrument's simplicity, one may combine the QCM with imaging techniques, with optical spectroscopy (even in transmission), and with electrical impedance spectroscopy. The situation is open. Probing whole cells and cell layers with a QCM is already a robust and reliable technique. A better understanding of data interpretation will expand the scope of possible applications.

Keywords Acoustic sensors · Biocolloids · Quartz crystal microbalance · Whole-cell-based biosensing

1 Introduction

Chemical and biological sensors are commonly portrayed as a combination of a "receptor" with a "transducer" [1]. Today, the transducer is the easier part. It is a device which transforms the primary signal (usually a concentration) into a signal suitable for data processing (often a voltage or a current, sometimes a frequency). A number of technologies (optical, electrochemical, acoustic, and others) have matured. Of course one always strives for further improvement, but far more critical than the device producing the electronic signal at the sensor's rear-end is the receptor layer. The receptor layer is supposed to endow the device with specificity; specificity usually requires chemical recognition. While one would hope to imitate the natural organs (the nose and the tongue [2]), this goal is far away and the bottlenecks in the process of sensor improvement are better receptors.

Hoping to imitate nature, people have long ago proposed to not use molecules or assemblies of molecules for the purpose of recognition but rather to harness the potential of entire cells to that end [3]. One problem of course is to reliably position cells on a transducer and keep them alive and responsive there. A second question is which cells to use and how to possibly modify them genetically, so that they respond to the targets of interest. Both questions are outside the scope here.

A third problem is more interesting to the physicist. Whole-cell-based biosensing challenges the traditional way of dividing tasks between physicists (in charge of transduction) and chemists (in charge of receptors), because the signals, which the cells send out in response to changing environmental conditions, again are complex and require adapted technology for optimal translation into digital output. A second layer of sensing is introduced between the primary receptor layer (the cell) and the physical transducer. Expressing it as anthropomorphism, we need to listen to the cell carefully.

The two established techniques for interrogation of a cell's state are fluorescence and electric impedance spectroscopy, the latter sometimes being called "electrochemical cell-substrate impedance spectroscopy," abbreviated ECIS in that context [4, 5]. Fluorescence-based approaches often target one specific substance as, for instance, in reporter gene assays. Here, the cell is designed to produce a fluorescent molecule, if triggered to do so by the target. The limits of this approach are given by its specificity. The cell only responds to the substance it was designed to respond to. Should there be other dangers to the cell, it will not report it. ECIS is more responsive to the state of the cell layer as a whole than the fluorescence-based technologies. That is beneficial if the sensing target is not one chemical but rather the environmental conditions in a wider sense. Acoustic resonators can also perform this task, and they have unique capabilities in this regard. Cells carefully regulate their internal network of filaments and membranes. In concerted action, this machinery provides for mechanical stability, mechanical activity, and mechanical responsiveness. The QCM probes this viscoelastic network and thereby can provide relevant information on the cell's state.

The possibility of whole-cell-based sensing using a QCM crossed researcher's mind soon after the QCM had been immersed into liquid [6–12]. Of course whole-cell-based sensing is challenging when it comes to a detailed physical understanding. But a detailed understanding might not actually be needed; data analysis might rest on characteristic features and empirical relations. After all, large parts of biology and even of engineering are not quantitatively connected to the first principles of physics. When living cells are part of the transduction chain, the depth and the detail of the information obtained will not primarily rely on a detailed and accurate physical model of the acoustic situation. It will rather rely on cells' ability to specifically recognize and amplify signals from the environment. If these abilities can be exploited for sensing, compromises on the side of acoustics may be worthwhile.

There is a solid body of literature on the QCM being employed for the study of cells and cell layers. A good review is provided in Ref. [13]. This chapter is not intended as an update on the research field as such. Rather, it attempts to outline further options and possibilities from a technical point of view. "Technical" here means both instrumentation and data analysis. The author stays away from all questions concerning bioengineering. Quite obviously, identifying questions, which can be answered with a QCM/cell layer combination is equally important – if not more important – than the technical discussion. Put differently, this chapter addresses the two last steps in the transduction chain, which are the acoustic detection and the data processing. It does not address the question of how to find, engineer, and optimize cell layers and even less of how to understand the cell layer's behavior. The limited scope clearly in view, the technical discussion hopefully can make a contribution.

2   Instrumental Aspects

Before addressing cells and how they are seen (better: "heard" [14]) by the QCM, a few brief comments on instrumental issues shall provide context:

- The author works with impedance analysis [15]. The price of impedance analyzers has come down enormously. Good analyzers are now available from different sources for a few thousand USD. These will make the second-generation QCM available to even more laboratories than today.
- The noise level in $\Delta f/n$ (n the overtone order) can be as low as 30 mHz for a 5 MHz crystal, if the crystal is probed in transmission rather than reflection [16], which requires only a rather simple rewiring. A noise level in frequency of 30 mHz corresponds to a noise level in layer thickness of 0.006 nm upon conversion with the Sauerbrey equation (using $\rho = 1$ g/cm^3).
- Even with good electronics in place, quartz crystals often have somewhat of an individual character. Systematic errors are superimposed onto the statistical noise. These are recognized when, for instance, one overtone behaves differently from all others. All crystals contain defects, and these migrate inside the resonator in response to changes of temperature and stress. High-quality crystals are needed, but even with these, one finds "bad ones" and "good ones." For the time being, the community has to live with that situation.
- In complicated situations, where artifacts are to be expected and quantitative analysis is out of reach anyway, the bandwidth – to be used as an empirical indicator with some correlation to the state of the sample – is often more robust than the frequency. Bandwidth is equivalent to the dissipation factor describing energy losses during oscillation. Static stresses and fluctuations of temperature affect the bandwidth less than the frequency.

At this time, the QCM is the workhorse of the acoustic sensing. A typical fundamental frequency is 5 MHz. Over the years, numerous other acoustic devices have been designed and tested. That includes *surface acoustic wave* (SAW) devices [17], Love wave devices [18], QCM's with a base frequency much higher than 5 MHz [19, 20], and the thin *film bulk acoustic resonator* (FBAR) [21]. Looking at the underlying equations, these should outperform the QCM at least in some regards, but these devices all are considerably more complicated than the QCM. When all is said and done, most practically minded researchers today end up sticking with the QCM. That is not meant to be a statement about the future. It will be interesting to see how the field evolves.

3   Background: The Non-gravimetric QCM

The classical quartz crystal microbalance (QCM), as for instance described by Sauerbrey in 1959 [22], measures *inertial* loads. A film deposited onto the resonator surface exerts an oscillatory stress onto the surface, which is proportional to the film's mass. The frequency shift, Δf, therefore is also proportional to the mass (more precisely: to the mass per unit area), which motivates the name "microbalance." The relation between frequency and mass is the content of the famous Sauerbrey equation. Since the layer's density is often known, the mass per unit area is easily converted to a film thickness, and these devices therefore basically are film-thickness monitors. When operated in air, Δf depends on the mass alone for films with a thickness of up to about 50 nm. For thicker films, viscoelastic effects increase the *apparent* mass.

When the QCM was first immersed into a liquid, people were still primarily interested in measuring the thickness of an experimentally deposited layer [23, 24]. These researchers were electrochemists. The goal was (and still is) to compare the mass deposited during some electrochemical process to the total charge passed through the electrode surface in that same process. The two should be closely related to each other according to Faraday's law. Data analysis is in this context often also carried out with the Sauerbrey equation. If one looks at the matter a little closer, one finds that Δf is not strictly proportional to the mass per unit area even in the thin-film limit. Precise analysis requires a viscoelastic correction [25, 26]. The oscillatory stress exerted by the electrodeposited sample should be calculated with an acoustic multilayer formalism [26]. Strictly speaking, the QCM does not measure a mass per unit area but rather a quantity, which is not easily defined in one line and which depends on the film's thickness *and* its viscoelastic compliance. Only for sufficiently rigid films this quantity is equivalent to the mass per unit area. Otherwise, one has to include the sample's finite compliance in the analysis.

As far as the interpretation of the frequency shifts obtained with an electrochem-ical QCM (i.e., a QCM setup in which the surface electrode facing the solution is used for electrochemistry, abbreviated EQCM) is concerned, a number of further complications were soon recognized [27]. Δf is also affected by roughness, by a variable viscosity of the bulk liquid, by nanobubbles, by slip, and by subtle effects related to piezoelectric stiffening [28]. Accordingly, the quantitative interpretation of EQCM results remains somewhat of a challenge. We will come back to challenges of this kind when discussing living cells growing on a QCM surface. For the EQCM, quantitative analysis is difficult but definitely feasible. Depending on the system (its thickness, its roughness, etc.), the parameters derived from the experiments are robust and worth further discussion.

Once the QCM had been established as an instrument working in the liquid phase, people started to apply it to life-science problems [29]. That, for instance, included and still includes protein adsorption to man-made surfaces [30]. To this day, the QCM is one of the tools of label-free biosensing [31]. It is not used quite as widely as surface plasmon resonance (SPR) spectroscopy because the limit of detection is not as good [32]. Also, SPR spectroscopy has better baseline stability. The QCM has certain advantages compared to SPR, but these have not turned the QCM into a routine tool to the same extent as this has happened for SPR.

Importantly, the modern QCMs have evolved to be used in applications beyond micro-weighting [33, 34]. Accordingly, these instruments are sometimes called "second-generation QCMs." They determine the resonance bandwidth in addition to the resonance frequency, and they do so for a number of different overtones. Typical frequencies of the measurement are 15 MHz ($n = 3$), 25 MHz ($n = 5$), 35 MHz ($n = 7$), 45 MHz ($n = 9$), and 55 MHz ($n = 11$) using 5 MHz resonators at their nth overtone. The resonance bandwidth is sometimes converted to an inverse Q-factor, also called "dissipation factor" [35]. This added information (a set of shifts in frequency and bandwidth, rather one single frequency shift) can be exploited to infer certain properties of the sample in addition to its thickness. Often, it is used to infer the sample's *softness*.

Along with the second-generation QCMs, a modified frame of data analysis has been developed, based on what the author calls the *small-load approximation* (SLA) [29]. In a first step, the resonance frequency and shifts thereof are defined as *complex parameters*, $f + i\Gamma$ and $\Delta f + i\Delta\Gamma$. Γ is the half bandwidth at half maximum when the conductance is plotted as a function of frequency. $\Delta\Gamma$ is equivalent to the shift in dissipation factor, ΔD. Researchers used to ΔD can avoid the use of a new variable by writing the complex frequency shift as $\Delta f + i\Delta D(f/2)$. To illustrate that the concept of a complex resonance frequency makes sense, consider the freely decaying oscillation. The freely decaying oscillation depends on time as $\exp(2\pi i)$ $(f + i\Gamma) = \exp(2\pi i f) \exp(-2\pi\Gamma)$. The fact that Γ (the decay rate divided by 2π) is equal to the half bandwidth (for narrow resonances) is proven in Ref. [29]. The second step contained in the small-load approximation is a relation between the shift of the complex resonance frequency, $\Delta f + i\Delta\Gamma$, and the area-averaged oscillatory stress at the resonator surface. The two are connected as

$$\frac{\Delta f + i\Delta\Gamma}{f_0} \approx \frac{i}{\pi Z_q}\langle Z_L \rangle = \frac{i}{\pi Z_q}\frac{\langle \sigma_S \rangle}{v_S} \tag{1}$$

f_0 is the resonance frequency at the fundamental (often 5 MHz), Z_q is the shear-wave acoustic impedance of the resonator crystal (equal to 8.8×10^6 kg m^{-2} s^{-1} for AT-cut quartz), σ_S is the tangential stress at the resonator surface, v_S is the tangential velocity at the resonator surface, and angle brackets denote averaging over the resonator's active area with the suitable statistical weight applied [27]. Both σ_S and v_S are complex amplitudes of oscillatory quantities. If the tangential stress is laterally homogeneous, the ratio of stress and velocity is called *load impedance*, Z_L. Z_L has the same units as the resonator's shear-wave impedance, but it is not a material's parameter but rather depends on the geometry of the sample. For instance, the *load impedance* is $Z_L = i\omega m$ (m the mass per unit area) for the Sauerbrey film.

The SLA's range of applicability is tremendously wide. It covers films, liquids, viscoelastic media, acoustic multilayers, nanobubbles, vesicles, sand-piles, and just

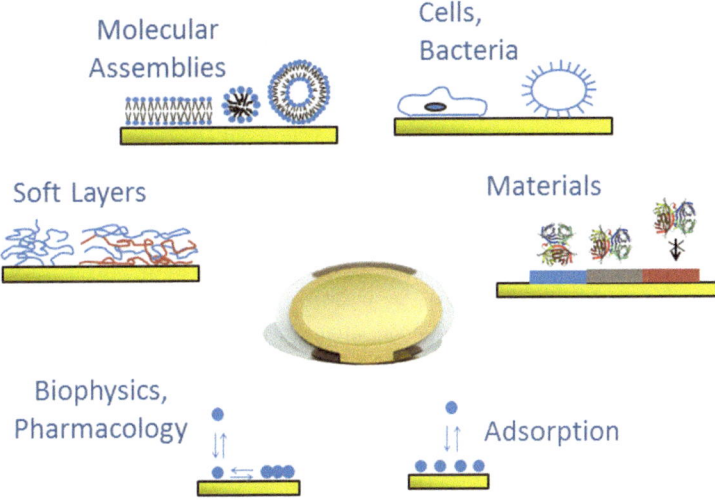

Fig. 1 Samples which can be studied with a QCM. As long as the stress distribution at the resonator surface for the various systems is accessible to modeling, the QCM response can be predicted for different load situations

about any other samples that might come to mind (Fig. 1). If the area-average of the oscillatory tangential stress at the resonator surface can be computed in one way or another, one may predict Δf and $\Delta\Gamma$ from the material's properties and dimensions. One may compare these predictions with experiment and thereby test whether the assumptions used in the calculation of the stress field are correct. However, this approach only runs forward. Assumptions are made; they are used to predict and Δf and $\Delta\Gamma$ at the different overtones; they are tested against experiment; and they are discarded in case they disagree with experiment. Unfortunately, there is no way to run this procedure backwards, that is, there is no way to explicitly compute geometric parameters from the experimental values of Δf and $\Delta\Gamma$ at the different overtones.

3.1 Full-Fledged Viscoelastic Modeling Is Difficult for Biological Cells

Before laying out the difficulties with quantitative models for biofilms and eukaryotic cells on QCM surfaces, it needs to be emphasized that the obstacles on the way to quantitative understanding are not of fundamental nature. Pessimism will be expressed below, but the pessimism may, at least in principle, be unjustified. If the geometry and the viscoelastic parameters of a given sample are known, a PDE-solver can compute the oscillatory stress field. The *partial differential equation* (PDE) to be solved is the frequency-domain version of the Stokes equation, which is equal to the Navier-Stokes equation with the nonlinear term omitted [36].

$$i\rho\omega\mathbf{v} = \eta\nabla^2\mathbf{v} - \nabla(\delta p)$$
$$\delta p = \frac{-K}{i\omega}(\nabla\cdot\mathbf{v}) \qquad (2)$$

where ρ is the density, ω is the frequency, η is the viscosity, $\mathbf{v}$ (a vector) is the velocity, δp is the deviation of the local pressure from equilibrium, and K is the modulus of compression. If K is infinite ($\nabla\cdot\mathbf{v} = 0$), the flow is incompressible. δp and $\mathbf{v}$ are complex amplitudes of corresponding oscillatory quantities. Viscoelasticity is accounted for by making η complex: $\eta = \eta' - i\eta'' = (G' + G'')/i\omega$ with the shear modulus G. From the flow fields and the viscosity, the stress field is calculated. Once the stress field is known, it is trivial to calculate Δf and $\Delta\Gamma$ from the stress field applying the small-load approximation. *The partial differential equation* in Eq. (2) is not particularly difficult to solve. Two commonly used techniques are the finite-element method and the finite-volume method.

Again: On a fundamental level, the road to quantitative modeling of cells adhering on a QCM resonator is open. The trouble is in the details. Here is a list of problems:

- A robust hypothesis for the geometry and the viscoelastic parameters of the load material is needed as input to the calculation. Such a detailed hypothesis is typically not available. Usually, the various candidate geometries will have so many free parameters that the limited set of experimental output parameters (Δf and $\Delta\Gamma$ for about five overtones) does not suffice to identify a unique solution.
- The viscoelastic parameters (G' and G'') in soft matter all depend on frequency. Soft matter undergoes a wide variety of relaxation processes, some of them with rates in the MHz range. These relaxation processes give rise to viscoelastic dispersions. For realistic modeling, it is mandatory to know the complex shear moduli and the complex moduli of compression at the frequencies of the different overtones. These are not usually known. With regard to compression, one might assume incompressibility so that $|K| \gg |G|$. With regard to shear, similar simplifications or even reasonable assumptions are not easily identified.
- Biological samples often have internal surfaces with nonzero surface tension. Also, they contain membranes with a finite bending stiffness. Interfaces with a dynamical behavior of their own are not accounted for in Eq. (1). Of course they can be accounted for, but this requires a separate effort.
- Most samples of practical relevance will display significant variability in their properties over the active area of the resonator. To account for this situation, it takes many simulation runs, even if all (statistical and structural) parameters are known, to apply the appropriate statistics.

If we accept that detailed viscoelastic modeling is not feasible at this time, we basically have two options: Either, we apply simplified models, or we try to extend the capabilities of the instrument and search for well-controlled experimental parameters, which do not require full-fledged viscoelastic modeling for interpretation. The two following chapters deal with these two options. We finish with a few remarks on combined instruments in the final chapter of this article.

4 Simple Models and Situations: Where These Can Be Applied

The three simple models in the physics of the QCM are the Sauerbrey model (treating the load material as a rigid thin film), the semi-infinite viscoelastic liquid, and the coupled resonance. The equation predicting the frequency shift induced by a liquid is sometimes attributed to Gordon and Kanazawa [37], but similar relations have been published before [38–41]. That $(-\Delta f)$ is proportional to the square root of the viscosity-density product follows straightforwardly from Mason's work from the first half of the last century [42]. The model of the coupled resonance goes back to Dybwad [43]. The coupled resonance model stands a bit distant to the two other models because it describes particulate samples. Also, the coupled resonance looks simple only at first glance.

The Sauerbrey model – when expanded by a viscoelastic correction [26] – contains five free parameters (the thickness of the film, its complex shear modulus, $G = G' + iG''$, and two power-law exponents quantifying viscoelastic dispersion [26]), while the semi-infinite viscoelastic medium needs four (G', G'' and two power-law exponents) [26]. The second-generation QCMs report 8–12 parameters, depending on the number of overtones measured. Given that the problem is over-determined by the experimental parameters, the model can be fitted to the data, and the agreement between the fit and the data can serve to assess the degree, to which these simple models are indeed adequate.

4.1 Cell Layers as Semi-infinite Viscoelastic Media

For QCMs working in the MHz range, biological cells never form Sauerbrey layers. They are too thick and too soft. On the other end of the scale is the semi-infinite medium. The shear wave enters the medium and propagates forward, until it eventually decays. The decay length of the shear wave is a few microns, at most. In mathematical terms, the shear wave $u(z,t)$ takes the form

$$u(z,t) = \mathrm{Re}(u_S \exp(i(\omega t - kz)))$$
$$= u_S \cos(\omega t - k'z)\exp(-k''z) \tag{3}$$

where u_S is the amplitude of the displacement, to be evaluated at the surface. A typical value is a fraction of a nanometer. $k = k' - ik''$ is the wavenumber; z denotes the distance from the surface. For the general case of viscoelastic media, the depth of penetration is

Fig. 2 For typical biological cells, the shear wave launched by the resonator surface does not reach to the top. The QCM only senses the viscoelastic properties of the material at the bottom

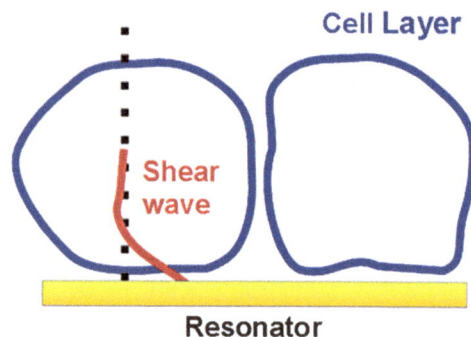

$$\delta = k'^{-1} = -\frac{1}{\omega\sqrt{\rho_{\text{liq}}}}\frac{\left|G' + iG''\right|}{\text{Im}\left(\sqrt{G' + iG''}\right)} \qquad (4)$$

If the medium is a *Newtonian* liquid ($G' = 0$, $G'' = \omega\eta$), this relation simplifies as

$$\delta = \sqrt{\frac{2\eta}{\rho\omega}} \qquad (5)$$

In water, the penetration depth at 5 MHz is 250 nm and scales as $n^{-1/2}$ (n the overtone order). Biological samples are never *Newtonian* liquids, but this simple equation can still serve to estimate the order of magnitude. δ comes out to be in the range of a few hundred nanometers, which is less than the thickness of the cell. Accordingly, the QCM senses predominantly the near-surface part of the sample (Fig. 2).

The complex viscosity is derived from Δf and $\Delta\Gamma$ as

$$\begin{aligned}
\eta'\rho &= \frac{G''\rho}{\omega} = -\frac{\pi Z_q^2}{f_r}\frac{\Delta f\Delta\Gamma}{f_0^2}\\
\eta''\rho &= \frac{G'\rho}{\omega} = \frac{1}{2}\frac{\pi Z_q^2}{f_r}\frac{\left(\Delta\Gamma^2 - \Delta f^2\right)}{f_0^2}
\end{aligned} \qquad (6)$$

where $\eta = \eta' - i\eta''$ and ρ are the viscosity and the density of the medium, respectively. The density, ρ, was moved to the left-hand side in order to emphasize that the QCM determines the product of G and ρ. The latter is often assumed to be about 1 g/cm^3. Note: *G is expected to differ between overtones.* Using Eq. (6), a viscoelastic spectrum in a frequency range of about a decade can be explicitly obtained.

In case the sample is heterogeneous, the *medium* is an *effective medium*. Effective medium theories make sense when the feature size is small compared to the wavelength of sound. Here, the latter can be replaced by the penetration depth. If, further, the sample is much thicker than δ, modeling the layer as semi-infinite is reasonable. If the effective near-surface shear modulus of a cell layer, as calculated with Eq. (6), is changed by some event such as the addition of a drug, cell death, or cell proliferation, the QCM will catch this event.

A few comments:

- The complex shear modulus of a cell layer as reported by the QCM is related to the properties of the cytoskeleton.
- If a cell can be made to modulate its viscoelastic properties close the substrate, the QCM will respond to such changes particularly well. For instance, there is a gap between the cell and the resonator surface. If the viscoelastic properties of the material in these gaps vary, the QCM will presumably report that.
- A cell's death can be expected to result in a softening, that is, in a decrease of $G' + iG''$. For expanded information on this topic, see Ref. [44].
- It is emphasized below that kHz resonators can be used to determine the thickness of a biofilm. These resonators have also been used to measure the viscoelastic properties of tissues [45–47]. When used in this mode, an equation similar to Eq. (6) is applied.
- If a biofilm has formed, the QCM cannot determine its thickness. Of course it can always detect the onset and expansion of biofilm formation [48].
- There have been a number of attempts to monitor blood clotting with a QCM [49–51].

4.2   Torsional Resonators Can Measure the Thickness of Biofilms

Since the QCM only senses the bottom of a cell layer, it cannot determine the cell layer's thickness. The QCM's dynamic range does not extend to film thicknesses exceeding 10 μm. The limit depends on the layer's softness. In air, the drive to build less sensitive devices is moderate because conventional gravimetric balances – if operated with care – are sensitive enough to be used in this mass region. However, conventional balances cannot be used in liquids because of buoyancy. In consequence, there *is* drive to build acoustic devices with low mass sensitivity for weighing *in the liquid phase*.

A side remark is worthwhile with regard to optical thickness determination in the liquid phase and the corresponding acoustic techniques: The eminent role of the refractive index for optical techniques is taken over by the shear-wave acoustic impedance for the QCM [52]. The shear-wave acoustic impedance of typical soft matter (gels, polymer brushes, biofilms) is much larger (at least by a factor of three) than shear-wave acoustic impedance of water. In contrast, the refractive indices of the sample and the ambient liquid are often similar. They may even be so similar that there is only poor optical contrast. Poor contrast with shear waves is rare because the acoustic impedance of the sample is usually much larger than the impedance of

Fig. 3 Left: Principle of operation of a torsional resonator. Since there is a nodal plane at the waist, the resonator can be mounted with an O-ring there. There also is a nodal line along the cylinder's symmetry axis, which can be used to further stabilize the resonator with a pin at the top. The electrodes are arranged in quadrupolar geometry. The image on the right-hand side shows the unit TQ56 supplied by flucon, Osterode, Germany (www.flucon.de)

the solvent. This is one of the advantages of the thickness-shear devices in addition to simplicity.

In order to acoustically measure the thickness of a biofilm, one would have to lower the QCM's sensitivity. However, sensitivity cannot be lowered easily because pure thickness-shear modes require resonators to have a width-to-thickness ratio of at least 10. Even with such thin plates, the thickness-shear modes are never perfect. There are always *flexural contributions* to the shear modes, which give rise to compressional waves [53]. One can live with these if the width-to-thickness ratio is larger than 10; otherwise compressional waves become detrimental.

Torsional resonators may serve as an alternative to thin plates [54, 55]. The torsional resonator shown in Fig. 3 is made of x-cut quartz and oscillates at a frequency of about 56 kHz. It is commercially used to monitor the viscosity of engine oil [56]. The lowered frequency compared to the conventional QCM is helpful for this application for two reasons: (1) It brings the frequency of the measurement closer to the technically relevant range. (2) More importantly in the biosensing context, the torsional resonator is less sensitive to adsorbed mass than the thickness-shear QCM. Adsorbed mass mostly consists of debris or contaminants in the case of the engine oil monitor, and it is favorable if the instrument is blind to such adsorbates. But in cell-based biosensing, however, the adsorbed mass consists of a biofilm or a cell layer, both of which are the core of the sensing target.

A sketch and an image of a torsional resonator are shown in Fig. 3. A typical length is 35 mm, and a typical diameter is 12 mm. Torsional resonators do not easily lend themselves to miniaturization; they are macroscopic devices. This also has advantages. Composite torsional resonators have been designed, the shape of which has been adjusted to the application [47]. For an overview of the use of kHz resonators in tissue characterization, see Reference [46].

As far as thickness determination on biofilms is concerned, torsional resonators do the trick. Figure 4 shows an application example.

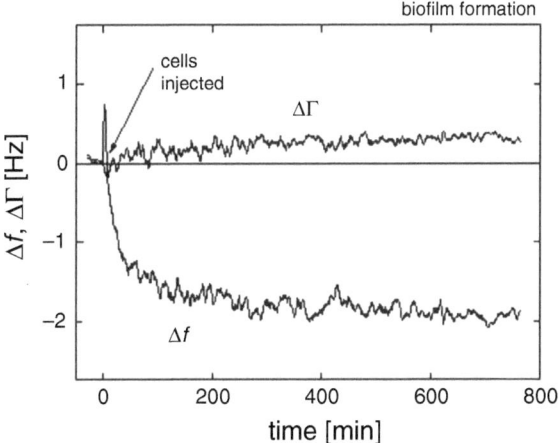

Fig. 4 Time course of bacterial biofilm formation monitored with a torsional resonator. Suspended *Pseudomonas fluorescens* were applied at time zero. The final film thickness is around 0.5 μm. The increase in bandwidth is smaller than the decrease in frequency, as typical for Sauerbrey films. Note: The area-averaged thickness of 0.5 μm indicates that the resonator was incompletely covered since the individual cells are known to be thicker than 0.5 μm. Adapted from Ref. [54]

At $t = 0$, cells from a culture of *Pseudomonas fluorescens* were injected into the sample compartment. Over the course of a few hours, they formed an adsorbate. As expected, the frequency decreased. The increase in bandwidth, $\Delta\Gamma$, was smaller than the decrease in frequency, which implies that effects of finite softness are small. Viscoelastic effects did exist, though. From the ratio of $\Delta\Gamma/(-\Delta f)$, one may obtain an estimate of the layer's elastic shear compliance [26]. The final average film thickness was calculated to be 0.5 μm, which indicates that coverage was incomplete in this particular case.

4.3 Coupled Resonances

A finding, which has repeatedly irritated researchers using the QCM for studies of biocolloids, is the positive frequency shifts which occur occasionally [57–59]. Such shifts are of particular relevance in the context of bacterial adhesion [60]. Positive Δf cannot be explained by inertial loads because these increase the resonator's mass, thereby decreasing the frequency. Inertial loads contrast the so-called elastic loads, which increase the resonator's effective stiffness, κ_R, rather than its effective mass, M_R. Elastic loads are objects, which are too heavy to follow the resonator's MHz motion [43]. They are clamped in space by inertia. In the simple models, they make contact with the resonator surface by a small link considered as a *point contact*. Such links can be modeled as Hookean springs. The external object exerts a restoring force onto the resonator surface via the link. It thereby increases the composite

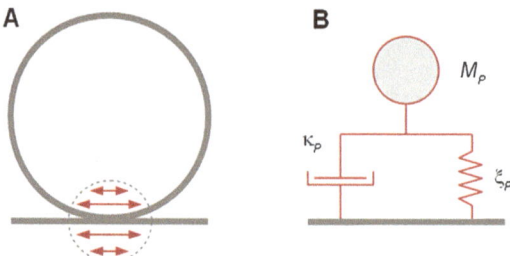

Fig. 5 (**a**) Sketch of a contact between a particle and the resonator surface. If the contact is smaller than both the sphere and the wavelength of sound, it may be represented as a Hookean spring. (**b**) Equivalent mechanical circuit representing the situation in A as a *lumped element representation*. Viscous dissipation is accounted for by the dashpot. An arrangement of this type gives rise to the coupled resonances as depicted in Fig. 6

resonator's stiffness and, in consequence, increases its resonance frequency, following the relation $\omega = (\kappa_R/M_R)^{1/2}$. Δf is positive, proportional to the stiffness of the contact and inversely proportional to the overtone order ($\Delta f \propto 1/n$). If this inverse-n-scaling (also called *Dybwad scaling*) is observed, the interpretation is straightforward.

Many biological colloids are not large enough to be clamped in space by inertia. One does find $\Delta f > 0$, but no $1/n$-scaling. Sometimes Δf crosses from negative values at low overtone orders to positive values for the high-order overtones. The interpolated frequency, at which Δf would cross the abscissa (if there was an overtone with this exact frequency), is the *frequency of zero crossing*, f_{ZC} is calculated by interpolation. Zero crossing of Δf is explained by the coupled resonance. Such loads are neither of the inertial type nor of the elastic type; they are an intermediate between the two. The particle together with its link to the surface forms a resonator of its own, as shown in Fig. 5a. Figure 5b shows the mechanical equivalent circuit. The simple models of a coupled resonance predict the complex frequency shift as

$$\Delta f + i\Delta\Gamma = -f_{OS}\frac{f_0}{\pi Z_q A}\omega M_P\frac{\omega_P^2}{\omega_P^2 - \omega^2} \tag{7}$$

where M_P is the modal mass of the particle and ω_P is the (complex) particle resonance frequency. If the damping of the coupled resonance frequency is sufficiently small, the value of its real part, ω_P', is in the range of $2\pi f_{ZC}$. f_{ZC} can be easily read from the experimental data; fitting is not required. The imaginary part is related to the bandwidth of the coupled resonance. It is not strictly equal to the half bandwidth divided by 2π because the coupled resonance rarely is sharp enough to let this approximation be applicable. There is a pre-factor, f_{OS} (the oscillator strength), which accounts for a less-than-perfect coupling between the motion of the resonator plate and the motion of the particle.

Fig. 6 When a sample gives rise to a coupled resonance, Δf and $\Delta\Gamma$ display some characteristic features: (1) Δf crosses from negative to positive at the zero-crossing frequency, f_{ZC}. $2\pi f_{ZC}$ is close to the undamped resonance frequency of the coupled resonance $\omega_P' = (\kappa_P/M_P)^{1/2}$. (2) At the zero-crossing frequency, $\Delta\Gamma$ goes through a maximum. (3) Plotting $\Delta\Gamma$ versus Δf, one finds a circle (more generally: a spiral). The radius of this circle is a measure of the contact stiffness. The open symbols were inserted to remind the reader that the frequency, ω, can only take discrete values set by the overtone orders

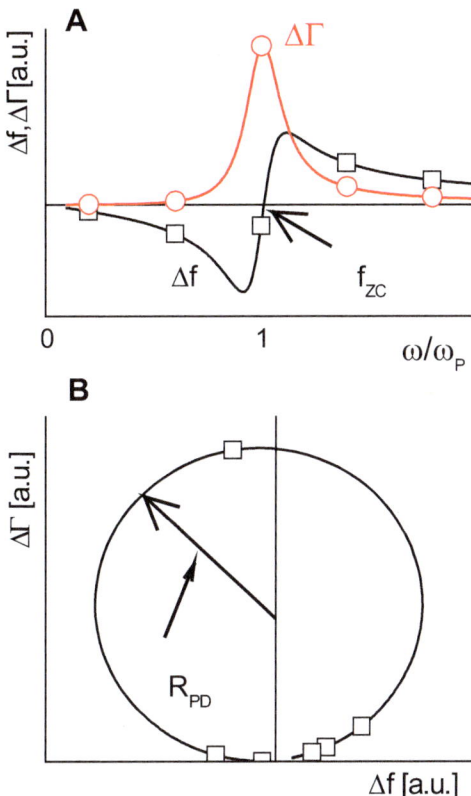

For more background on f_{OS} and also on the differences between the modal mass and the mass of the particle, the interested reader is referred to Refs. [28, 61]. It turns out that quantitative fitting of Eq. (7) to the experimental data obtained with bacteria is difficult. Even with silica spheres, the model has to be extended. The principal problem is that the different particles (or different cells) adhering to the resonator surface show a variable strength by which they adhere. There is a distribution in values of ω_P', which smoothens out the coupled resonance.

Given that quantitative fitting is not easily achieved, one can still search for robust parameters, which are easily derived from the experimental data and which can be compared between experiments. The first parameter of that kind is the frequency of zero crossing, f_{ZC}. A second one is the radius of the circle in a polar diagram of $\Delta\Gamma$ versus Δf. The radius is called R_{PD} in Ref. [63]. Whether such a circle or a spiral is actually observed is a check of consistency. Often, the range of accessible frequencies only covers a fraction of a circle. If a full circle is observed, its radius is an estimate for the following combination of parameters:

$$R_{PD} \approx \frac{N_P f_0}{2\pi Z_q} f_{OS} \frac{\kappa_P}{\gamma_{het}} \qquad (8)$$

where γ_{het} is the heterogeneous bandwidth of the resonance. It usually differs from the homogeneous bandwidth, $\omega_P''/2\pi$, because the different contacts vary in their individual stiffness. N_P is the number of particles per unit area. Given the numerous parameters involved, quantitative interpretation based on Eq. (8) is problematic, but Eq. (8) may still be the basis for a comparison between experiments [64].

5 Other Dimensions of Sensing

5.1 Variation of Amplitude

Searching for other dimensions that are useful for QCM-based sensing, variation of oscillation amplitude is an interesting approach. The idea is old [65–67]. A few general remarks before we go into the details:

- The quartz plate itself has an intrinsic dependence of the resonance frequency on amplitude. It is termed *drive level dependence* (DLD) in the field of time and frequency control [68]. The DLD mostly goes back to a slightly nonlinear elasticity of the crystal. Changes in temperature induced by high-oscillation amplitudes may also play a role [69]. The drive level dependence is hysteretic. After running a resonator at high amplitude (high drive level) and returning to low amplitudes, the resonance frequency is displaced from where it was during initial low amplitude runs. The DLD is usually considered to be a problem, and the simplest way to stay away from it is by choosing the driving voltage low enough. A *drive level dependence* is less of a problem for crystals immersed in a liquid than for crystals in air because the liquid lowers the oscillation amplitude via damping. Also, the crystal warms up to a smaller extent than in air because of the good thermal contact to the liquid.
- Amplitude effects originating from the sample require amplitudes beyond what is applied under normal operating conditions. Under normal conditions, linear response holds: Stress is proportional to displacement. The load impedance therefore does not depend on amplitude, and neither does the frequency shift, following Eq. (1).
- There is a limited range of amplitudes, which are large enough to let the non-linearities from the sample be visible but are still small enough to let the electronics run well. At displacement amplitudes beyond 10 nm, the equipment available to the author runs into saturation and displays nonlinearities of its own. An amplifier (such as the unit ZHL-1-2W+ from Mini-Circuits) may overcome these problems.

With regard to nonlinearities originating from the sample, two separate types of nonlinearities need to be distinguished. They are of *mechanical* and of *hydrodynamic* origin, respectively. The *mechanical* nonlinearities are the more interesting ones. These depend on the properties of the link between the cell and the resonator surface. The *hydrodynamic* effects, on the contrary, are universal. They are present in all liquids, and they depend on the density and the viscosity of the liquid, only. They do affect QCM readings of cells in contact to the resonator, but they do not necessarily tell us something about this cell layer.

Mechanical nonlinearities are most prominent at localized contacts, for instance between a sphere (or a virus) and the resonator plate [70]. At localized contacts, there is a stress concentration, which makes the observation of nonlinear effects easier. Prominent features are partial slip (slip at the edge of the contact only), followed by gross slip at very large amplitudes [71]. At the largest amplitudes accessible, the object under study can be seen to start sliding if the resonator surface is tilted [72]. Partial slip and gross slip have been studied in some detail in the context of contact mechanics [67, 69, 73–75]. They have so far not been exploited systematically in the context of cell-based sensing. However, slip may be involved in *rupture event scanning* (REVS) as discussed below.

Hydrodynamic nonlinearities, which are less interesting than mechanical nonlinearities, come about by the advection term in the Navier-Stokes equation. The importance of the advection term is sometimes quantified by the Reynolds number [76]. Generally speaking, QCM experiments occur in the low Reynolds number regime. There never is turbulence above a QCM surface. At least, it is never caused by the resonator's motion. However, there is a more subtle effect, which is *steady streaming*. Steady streaming denotes a steady flow of material caused by a high-frequency periodic oscillation [77]. Steady streaming is to be distinguished from acoustic streaming [78], with the latter being caused by compressional waves. Acoustic streaming drives dust particles in a Kundt's tube toward the nodes of the standing acoustic wave. It has numerous applications in microfluidics [79]. Steady streaming, on the contrary, does not necessarily require pressure waves as produced in conventional ultrasound. Steady streaming is observed when immersing a resonator into a liquid and raising the amplitude to a few nanometers [80]. There is a steady flow of liquid along the direction of shear, directed toward the center of the crystal. The mechanism, by which the flow is generated, requires flexural contributions to the pattern of vibration. These are always present because of energy trapping. For the mathematical details, see Ref. [80].

There are a number of reports on amplitude effects in the liquid phase. It is not always clear whether they go back to *mechanical* or *hydrodynamic* effects. In some cases, the systems under study contain narrow contacts subjected to large oscillatory stresses, which implies that the partial slip and gross slip may have been of importance. That concerns, for instance, the class of experiments called *rupture event scanning* (REVS) [61]. In rupture event scanning, the amplitude of the resonator is ramped up until the particles in question detach from the surface. Detachment can be followed optically but is also indicated by a sudden spike at the third harmonic of the exciting signal. Particle detachment amounts to a strong nonlinearity. Third-harmonic generation evidences. Third-harmonic generation has later been used to probe nonlinearities in situations without particle detachment [81].

Another type of experiment explores whether high-amplitude excitation of the resonator can prevent adsorption of particles, liposomes, or living cells [66, 67]. That is the case. Note that prevention of particle adsorption is different from particle detachment [61].

In all these cases, *steady streaming* comes into play, at least potentially. Particle detachment may have been caused by partial slip and gross slip at the point of contact, but it may just as well have been the consequence of a transverse flow of liquid at the resonator surface. In the latter case, the underlying physics would have been the same as in more conventional detachment experiments using shear flow [82].

5.2 Piezoelectric Stiffening

Piezoelectric stiffening in the context of QCM experiments is typically considered as a problem rather than a novel dimension of sensing. Effects of piezoelectric stiffening are usually avoided by grounding the front electrode well [83]. If the front electrode is grounded well, all effects of piezoelectric stiffening are confined to the back electrode, where they can be controlled. One cannot avoid piezoelectric stiffening completely when electrically exciting the resonance. Applying experimental means to control piezoelectric stiffening is the best one can do.

A reminder on piezoelectric stiffening might be useful at this point. A good start is the analogy to the heat capacity of gases. When the temperature of a gas is raised by some small amount dT, the heat needed to do that, δQ, depends on whether heating occurs at constant volume, at constant pressure, or at some condition in-between. Defining the specific heat capacity as $c = dQ/dT$ is not meaningful as there are two specific heat capacities: one at constant volume, $c_V = (dQ/dT)_V$, and one at constant pressure, $c_p = (dQ/dT)_p$. The reason for this difference is that a gas heated under conditions of constant pressure expands and thereby does work on its environment, given as $|\delta W| = pdV$. No such work is done under conditions of constant volume.

Likewise, the work done on a piezoelectric body while deforming it contains two contributions. After the deformation, there is both mechanical and electric energy contained in the system. Actually, some electrical energy is contained in the body, but some energy is also contained in the electric field around it. If the body is deformed with no electrodes present, the ratio of the electric energy and the mechanical energy is k_t^2 with k_t representing the *piezoelectric coupling coefficient*. k_t^2 is 0.8% for laterally infinite plates of AT-cut quartz. This is only true as long as the open-circuit condition holds. The opposing limit is the piezoelectric plate with electrodes on both sides and an electrical connection short-circuiting them. When we now deform the plate, the piezoelectrically induced polarization is compensated by charge flowing between the two electrodes. The charge in the electrodes screen the electric field, thereby reducing the electrical contribution to the work of deformation and, in consequence, reducing the total work of deformation. The effective stiffness of the plate, κ_R (the analogue to the specific heat of the gas), depends on whether the deformation is carried out at constant charge (i.e., the open-circuit condition) or at constant voltage (i.e., with short-circuited electrodes). The second stiffness is smaller than the first because the electric contribution is missing. Translated to the resonance frequency following $\omega = (\kappa_R/M_R)^{1/2}$ and compared to the frequency shifts obtained in gravimetric experiments, the effect is rather substantial. Grounding or not grounding electrodes can shift the resonance frequency by a kHz and more. This effect is actually exploited in time and frequency control. The frequency of a quartz-based clock can be "pulled" within certain limits by placing a capacitor in the circuit next to the resonator [84].

Piezoelectric stiffening may – at least in principle – be avoided by short-circuited the electrodes. With short-circuited electrodes, one cannot electrically interrogate the state of the crystal, though. One has to live with piezoelectric stiffening and control its magnitude. Above, we discussed the two limiting cases, which are the open-circuit condition and short-circuited electrodes. In practical situations, the two electrodes will be connected across a finite electrical impedance, Z_{ex}. As always, the electrical impedance may be complex. The electrical circuitry affects the external impedance but also the stray capacitances and currents inside the sample. Further complicating the matter, the resonator is not a laterally infinite plate, and it is not covered with electrodes across the entire surface. Calculating Z_{ex} for such complicated geometries is not easy (but possible with the finite-element method, as made available by COMSOL and other software suppliers).

In Fig. 7 (showing a "lumped element circuit"), we assume that whatever happens to the electrical fields around the resonator is represented by the complex impedance Z_{ex}. As shown in Ref. [28], a change in Z_{ex} results in a shift of the complex resonance frequency, given as

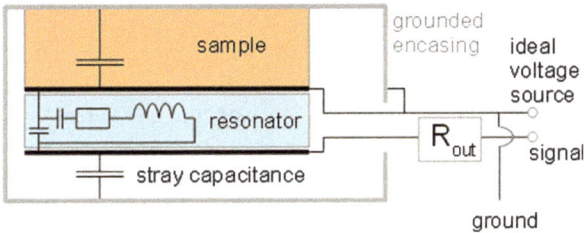

Fig. 7 Equivalent circuit diagram of a resonator emphasizing the output resistance of the driving circuitry and the stray capacitances between the resonator and its housing. The sample has been represented as a capacitor here, but it may require a more elaborate network, accounting for electric conductivity, as well

$$\frac{\Delta f_{PE} + i\Delta\Gamma_{PE}}{f_0} = \frac{4k_t^2}{\pi^2 n}\left(\left(1 + \frac{1}{Z_{ex}i\omega C_0}\right)^{-1} - \left(1 + \frac{1}{Z_{ex,ref}i\omega C_0}\right)^{-1}\right) \qquad (9)$$

This relation is simplified if $Z_{ex}(i\omega C_0) << 1$. This situation is to be expected in an electrochemical setting in which a supporting electrolyte is used to lower the bulk resistance to the extent that it does not influence the outcome of the measurement. If $Z_{ex}(i\omega C_0) << 1$ applies, Eq. (9) simplifies as follows:

$$\Delta f_{PE} + i\Delta\Gamma_{PE} \approx k_t^2 \frac{8i}{\pi} f_0^2 C_0 (Z_{ex} - Z_{ex,ref}) \qquad (10)$$

How do we put this insight to work for sensing purposes? If the front electrode is grounded and if the measurement cell has a grounded metal housing, the entire sample compartment is close to ground potential. Variations in conductivity and dielectric properties of the sample do not matter much. Z_{ex} is governed by the electrical conditions at the back electrode. This state can serve as the reference state. Alternatively, one might also ground the back electrode and apply the signal to the front. Assuming that the housing still is on ground potential, there now *are* electrical fields permeating the sample, and the sample's electric properties therefore do matter [85–87]. The frequency shifts are different between the two situations, and one can make use of this difference for sensing purposes. All that is needed is a switch in the driving circuitry, which applies the signal to either the front electrode or the back electrode [88]. Applying it to the back electrode and grounding the front electrode well is the conventional procedure. It avoids effects of piezoelectric stiffening originating from the sample. With the switch in place, it is possible to turn those effects on and off and analyze the differences in Δf and $\Delta\Gamma$ linked to switching the piezoelectric stiffening. That information is available with very moderate efforts.

A few more remarks:

- Applying the oscillating potential to the front is not an option if the sample is a high-concentration electrolyte. The electrolyte then short-circuits the excitation, and the resonance curve is too noisy to be analyzed. Most of the current passes through the electrolyte; only a small fraction passes through the crystal in the form of a displacement current driving the resonance.
- One might measure Z_{ex} electrically and obtain equivalent information. The motivation for relying on Δf and $\Delta\Gamma$ rather than the electrical measurement is that the electrical measurement requires careful calibration of the impedance analyzer, while the frequency-based measurement does not. (Some calibration is needed for operation of the resonator, but one may be lenient and account for imperfect calibration with correction factors in the fit function [28].)

– The electrical impedance, Z_{ex}, pertains to MHz frequencies. At these frequencies, inductances of cables have an impact. These complicate the matter. MHz frequencies are avoided for that reason in conventional electrochemical impedance spectroscopy (EIS).

Quantitative interpretation of frequency shifts caused by piezoelectric stiffening Δf_{PE} as determined with the switching procedure has been difficult in the past. Still: The effects are clearly noticeable; they differ between different samples. One might either put more effort into quantitative interpretation or simply use Δf_{PE} and $\Delta \Gamma_{PE}$ on the different overtones as empirical parameters.

5.3   Temporal Variations

Contrary to conventional adsorbates, cells adsorbed to a QCM surface display activity on their own. A trace of frequency and bandwidth versus time may display fluctuations, which are characteristic for the type of cell and its state. Particularly interesting are the periodic contractions of cardiac myocytes [89–91]. Unfortunately, the preparation of these samples entails a considerable effort. If prepared properly, these tissues start to contract and expand cooperatively with a frequency around 1 Hz. This motion is picked up easily with a QCM. Note: It is also picked up easily with a conventional microscope or by electrical means. The QCM is not necessarily the easiest way of monitoring their activity. In Ref. [90], it was reported that Δf increased transiently during the pulse, while $\Delta \Gamma$ decreased. However, within those assays neither the sign nor the magnitudes of these dependences matter. The focus is on the beat frequency and its change upon addition of certain drugs. In addition to the beat frequency, it is interesting to study the strength of higher harmonics within the data trace. This is achieved by computing the Fourier transforms of $\Delta f(t)$ and $\Delta \Gamma(t)$ and converting these to power spectral densities (PSDs).

Cell cultures other than cardiac myocytes display a less impressive activity, but they still do show activity. The time traces of the fluctuations can be converted to a power spectral density and the features of the PSD can be correlated with the state of the sample. This mode was demonstrated in Ref. [92], see also Ref. [93].

5.4   Temperature Sweeps

Cells are highly responsive to all kinds of stimuli. Temperature variation as a stimulus is always available because a temperature control is needed anyway. Of course one needs to stay inside the range of temperatures which the cells can tolerate. The author is not aware of a systematic study of temperature effects in QCM-based whole-cell

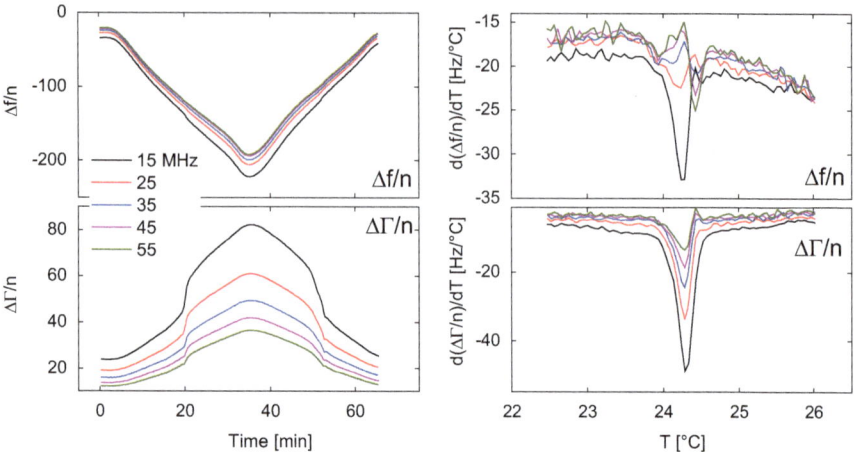

Fig. 8 Temperature sweeps on a multilayer of the phospholipid dimyristoyl-phosphatidylcholine (DMPC). The lipid's main phase transition is reflected as a kink. The right-hand side shows the temperature derivatives of the data on the left. There is a close analogy with differential scanning calorimetry

experiments. As a demonstration for temperature effects on biological samples, Fig. 8 shows temperature scans on multilayers of the phospholipid dimyristoyl-phospatidyl-choline (DMPC). The graphs on the right-hand side show the derivatives with respect to temperature. The lipid's main phase transition is clearly indicated in the data. The QCM data shows some similarity to data recorded by differential scanning calorim-etry and the analogies are indeed far-reaching [94, 95].

6 Combined Instruments

In search for more in-depth information on complicated samples, one can combine the QCM with other techniques. The QCM is well-suited for such combinations because it is so simple. Here are examples:

– Combination with optical microscopy is trivial and is routinely done in many labs.
– The QCM has been combined with optical reflectometry [56, 96–98]. There are technical difficulties, though. Also, cell layers are thicker than the wavelength of light, and they are heterogeneous. This complicates the analysis.
– Combination with electrochemistry is another classical approach. Of course one has to be careful in choosing potentials which the cells can tolerate. There is considerable interest in bioelectrochemically active cell layers [99, 100]. The

  envisioned application is the biofuel cell, but the academic interest reaches
  beyond that application [101].
- Combination with UV-Vis spectroscopy is possible in reflection and in transmission. It can be done in transmission when the crystals are coated with transparent ITO electrodes [102].
- The QCM can be combined with electrical impedance spectroscopy. The latter is called *electrochemical impedance spectroscopy* (EIS) in the context of electrochemistry and called *electrical cell-substrate impedance spectroscopy* (ECIS), when applied to biological cells [12, 91, 103]. In both cases, the front electrode of the resonator is subjected to a small-amplitude oscillatory voltage. The voltage-to-current ratio is recorded as a function of frequency, where a typical sweep extends from 1 Hz to 100 kHz. A caveat: EIS or ECIS do not run well on the conventional SiO_x-coated resonators because the SiO_x layer blocks the current. Even its capacitance is so low that the corresponding impedance equal to $(i\omega C)^{-1}$ dominates the overall impedance, so that the sample is masked. Special crystals with thin SiO_x layers are needed, if they are to be used for ECIS as a parallel readout technology.

7 Conclusions

With regard to bringing the QCM to whole-cell-based sensing, the following issues deserve consideration:

- Can cells or cell layers be identified, which are particularly well defined with regard to their viscoelastic behavior, thereby improving reproducibility?
- To what extent will it be possible to interpret apparent complex shear moduli, as extracted from the shifts of frequency and bandwidth, as a state of the cytoskeleton?
- To what extent is the coupled resonance model capable of explaining experimental data and allowing for quantitative analysis?
- Can the additional dimensions of sensing (amplitude and temperature, in particular) increase the depth of information?
- Can combined measurements (with ECIS, in particular) bring a synergistic benefit?

  The success of such efforts cannot be foreseen, but there is certainly room for further progress.

Acknowledgments The author has enjoyed a long-standing collaboration with Ilya Reviakine on the application of the QCM to biosystems, which has influenced this chapter in many ways. Astrid Peschel provided the data shown in Fig. 8.

References

1. Janata J (2009) Principles of chemical sensors. Springer, Berlin
2. Rock F, Barsan N, Weimar U (2008) Electronic nose: current status and future trends. Chem Rev 108(2):705–725
3. Bousse L (1996) Whole cell biosensors. Sens Actuators B Chem 34(1–3):270–275
4. Gu MB, Mitchell RJ, Kim BC (2004) Whole-cell-based biosensors for environmental biomonitoring and application. Adv Biochem Eng Biotechnol 87:269–305
5. Pancrazio JJ, Whelan JP, Borkholder DA, Ma W, Stenger DA (1999) Development and application of cell-based biosensors. Ann Biomed Eng 27(6):697–711
6. Gryte DM, Ward MD, Hu WS (1993) Real-time measurement of anchorage-dependent cell-adhesion using a quartz crystal microbalance. Biotechnol Prog 9(1):105–108
7. Wegener J, Janshoff A, Galla HJ (1998) Cell adhesion monitoring using a quartz crystal microbalance: comparative analysis of different mammalian cell lines. Eur Biophys J Biophys Lett 28(1):26–37
8. Zhou T, Marx KA, Warren M, Schulze H, Braunhut SJ (2000) The quartz crystal microbalance as a continuous monitoring tool for the study of endothelial cell surface attachment and growth. Biotechnol Prog 16(2):268–277
9. Wegener J, Seebach J, Janshoff A, Galla HJ (2000) Analysis of the composite response of shear wave resonators to the attachment of mammalian cells. Biophys J 78(6):2821–2833
10. Redepenning J, Schlesinger TK, Mechalke EJ, Puleo DA, Bizios R (1993) Osteoblast attachment monitored with a quartz-crystal microbalance. Anal Chem 65(23):3378–3381
11. Wegener J, Janshoff A, Steinem C (2001) The quartz crystal microbalance as a novel means to study cell-substrate interactions in situ. Cell Biochem Biophys 34(1):121–151
12. Heitmann V, Reiss B, Wegener J (2007) The quartz crystal microbalance in cell biology: basics and applications. In: Steinem C, Janshoff A (eds) Piezoelectric sensors. Springer, Berlin
13. Saitakis M, Gizeli E (2012) Acoustic sensors as a biophysical tool for probing cell attachment and cell/surface interactions. Cell Mol Life Sci 69(3):357–371
14. Reviakine I, Johannsmann D, Richter RP (2011) Hearing what you cannot see and visualizing what you hear: interpreting quartz crystal microbalance data from solvated interfaces. Anal Chem 83(23):8838–8848
15. Beck R, Pittermann U, Weil KG (1988) Impedance analysis of quartz oscillators, contacted on one side with a liquid. Ber Bunsen Phys Chem 92(11):1363–1368
16. Böttcher A, Peschel A, Johannsmann D (2015) A backing plate for quartz crystal resonators improves the baseline stability and the baseline reproducibility. Meas Sci Technol 26:035303
17. Länge K, Rapp BE, Rapp M (2008) Surface acoustic wave biosensors: a review. Anal Bioanal Chem 391(5):1509–1519
18. Gaso Rocha MA, Jiménez Y, Laurent FA, Arnau A (2013) Love wave biosensors: a review. In: Rinken T (ed) State of the art in biosensors – general aspects. Intech, London. https://doi.org/10.5772/53077. http://www.intechopen.com/books/state-of-the-art-in-biosensors-general-aspects/love-wave-biosensors-a-review. 17 Dec 2015
19. March C, Garcia JV, Sanchez A, Arnau A, Jimenez Y, Garcia P, Manclus JJ, Montoya A (2015) High-frequency phase shift measurement greatly enhances the sensitivity of QCM immunosensors. Biosens Bioelectron 65:1–8
20. Zimmermann B, Lucklum R, Hauptmann P, Rabe J, Büttgenbach S (2001) Electrical characterisation of high-frequency thickness-shear-mode resonators by impedance analysis. Sens Actuators B Chem 76(1–3):47–57
21. Wingqvist G, Bjurstrom J, Liljeholm L, Yantchev V, Katardjiev I (2007) Shear mode AlN thin film electro-acoustic resonant sensor operation in viscous media. Sens Actuators B Chem 123 (1):466–473
22. Sauerbrey G (1959) Verwendung von Schwingquarzen zur Wägung Dünner Schichten und zur Mikrowagung. Z Phys 155(2):206–222

23. Bruckenstein S, Shay M (1985) Experimental aspects of use of the quartz crystal microbalance in solution. Electrochim Acta 30(10):1295–1300

24. Nomura T, Okuhara M (1982) Frequency-shifts of piezoelectric quartz crystals immersed in organic liquids. Anal Chim Acta 142:281–284

25. Johannsmann D (1999) Viscoelastic analysis of organic thin films on quartz resonators. Macromol Chem Phys 200(3):501–516

26. Voinova MV, Jonson M, Kasemo B (2002) 'Missing mass' effect in biosensor's QCM applications. Biosens Bioelectron 17(10):835–841

27. Tsionsky V, Daikhin L, Zilberman G, Gileadi E (1997) Response of the EQCM for electrostatic and specific adsorption on gold and silver electrodes. Faraday Discuss 107:337–350

28. Johannsmann D (2014) The quartz crystal microbalance in soft matter research: fundamentals and modeling. Springer, Berlin

29. Marx KA (2003) Quartz crystal microbalance: a useful tool for studying thin polymer films and complex biomolecular systems at the solution-surface interface. Biomacromolecules 4 (5):1099–1120

30. Ward MD, Buttry DA (1990) In situ interfacial mass detection with piezoelectric transducers. Science 249(4972):1000–1007

31. Cooper MA (2003) Label-free screening of bio-molecular interactions. Anal Bioanal Chem 377(5):834–842

32. Homola J (2003) Present and future of surface plasmon resonance biosensors. Anal Bioanal Chem 377(3):528–539

33. Lucklum R, Hauptmann P (2006) Acoustic microsensors-the challenge behind microgravimetry. Anal Bioanal Chem 384(3):667–682

34. Steinem C, Janshoff A (2007) Piezoeletric sensors. Springer, Heidelberg

35. Rodahl M, Hook F, Fredriksson C, Keller CA, Krozer A, Brzezinski P, Voinova M, Kasemo B (1997) Simultaneous frequency and dissipation factor QCM measurements of biomolecular adsorption and cell adhesion. Faraday Discuss 107:229–246

36. Johannsmann D, Reviakine I, Rojas E, Gallego M (2008) Effect of sample heterogeneity on the interpretation of QCM(−D) data: comparison of combined quartz crystal microbalance/ atomic force microscopy measurements with finite element method modeling. Anal Chem 80 (23):8891–8899

37. Kanazawa KK, Gordon JG (1985) Frequency of a quartz microbalance in contact with liquid. Anal Chem 57(8):1770–1771

38. Borovikov AP (1976) Measurement of viscosity of media by means of shear vibration of plane piezoresonators. Instrum Exp Tech 19(1):223–224

39. Tabidze AA, Kazakov RK (1983) High-frequency ultrasonic unit for measuring the complex shear modulus of liquids. Meas Tech USSR 26(1):24–27

40. Stockbridge CD (1966) In: Beckum KH (ed) Vacuum microbalance techniques, vol 5. 4th edn. Plenum Press, New York

41. Glassford APM (1978) Response of a quartz crystal microbalance to a liquid deposit. J Vac Sci Technol 15(6):1836–1843

42. Mason WP (1948) Piezoelectric crystals and their applications to ultrasonics. Van Nostrand, Princeton

43. Dybwad GL (1985) A sensitive new method for the determination of adhesive bonding between a particle and a substrate. J Appl Phys 58(7):2789–2790

44. Scholz M, Noack V, Pechlivanis I, Engelhardt M, Fricke B, Linstedt U, Brendel B, Schmieder K, Ermert H, Harders A (2005) Vibrography during tumor neurosurgery. J Ultrasound Med 24(7):985–992

45. Hemsel T, Stroop R, Uribe DO, Wallaschek J (2007) Resonant vibrating sensors for tactile tissue differentiation. J Sound Vib 308(3–5):441–446

46. Stroop R, Uribe DO, Martinez MO, Brokelmann M, Hemsel T, Wallaschek J (2008) Tactile tissue characterisation by piezoelectric systems. J Electroceram 20(3–4):237–241

47. Valtorta D, Mazza E (2006) Measurement of rheological properties of soft biological tissue with a novel torsional resonator device. Rheol Acta 45(5):677–692
48. Bressel A, Schultze JW, Khan W, Wolfaardt GM, Rohns HP, Irmscher R, Schoning MJ (2003) High resolution gravimetric, optical and electrochemical investigations of microbial biofilm formation in aqueous systems. Electrochim Acta 48(20–22):3363–3372
49. Tessier L, Patat F, Schmitt N, Lethiecq M, Frangin Y, Guilloteau D (1994) Significance of mass and viscous loads discrimination for an at-quartz blood-group immunosensor. Sens Actuators B Chem 19(1–3):698–703
50. Bandey HL, Cernosek RW, Lee WE, Ondrovic LE (2004) Blood rheological characterization using the thickness-shear mode resonator. Biosens Bioelectron 19(12):1657–1665
51. Muller L, Sinn S, Drechsel H, Ziegler C, Wendel HP, Northoff H, Gehring FK (2010) Investigation of prothrombin time in human whole-blood samples with a quartz crystal biosensor. Anal Chem 82(2):658–663
52. Plunkett MA, Wang ZH, Rutland MW, Johannsmann D (2003) Adsorption of pNIPAM layers on hydrophobic gold surfaces, measured in situ by QCM and SPR. Langmuir 19(17):6837–6844
53. Eggers F, Funck T (1987) Method for measurement of shear-wave impedance in the Mhz region for liquid samples of approximately 1 Ml. J Phys E Sci Instrum 20(5):523–530
54. Bücking W, Du B, Turshatov A, Konig AM, Reviakine I, Bode B, Johannsmann D (2007) Quartz crystal microbalance based on torsional piezoelectric resonators. Rev Sci Instrum 78 (7):074903
55. Sievers P, Moss C, Schroeder U, Johannsmann D (2018) Use of torsional resonators to monitor electroactive biofilms. Biosens Bioelectron 110:225–232
56. Bode B (1984) Entwicklung eines Quarzviskometers für Messungen bei hohen Drücken. Clausthal University of Technology, Clausthal-Zellerfeld
57. Vaughan RD, O'Sullivan CK, Guilbault GG (2001) Development of a quartz crystal micro-balance (QCM) immunosensor for the detection of Listeria monocytogenes. Enzym Microb Technol 29(10):635–638
58. Molino PJ, Hodson OA, Quinn JF, Wetherbee R (2008) The quartz crystal microbalance: a new tool for the investigation of the bioadhesion of diatoms to surfaces of differing surface energies. Langmuir 24(13):6730–6737
59. Poitras C, Fatisson J, Tufenkji N (2009) Real-time microgravimetric quantification of Cryptosporidium parvum in the presence of potential interferents. Water Res 43(10):2631–2638
60. Wang Y, Narain R, Liu Y (2014) Study of bacterial adhesion on different glycopolymer surfaces by quartz crystal microbalance with dissipation. Langmuir 30(25):7377–7387
61. Peschel A, Langhoff A, Johannsmann D (2015) Coupled resonances allow to study the aging of adhesive contacts between a QCM surface and single, micrometer-sized particles. Nanotechnology 26(48):484001–484009
62. Johannsmann D (2016) Towards vibrational spectroscopy on surface-attached colloids performed with a quartz crystal microbalance. Sens Biosens Res 11:86–93
63. Olsson ALJ, van der Mei HC, Johannsmann D, Busscher HJ, Sharma PK (2012) Probing colloid-substratum contact stiffness by acoustic sensing in a liquid phase. Anal Chem 84 (10):4504–4512
64. Olsson ALJ, Arun N, Kanger JS, Busscher HJ, Ivanov IE, Camesano TA, Chen Y, Johannsmann D, van der Mei HC, Sharma PK (2012) The influence of ionic strength on the adhesive bond stiffness of oral streptococci possessing different surface appendages as probed using AFM and QCM-D. Soft Matter 8(38):9870–9876
65. Cooper MA, Dultsev FN, Minson T, Ostanin VP, Abell C, Klenerman D (2001) Direct and sensitive detection of a human virus by rupture event scanning. Nat Biotechnol 19(9):833–837
66. Edvardsson M, Rodahl M, Hook F (2006) Investigation of binding event perturbations caused by elevated QCM-D oscillation amplitude. Analyst 131(7):822–828
67. Heitmann V, Wegener J (2007) Monitoring cell adhesion by piezoresonators: impact of increasing oscillation amplitudes. Anal Chem 79(9):3392–3400

68. Nosek J (1999) Drive level dependence of the resonant frequency in BAW quartz resonators and his modeling. IEEE Trans Ultrason Ferroelectr Freq Control 46(4):823–829
69. http://www.am1.us/Local_Papers/U11625%20VIG-TUTORIAL.pdf. Accessed 18 June 2014
70. Berg S, Johannsmann D (2003) High speed microtribology with quartz crystal resonators. Phys Rev Lett 91(14):145505
71. Mindlin RD, Deresiewicz H (1953) Elastic spheres in contact under varying oblique forces. Trans ASME J Appl Mech 20(3):327–344
72. Leopoldes J, Conrad G, Jia X (2013) Onset of sliding in amorphous films triggered by high-frequency oscillatory shear. Phys Rev Lett 110(24):248301
73. Hanke S, Petri J, Johannsmann D (2013) Partial slip in mesoscale contacts: dependence on contact size. Phys Rev E 88(3):032408
74. Vlachová J, König R, Johannsmann D (2015) Stiffness of sphere–plate contacts at MHz frequencies: dependence on normal load, oscillation amplitude, and ambient medium. Beilstein J Nanotechnol 6:845–856
75. Borovsky B, Booth A, Manlove E (2007) Observation of microslip dynamics at high-speed microcontacts. Appl Phys Lett 91(11):114101
76. Batchelor GK (1967) An introduction to fluid dynamics. Cambridge University Press, Clausthal-Zellerfeld
77. Riley N (2001) Steady streaming. Annu Rev Fluid Mech 33:43–65
78. Riley N (1998) Acoustic streaming. Theor Comput Fluid Dyn 10(1–4):349–356
79. Friend J, Yeo LY (2011). Rev Mod Phys 83:647
80. König R, Langhoff A, Johannsmann D (2014) Steady flows above a quartz crystal resonator driven at elevated amplitude. Phys Rev E 89(4):043016
81. Ghosh SK, Ostanin VP, Johnson CL, Lowe CR, Seshia AA (2011) Probing biomolecular interaction forces using an anharmonic acoustic technique for selective detection of bacterial spores. Biosens Bioelectron 29(1):145–150
82. Salazar-Banda GR, Felicetti MA, Goncalves JAS, Coury JR, Aguiar ML (2007) Determination of the adhesion force between particles and a flat surface, using the centrifuge technique. Powder Technol 173(2):107–117
83. Rodahl M, Hook F, Krozer A, Brzezinski P, Kasemo B (1995) Quartz-crystal microbalance setup for frequency and Q-factor measurements in gaseous and liquid environments. Rev Sci Instrum 66(7):3924–3930
84. Driscoll MM, Healey DJ (1971) Voltage-controlled crystal oscillators. IEEE Trans Electron Dev ED18(8):528
85. Shana ZA, Zong H, Josse F, Jeutter DC (1994) Analysis of electrical equivalent-circuit of quartz-crystal resonator loaded with viscous conductive liquids. J Electroanal Chem 379(1–2):21–33
86. Shana ZA, Josse F (1994) Quartz-crystal resonators as sensors in liquids using the acoustoelectric effect. Anal Chem 66(13):1955–1964
87. Zhang C, Vetelino JF (2003) Chemical sensors based on electrically sensitive quartz resonators. Sens Actuators B Chem 91(1–3):320–325
88. Peschel A, Boettcher A, Langhoff A, Johannsmann D (2016) Probing the electrical impedance of thin films on a quartz crystal microbalance (QCM), making use of frequency shifts and piezoelectric stiffening. Rev Sci Instrum 87:115002
89. Vidarsson H, Hyllner J, Sartipy P (2010) Differentiation of human embryonic stem cells to cardiomyocytes for in vitro and in vivo applications. Stem Cell Rev Rep 6(1):108–120
90. Pax M, Rieger J, Eibl RH, Thielemann C, Johannsmann D (2005) Measurements of fast fluctuations of viscoelastic properties with the quartz crystal microbalance. Analyst 130 (11):1474–1477
91. Tymchenko N, Kunze A, Dahlenborg K, Svedhem S, Steel D (2013) Acoustical sensing of cardiomyocyte cluster beating. Biochem Biophys Res Commun 435(4):520–525
92. Sapper A, Wegener J, Janshoff A (2006) Cell motility probed by noise analysis of thickness shear mode resonators. Anal Chem 78(14):5184–5191

93. Gutman J, Walker SL, Freger V, Herzberg M (2013) Bacterial attachment and viscoelasticity: physicochemical and motility effects analyzed using quartz crystal microbalance with dissipation (QCM-D). Environ Sci Technol 47(1):398–404
94. Wargenau A, Tufenkji N (2014) Direct detection of the gel-fluid phase transition of a single supported phospholipid bilayer using quartz crystal microbalance with dissipation monitoring. Anal Chem 86(16):8017–8020
95. Losada-Perez P, Khorshid M, Yongabi D, Wagner P (2015) Effect of cholesterol on the phase behavior of solid-supported lipid vesicle layers. J Phys Chem B 119(15):4985–4992
96. Domack A, Prucker O, Ruhe J, Johannsmann D (1997) Swelling of a polymer brush probed with a quartz crystal resonator. Phys Rev E 56(1):680–689
97. Edvardsson M, Svedhem S, Wang G, Richter R, Rodahl M, Kasemo B (2009) QCM-D and reflectometry instrument: applications to supported lipid structures and their biomolecular interactions. Anal Chem 81(1):349–361
98. Laschitsch A, Menges B, Johannsmann D (2000) Simultaneous determination of optical and acoustic thicknesses of protein layers using surface plasmon resonance spectroscopy and quartz crystal microweighing. Appl Phys Lett 77(14):2252–2254
99. Babauta JT, Beasley CA, Beyenal H (2014) Investigation of electron transfer by geobacter sulfurreducens biofilms by using an electrochemical quartz crystal microbalance. ChemElectroChem 1(11):2007–2016
100. Liu Y, Berna A, Climent V, Miguel Feliu J (2014) Real-time monitoring of electrochemically active biofilm developing behavior on bioanode by using EQCM and ATR/FTIR. Sens Actuators B Chem 209:781–789
101. Rabaey K, Angenent L, Schroder U (eds) (2009) Bioelectrochemical systems: from extracellular electron transfer to biotechnological application. IWA Publishing, London
102. ITO coated resonator crystals are available from microvacuum: http://www.microvacuum.com/
103. Giaever I, Keese CR (1991) Micromotion of mammalian-cells measured electrically. Proc Natl Acad Sci U S A 88(17):7896–7900

BIOREV (2019) 2: 163–188
https://doi.org/10.1007/11663_2018_2
© Springer International Publishing AG 2018
Published online: 6 February 2018

Microphysiometry

Martin Brischwein and Joachim Wiest

Contents

M. Brischwein
Technische Universität München (TUM) – Heinz Nixdorf Lehrstuhl für Biomedizinische
Elektronik, Munich, Germany
e-mail: brischwein@tum.de

J. Wiest (✉)
cellasys GmbH, Kronburg, Germany
e-mail: wiest@cellasys.com

Abstract Time-resolved monitoring of metabolic activities in vitro (microphysiometry) is necessary to study the dynamic regulations of core biochemical pathways in cellular disease models. Following a brief review of recent developments in the field, this contribution presents the most important branches of current sensor-based methods of microphysiometry. The primary parameters assessed by microphysiometry are extracellular pH changes and the concentration of dissolved oxygen to conclude on extracellular acidification (EAR) and oxygen uptake rates (OUR) as quantitative measures for cell metabolism. Direct sensing of selected small molecule metabolites or metabolic heat are alternative assay strategies. The major physical transduction principles encompass electrochemical and optochemical sensing complemented by less popular approaches for highly specific applications. All microphysiometric devices include tissue culture maintenance systems that have to guarantee physiological conditions and that must be functionally aligned with the sensing component. The interplay of cell metabolic activity and sensors in microscaled reaction volumes can be simulated with appropriate numerical models describing the physical processes of reaction and diffusion. While aspects of automation and throughput belong more to the engineering side of microphysiometry, both form the basis for the inevitable statistical data acquisition and analysis. This paper concludes with a description of two selected applications of microphysiometry, one in toxicology and the other one in clinical cancer research.

Keywords Cell culture · Cellular metabolism · Cellular respiration · Extracellular acidification · Microphysiometry · Oncology · Oxygen sensing · pH sensing · Toxicology

1 Introduction

1.1 What Is Microphysiometry?

The most prevalent human diseases in industrialized countries include (type 2)-diabetes, obesity, cancer, and cardiovascular diseases. A common feature of these ailments is dysregulation or dysfunction of cellular energy metabolism. To study the mechanisms behind such diseases on a cellular level, we need to analyze the dynamic activity of the cells' core metabolic pathways. This basic idea is fundamental to sensor-based microphysiometric methods or as stated by Alajoki et al. "The concept is that it is possible to detect receptor activation and other physiological changes in living cells by monitoring the activity of energy metabolism" [1].

It is a well-known fact from systems theory that the behavior of a dynamic system can only be understood if it is systematically perturbed. This implies that we have to be in a position to define input variables, keep others constant while observing how output signals evolve over time. The primary parameters assessed in microphysiometry comprise pH and the concentration of dissolved oxygen, glucose, and lactate, with a clear emphasis on the first two. Measuring these parameters experimentally in combination with a fluidic system for cell culture maintenance and the well-defined application of drugs or toxins provides the quantitative output parameters extracellular acidification rates (EAR), oxygen consumption rates (OUR), and rates of glucose consumption or lactate release to characterize the metabolic situation. Thanks to the label-free nature of sensor-based measurements, dynamic monitoring of cells or tissues for several days or even longer is feasible. On an extended timescale, a dynamic analysis of a cell's metabolic response to an experimental treatment can distinguish acute effects (e.g., 1 h after a treatment), early effects (e.g., at 24 h), and delayed, chronic responses (e.g., at 96 h).

Despite the microscale dimensions of many planar sensors, a single-cell microphysiometer is usually below the limits of sensitivity. Nevertheless, a statistical assessment of single cell metabolic rates might indeed be valuable to reveal phenomena like cell-to-cell variability or a potential cell-cycle dependency of metabolism. Scanning electrochemical microscopy [2], enzyme-modified electrodes [3], and pH-ISFETs [4–6] have been described as tools for metabolic monitoring of either isolated single cells or small cell groups growing within extended cell networks. In the context of biological relevance, any physiological characterization of isolated single cells is controversial since cell–cell contacts are vitally important for most, normally differentiated cell types – presumably except for cells of lymphatic origin.

An important distinction should be made with respect to the *intention* of microphysiometry (similar to other cell-based assays): On the one hand, it is the objective of microphysiometry to analyze the basal metabolic status of a cell population and its response to a defined treatment or exposure in vitro. This situation typically applies to all biomedical cell-based assays and also to techniques used in bioprocess engineering. It is an entirely physiological intention. In vivo microphysiometry based on active, implanted devices for monitoring physiological parameters in the body is a related issue which is, however, not within the focus of this chapter. On the other hand, using a distinct cell type with well-defined expression of cell-surface receptors and signal transduction cascades as the biological sensing element within the so-called whole cell biosensors to study the bioresponse to samples with unknown composition is, however, a more analytical intention. Such devices have the potential to become useful elements in an *effect-directed analysis* (EDA) of environmental samples [7]. Certainly, all devices and technical components in contact with living cells have to be fully biocompatible, implying the absence of any damaging effects emerging from the technical device and the procedures to the living matter and in the reverse direction.

In a typical microphysiometric experiment, a cell ensemble is subjected to an experimental stimulation (which may be physical, chemical, or biological) in a way that depends on both, the type of stimulus and the type and differentiation of the cells (Fig. 1). The cellular signaling cascades, often triggered by specific receptors, cause a cellular response that is most frequently coupled to a change in the activity of cellular core metabolism. This in turn leads to a changing rate, e.g., of extracellular acidification and oxygen consumption which are relayed into the extracellular microenvironment and become assessable by the corresponding sensors. Notably, the physicochemical parameters of the cell microenvironment itself may have the character of input signals, a factor that is evident, e.g., in hypoxia signaling [8].

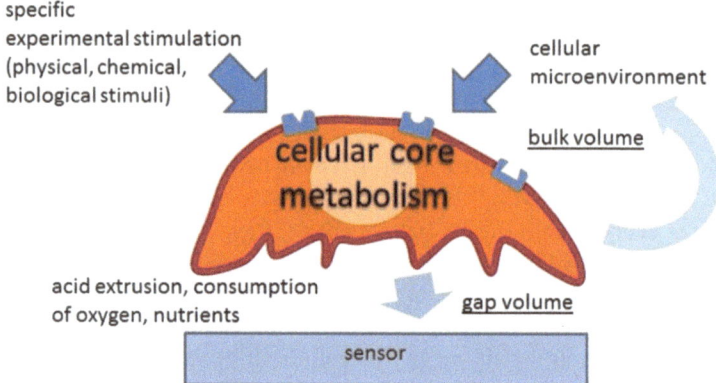

Fig. 1 Signaling scheme in a typical microphysiometric experiment. The quality of the sensor signals depends on tightness of the cell–sensor coupling and the performance of the transducer. Cell–sensor coupling may be very close – in cases when cells grow immediately on the surface of a microscale sensor and metabolic output is recorded in the small volume underneath the basal cell membrane – or it may be looser or more distant so that there is a larger diffusion distance and the metabolic output is detected in the bulk volume. The sensing performance is characterized by sensitivity and noise level, response time, and sampling frequency (Copyright: TUM)

1.2   A Brief Review of Microphysiometry

The term *microphysiometry* was mentioned for the first time in a paper describing a novel way of detecting the EAR of cultured cells grown directly on the surface of a silicon sensor chip [9, 10]. Microphysiometers are considered as a subgroup of different microscale sensor devices that are used to monitor cell-based assays [11]. The planar surfaces of those sensor chips are well suited for cell growth and cell culturing. The immediate contact between sensor and the adhered cell population allows high sensitivity measurements. Moreover, the miniaturized dimensions of those "cell-on-a-chip" systems help to save primary cell or tissue material and expensive chemical compounds that might be needed to perform the assay.

As early as 1972, Thomas et al. [12] introduced a microelectrode array (MEA) to monitor the bioelectric activity of cardiomyocytes. In 1984, Giaever and Keese [13] published the concept of *electric cell-substrate impedance sensing* (ECIS) to monitor morphology and motion of cells by measuring impedance via gold film electrodes. In the aftermath of these milestones in label-free live cell monitoring, the term *microphysiometry* was introduced by Parce and Owicki in 1989 [10, 14]. The instrument used for the original work is the *cytosensor microphysiometer* which monitors – in four channels – the EAR of living cells employing a miniaturized pH sensor [15]. Soon after its development, the device was commercially available from Molecular Devices (Sunnyvale, CA, USA). A different multisensor device for cell monitoring was developed by Wolf et al. that additionally included sensors for dissolved oxygen and electric impedance [16]. This multiparametric approach was later commercialized as the *Bionas 2500* system (Bionas GmbH: Rostock,

Germany), the *Intelligent Mobile Lab for in-vitro diagnostic* (IMOLA-IVD) technology (cellasys GmbH: Kronburg, Germany), and the *Intelligent Multiplate Reader* (IMR) platform (HP Medizintechnik GmbH: Oberschleißheim, Germany). In the USA, the company Seahorse Biosciences (North Billerica, MA, USA) developed the *XF Analyzer* system with optochemical sensors for pH and dissolved oxygen. Alongside these commercial developments, some research groups advanced the field, e.g., by introducing enzyme sensors to monitor glucose and lactate concentrations in the cellular microenvironment [17, 18].

2   Parameters and Sensors

The core metabolism of living cells goes along with an active or passive transport of metabolites across the cell membrane from the extracellular environment into the cells or vice versa. The extracellular microenvironment of the cells serves as source or sink of substances such as dissolved oxygen, nutrients such as sugars, amino acids, fatty acids, and glycerol, or metabolic end products. The extracellular concentration of these metabolites is analyzed by methods of microphysiometry. It provides, for instance, the oxygen uptake rate (OUR) or the EAR as key parameters for cellular metabolism. Since the OUR mainly reflects mitochondrial, oxidative pathways of cellular energy metabolism and the EAR is dominated by the activity of non-oxidative pathways, the ratio of both is a measure of the metabolic status of a cell population independent of the absolute rates [19].

2.1   *Extracellular Acidification and pH Sensors*

The physical chemistry and cell biology of extracellular acidification was excellently reviewed in a paper by Owicki and Parce [20]. Both, anaerobic and aerobic pathways contribute to the apparent rate of extracellular acidification, but the contribution of the non-oxidative, glycolytic pathway in terms of moles H^+ secreted per moles of glucose consumed is much higher. Typically, proton extrusion rates in the order of 10^{-8} mol/cells are observed. Acid release out of the cells occurs either passively by diffusion of dissolved CO_2 or by active transport via carrier proteins (ion exchangers) localized in the plasma membrane. Notably, the measurement of changing pH in the extracellular environment is only an indirect measure of the cellular acid extrusion rate: Buffering species will bind a large portion of the released protons in equilibrium reactions, reducing the number and concentration of free protons that are measured by pH sensors. In addition, the effective diffusion constant of protons is reduced by their binding to other chemical species with a relatively high molecular mass. Therefore, the measurement of extracellular acidification is usually performed in culture medium with reduced buffer capacity, i.e., without adding species like sodium bicarbonate and in ambient air.

Established microscale electrochemical pH sensors are: (1) the Ion Sensitive Field Effect Transistor (pH-ISFET) [21, 22], (2) the Light Addressable Potentiometric Sensor [23], and (3) simple potentiometric sensors based on oxides of transition metals such as ruthenium or iridium [24]. The various types of electrochemical pH microsensors have been reviewed recently [25]. pH-ISFETs and LAPS are active, silicon-based semiconductor elements while the potentiometric sensors simply measure a voltage passively. A common component in all devices is the pH-sensitive oxide–metal layer. Briefly, a (pH-dependent) potential difference across the oxide layer is: (1) amplified with the ISFET, (2) is made light addressable (or accessible) with the LAPS to gain spatial resolution, or (3) is measured directly with the potentiometric sensors. The benefit of using ISFETs is the easy scale-up to arrays and the immediate inclusion of signal processing circuitry into the CMOS-based chip. All electrochemical pH sensors provide quick response times and powerful pH resolution of at least 10^{-3} pH as long as the first input amplifier is near to the sensor. In addition to localized pH measurements, a spatial distribution of pH values can be recorded and imaged with the LAPS technology [26, 27] or with a CMOS-based multielectrode arrays in which every ISFET gate contributes a pixel to the overall image [28, 29]. A disadvantage of metal oxide sensors is signal drift that complicates the recording of absolute pH values on extended timescales [30].

Irrespective of the readout technology, all potentiometric electrochemical pH sensors require a reference electrode providing a constant potential with respect to the electrolyte solution. This reference electrode is either placed in an electrolyte bath at a distance from the cell culture within the fluidic system [31] or it is an integrated on-chip structure [32]. Since Ag/AgCl electrodes have to be handled with care so that no toxic silver ions leak into the cell culture medium, the use of planar iridium oxide (quasi-) reference electrodes has become more widespread [33].

Optochemical pH sensors make use of pH sensitive-fluorescent dyes immobilized in a chemical matrix consisting of polymer nanospheres or hydrogels [34]. The fluorophore doped material is either deposited on a suitable, transparent surface as an individual spot for integral measurements or a sensor foil is fabricated to allow for pH imaging in combination with a digital camera (www.presens.de). A very robust method is based on two different fluorophores with different lifetimes of the excited state: one with a pH-dependent fluorescence and a second reference dye whose fluorescence is independent of pH [34]. By using an amplitude-modulated excitation light for both dyes, the total of the fluorescent light adopts a phase shift that is a nonlinear function of the pH. This principle named "dual lifetime referencing" helps to avoid artifacts caused by dye bleaching or instable light sources. According to the manufacturer, the pH resolution is 5×10^{-2} pH (near the dye's pk_a value), the absolute precision (with batch calibration) is 0.2 pH, and the typical response time $(\tau_{63})^1$ is 30 s (www.presens.de). Although this lags behind the performance of electrochemical sensors, the signal drift is only ~0.1 pH/week.

---

[1]The time for the system's step response to reach $(1 - (1/e)) -$ of its final (asymptotic) value.

2.2 Cellular Oxygen Uptake and Sensors for Dissolved Oxygen

The concentration of dissolved oxygen is usually given either in percent of air saturation or in units of the partial pressure (in Pascal, Pa) in the gas phase that is in equilibrium with the liquid phase. The concentration of dissolved oxygen in liquids is linked to its partial pressure by Henry's law. The solubility of oxygen in water is relatively low (240 µM at 25°C), and even lower in culture medium at 37°C (about 200 µM) [36].

Typically, the vast majority of cellular oxygen consumption is due to the activity of the respiratory chain in the mitochondrial membranes. However, there are deviations from that rule [35]. The respiratory consumption of dioxygen from the surrounding medium by cultured cell monolayers shows a linear slope in time down to ≈ 20 µM. Compared to this, the apparent Michaelis–Menten constant K_M of cytochrome oxidase as the major oxygen converting enzyme is quite low and in the order of 1 µM. Thus, in a typical cell culture environment the respiratory chain will be active at its maximal rate V_{max} [36].

Similar to pH measurements, both electrochemical and optochemical sensors for dissolved oxygen have been described and used in microphysiometric applications. The basis for electrochemical oxygen sensing is the invention by Clark [37] who presented a useful amperometric sensor by coating an electrode surface with a selectively permeable gas diffusion membrane. The transfer of that general principle to planar geometries and micromachined sensors is far from trivial and it appears that only recently sensors with a sufficient reliability and long-term performance were described [18]. Figure 2 shows planar electrode structures for the amperometric sensing of dissolved oxygen placed within a fibroblast cell culture.

On the other hand, optochemical O_2 sensors have reached a mature stage of development. Molecular interaction of dissolved oxygen with a fluorophore integrated in or bound to a chemical matrix results in fluorescence quenching, i.e., a reduced lifetime τ of the excited state. The quantitative relationship between the concentration of the quencher (oxygen) and the shortening of the lifetime is given by the Stern–Volmer equation. Practically, the effect on lifetime is measured as a phase shift between the amplitude-modulated excitation light and the fluorescence light emitted by the sensor [38]. The specifications given by the manufacturer claim an effective resolution of $\pm 2\%$ O_2 at 100% air saturation (www.presens.de). The sensor drift is negligible. As with optochemical pH sensors, imaging of O_2 levels requires using sensor foils instead of sensor spots and a pixel-by-pixel readout with digital cameras (www.presens.de). The measurement range for O_2 reaches down to $<5\%$ air saturation or <10 µM, which is at the onset of hypoxia [8].

Fig. 2 Subconfluent monolayer of L929 cells on a chip with an amperometric sensor for dissolved oxygen (W: working-, R: reference-, and A: auxiliary-electrode arrangement). Cells at different stages of proliferation (1: adherent cells, 2: cell detaching, 3: cell division, and 4: divided cells adhere) are highlighted (Copyright: cellasys GmbH)

2.3 Sensors for CO_2

While microphysiometric sensors for pH reflect the total activity of all pathways contributing to extracellular acidification, sensors for dissolved CO_2 report more specifically on oxidative metabolism. Carbon dioxide either leaves the cell as dissolved gas or combines with water to form carbonic acid with a pseudo-first order rate constant of 0.062 s^{-1} for hydration [39], followed by a rapid dissociation into protons and bicarbonate anions (HCO_3^-) at physiological conditions (pH 7.4) and selective extrusion into the extracellular space by ion transport systems. The dissolved CO_2 as part of these equilibrium reactions can be detected by fluorescence-based pH sensors. As the working principle underlying these sensors is that of the Severinghaus electrode, the sensors are coated with a hydrophobic membrane to allow only dissolved gas to permeate and to prevent interference with the background acidity in the analyte solution [40]. Another, rather recent optochemical sensing approach for CO_2 is based on ion exchange polymer matrices with incorporated pairs of quaternary ammonium cation and a pH indicator anion dye [41].

2.4   Sensors for Glucose and Lactate

Obviously, the analysis of cellular core metabolism is strongly supported by sensors for key metabolites such as glucose and lactate. Monitoring of these species requires robust sensors with considerable long-term stability. Although enzyme-based amperometric (bio-)sensors remain commercially unchallenged, nonenzymatic sensors using materials that catalyze a direct oxidation of the substrate (e.g., glucose) have also been described [42]. Besides biomedical applications, real-time monitoring of glucose and lactate is one of the major analytical tasks in bioprocess technology and the introduction of micromachined sensors for these analytes allows an in situ process control inside small, analytical bioreactors [43]. Modified versions of the *cytosensor microphysiometer* including enzyme-based amperometric sensors have been established to monitor glucose and lactate metabolism in cell cultures [44]. A microscale cell culturing device including a dynamic real-time monitoring of glucose and lactate with biosensors (in addition to pH and dioxygen) was described recently [18]. The lifetime of the enzyme sensors while operating continuously is reported to be at least 42 days [45].

2.5   Sensing Metabolic Heat

A different detection concept for assaying cellular metabolism is based on the measurement of metabolic heat or the associated temperature shift. For mapping temperature at the single cell level, several microscopy techniques have been developed. Most of these techniques use the excitation of a fluorescent dye with selected fluorescence parameters (intensity, spectrum, and polarization) being dependent on temperature [46]. Infrared temperature mapping microscopy, in contrast, does not need any exogenous excitation sources but its sensitivity is limited to temperature variations in the order of 1 K [47]. Microcalorimetry measurements have shown that the typical heat power provided by a single cell's metabolism can reach $P \sim 100$ pW. The corresponding temperature increase in an aqueous environment has been estimated to be in the range of 10^{-5} K [47]. Thus, a cell aggregate containing $\sim 10^2$–10^3 cells providing a power of ~ 10–100 nW is well within the limit of detection of micromachined, planar calorimetric sensors [48].

The transducer principle of such sensors is based on the thermoelectric Seebeck effect, i.e., the generation of a voltage within an electric circuit that consists of two different conductors when the junctions between the two materials experience different temperatures. The material combinations selected for these types of devices (e.g., p$^+$ polysilicon/aluminium) have a particularly high Seebeck coefficient. The sensitivity is further increased by the arrangement of several of such thermocouples in series to form a thermopile structure and by limiting the heat conduction between thermopile junctions by a thermal resistance layer. If cultured cells [49] or active biofilms [50] are grown on one side of such a system, metabolic heat will cause a temperature difference and thus a measurable voltage difference.

A completely different, elaborate transducer principle has been described that is capable of measuring the metabolic heat generated by single cells [51]: The pico calorimeter comprises a sample stage (with individual brown fat cells in a liquid filled microchannel) that is separated from a cantilever resonator (oscillating within a vacuum chamber) by a glass wall, but thermally connected by a silicon heat guide. Temperature changes of the sample stage are monitored by changes in the resonance frequency of the cantilever in the vacuum chamber. The reported heat resolution of the device is 5.2 pJ.

3   Experimental Conditions and Automation

3.1   Climate Control and Fluidic Systems

Cells require a well-defined environment in order to establish a stable baseline activity and to respond reproducibly to signals from the outside. As a general rule, it is desirable to mimic the physiological conditions of the cells under study as close as possible. The major conditions that have to be maintained in cell cultures are: (1) temperature, (2) physicochemical properties of the culture medium, and (3) sterility [6].

In general, thermostatic environments (incubators) are used to keep mammalian cells at their physiological temperature of 37°C. Some systems allow controlling the concentration of carbon dioxide and/or oxygen in the gas phase to adjust either the basal pH of the medium or the availability of dissolved O_2 to the cells under study, respectively. The latter is required, for instance, in tumor hypoxia research [52]. It is still a matter of discussion which oxygen concentration is physiologically correct in cell cultures [53]. To prevent evaporation in open systems, humidifiers or membrane covered medium storage plates are used [54].

While microfluidic systems are readily integrated into cell culture devices to mimic physiological fluid flow and the associated shear forces (see, e.g., www.fluxionbio.com), the main function of liquid movement in microphysiometers is the exchange of culture medium and addition of chemical compounds. Classically, this is done with a peristaltic pump (CM, Bionas 2500, IMOLA). Alternative approaches use pipetting robots (IMR) or movable plungers (XF analyzer) to exchange the medium. It is a common principle to have a periodic sequence of medium exchange phases and static rest intervals with no medium flow. In these static phases, the microsensors measure the metabolic parameters within the so-called *microreaction volumes* as shown in Fig. 3.

Long-term monitoring of living cells or tissues in flow through systems triggers inevitably the economic issue of cost for cell culture medium. In view of the figures given in Table 1, the average throughput of culture medium is typically in the order of 100 μL/min per channel. Thus, depending on the system's design, the continuous exposure of cell cultures to fresh medium potentially supplemented with expensive chemical or biological compounds may be limited by cost considerations.

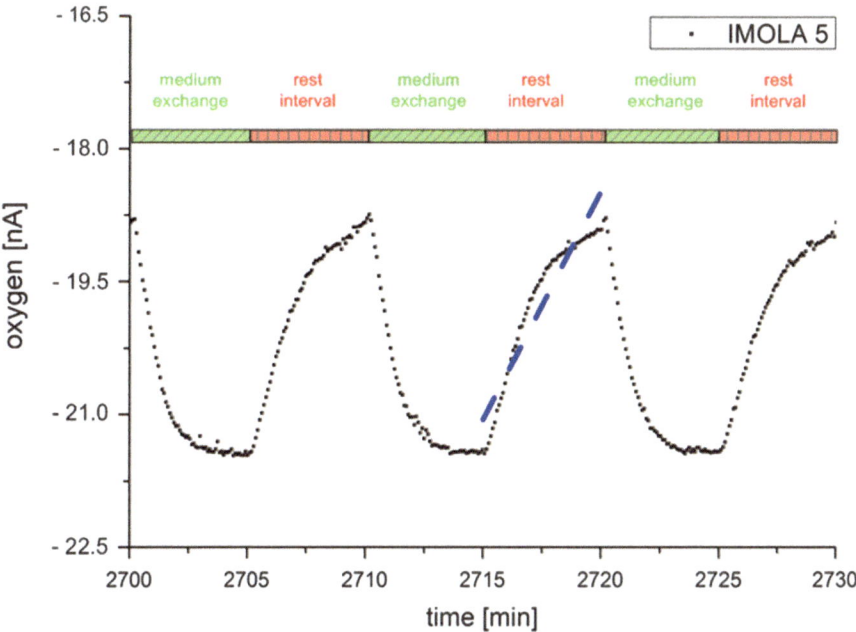

Fig. 3 Time-resolved raw data of a dissolved oxygen (DO) microsensor (measured amperometrically as current in nA) in a microscale cell culture chamber, illustrating the regular fluid cycles: Fresh cell culture medium is transported into the chamber during the exchange intervals calibrating the system to 100% saturation. In the following rest interval, cellular oxygen consumption leads to a decline of DO saturation (a higher DO saturation results in a lower oxygen sensor signal). The dashed blue line is a linear fit to the sensor raw data in one of the rest intervals to calculate a rate value dDO/dt to describe the oxygen consumption rate (OUR) (Copyright: cellasys GmbH)

Table 1 Volume of microchambers and flow rates in different microphysiometric systems

Microphysiometric system	Chamber volume (μL)	Flow rate
Cytosensor microphysiometer [23]	4	~100 μL/min
Silicon microphysiometer [14]	~10	10–50 μL/s
Bionas 2500 [55]	~15	~56 μL/min
XF analyzer [56]	7	Dilution system
IMOLA	6	~60 μL/min
IMR [19]	23	Hydrostatic pressure difference, ~100 μL/min

3.2 Microreaction Volumes

Generally, the culture system in microphysiometry confines the cellular specimen in a small microvolume to attain a high ratio of active cell biomass to the surrounding volume of culture medium. Therefore, microphysiometric platforms are usually

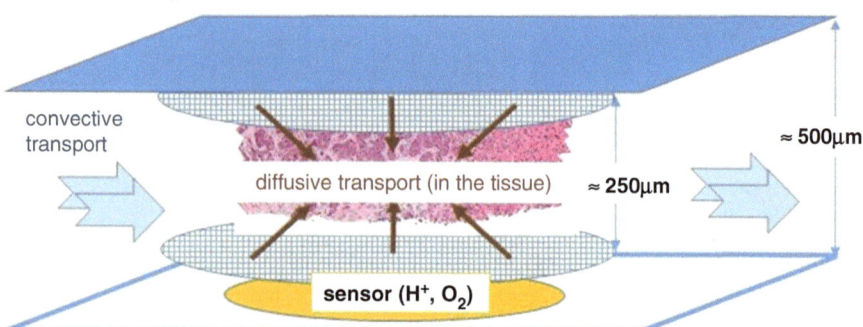

Fig 4 A flow of culture medium passes a tissue slice of ≈250 μm thickness, sandwiched between thin nylon meshes. Intensive convective flow increases the concentration gradients for oxygen, nutrient, and waste compounds towards the interior of the slice and thus enhances diffusional transport (Copyright: TUM)

closed culture systems, enabling a tight control of the culture environment. However, for loading the system with the cellular sample prior to the assay, the vessels have to be open and accessible.

Typical microreaction volumes in microphysiometer systems are in the lower μL range with chamber heights of about 200–500 μm. The medium in these volumes is transported in exchange intervals at flow rates high enough to achieve an even exchange and low enough to retain even loosely bound cells in the microchamber and to avoid undue shear stress. A reasoned geometry of the microreaction chamber will help to obtain a fast and uniform medium exchange at moderate flow rates [57]. Table 1 summarizes some features of microreaction volumes and flow systems found in some commercially available microphysiometer systems.

Figure 4 shows a schematic of a microreaction volume – as it is used in the IMR system for monitoring of primary tissue – in detail. In this particular case, the sensors on the bottom of the closed vessel clearly measure bulk properties of the medium at appreciable diffusional distances to the metabolically active tissue samples.

3.3 Automation

Automation is one of the most significant current topics of microphysiometry to reduce personal costs in laboratories and to increase reproducibility. This may concern both, the handling of viable cells and of liquid medium. Automation is also a prerequisite for an increased experimental throughput and thus, for the generation of data volumes needed to achieve statistical significance. This point is particularly relevant if the inherent heterogeneity of samples and/or the noise level of sensor data are high.

Until recently, the automation of cell seeding, realized by robotic handling of cell suspensions or printing of viable mammalian cells [58], was rarely implemented. In contrast, the automation of fluidics either with circuitries of pumps, valves, and air bubble traps or with liquid handling robotics has been at the core of microphysiometer developments from its beginnings. The on-chip microfluidic system is connected to storage vessels for culture medium. If compounds with a low half-life are under test, specific appliances for a controlled, automated addition of such compounds are required. Most microphysiometers also have the possibility to remove the compound under investigation in order to analyze the reversibility of effects [59].

4 Cell Culture

4.1 Cell Models

A crucial aspect to be considered for microphysiometry – as for any cell-based assay – is the selection of the cell model [60]. In general, the physiological relevance increases when established and immortalized cell lines are replaced by primary cells freshly isolated from the donor organism or defined cocultures of such primary cells.

However, those primary cell preparations may be of highly variable composition, quality, derived from different donors at different ages, with different underlying morbidity, environmental influences, and different isolation times. Therefore, physiological relevance is frequently in conflict with the reproducibility of an assay. Recently, the situation has been improved with the advent of (nonhuman) embryonic stem cells (ESCs) and human induced pluripotent stem cells (iPSCs). The identification and isolation of such stem cells, along with techniques for guiding their subsequent differentiation into terminal cell types, yields cell models that are both, relevant and reproducible enough to efficiently drive biomedical testing and *EDA* by cell-based assays [61].

4.2 Culture Systems

The details of cell growth and culturing are as important for the relevance of an assay as the cell model itself. The translational value of results from classical dish culture experiments is often very limited [62, 63]. In fact, the native environment of most cells comprises homotypic or heterotypic contact to neighboring cells, soluble factors, and components of the extracellular matrix. With regard to tumor biology, the outcome of an experimental treatment was shown to depend on the interaction between cancer cells and their microenvironment [64].

Soluble, biologically active compounds in serum, that is used to supplement culture medium, are among the most important active constituents of the microenvironment. For reasons of experimental standardization and quality control – but also for ethical reasons, the use of fetal calf serum should be avoided or reduced whenever possible. The state of the art in chemically defined cell culture medium has been reviewed recently [65].

The technical requirements for in vitro culture of mammalian cells are even more complex in the so-called tissue-on-chip or organ-on-chip devices [66]. For a review of three-dimensional culture systems with and without scaffolds, the reader is referred to [67]. 3D culture systems are a challenge with respect to quantitative analyses of cells. Until now, microarray scanners or confocal microscopy imaging in combination with fluorescence labeling approaches are the dominating methods [60]. However, microphysiometry platforms are beginning to incorporate 3D models [68–70]. While compartmentalized 3D coculture systems are synthetic models, there is a gradual transition towards models reflecting the native tissue structure more closely. For instance, spherical cell clusters such as tumor spheroids [71, 72], organoids [73], or tissue slices [74] show a more or less preserved tissue architecture. Although gradients of oxygen, nutrients, and metabolic waste products in such 3D structures are considered as an additional attribute of the organotypic situation, these gradients need to be controlled and kept within a defined range of tolerance to avoid unwanted physiological responses. This is both, a question of the diffusional distances that need to be overcome inside the samples and the steepness of concentration gradients maintained at the boundary surface by convection, i.e., by an efficient microfluidic system. Diffusion distances between cells and sensors should be as small as possible with a direct growth of 2D cell cultures on a planar sensor surface achieving the highest sensitivity and the fastest response in the detection of metabolic activity. The confinement of cells within a microvolume is easy for cells naturally adhering to the solid (sensor) substrate but is difficult to establish for suspension cultures, where immobilization techniques such as the encapsulation into agarose gels [75] may be required. For tissue slice cultures or spheroids, the use of porous layers (e.g., nylon meshes) allowing both, the nourishment of the tissue and protection against fluidic sheer stress, has been reported [68, 70].

Another issue is the longevity of on-chip cultures that must match the required assay duration. Long-term cell monitoring may be required to investigate the repeated dose toxicity of compounds, recovery effects, or subtle, slowly emerging effects on cell physiology. On-chip culture lifetimes of more than 30 days [76] and less than 6 months [77] have been reported, in applications that were designed to stabilize and to conserve mammalian cells on solid substrates for the preparation of portable cell-based biosensors with extended shelf lives.

5 Data Processing and Numerical Models

5.1 Dynamic Data Analysis

Dynamic analysis of microphysiometry data is advantageous in the field of toxicodynamics, i.e., the investigation of the dynamic interactions of a chemical with a biological target. A typical rationale of microphysiometry experiments uses a "black box" approach: Distinct "signal inputs" from the extracellular environment encounter the test organisms in a defined rest level and produce the "signal outputs", i.e., a change of metabolic activity measured by the sensors. This brings up the question for a meaningful evaluation of microphysiometry raw data. Common to all data sets derived from different sensor types is the structure of a time series of single measurements (e.g., $pH(t)$, $O_2(t)$) reflecting the cellular metabolic activity and its potential changes. As outlined above, this time series is subdivided into intervals by the periodic activity of a fluidic system.

The initial step of data evaluation should be an analysis of data quality, realized, e.g., by a Fourier transformation (Fig. 5) of the raw data to sort out low frequency interferences (e.g., sensor drift) and high frequency noise. A quantitative measure of data quality is the results of such a procedure and it may be used to introduce a weight factor for a subsequent statistical analysis that aims to test for the significance of differences between differently treated groups of cell cultures [78].

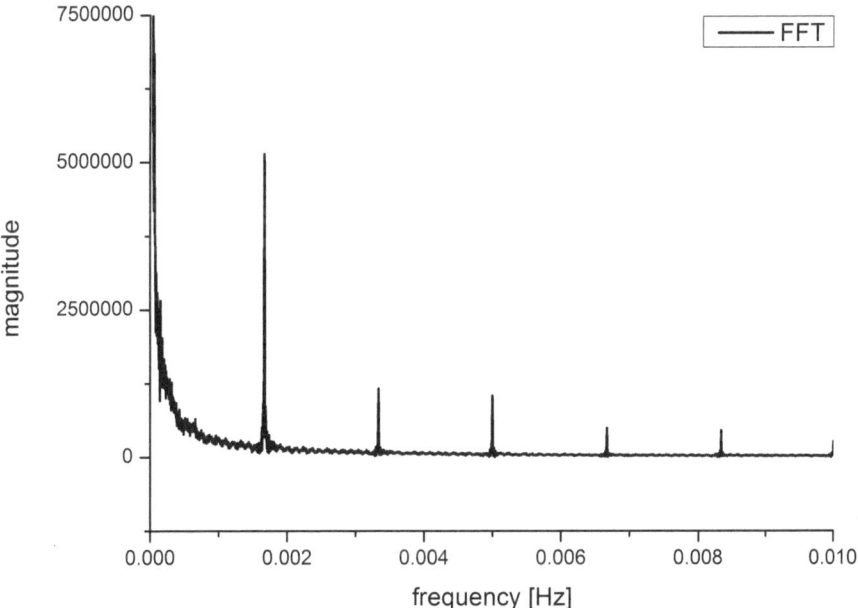

Fig. 5 Fast Fourier transformation of the time course of cellular respiration rates (L929 fibroblasts; cf. Fig. 3). The peak and its harmonics at 1.667E-3 Hz represent the oscillation period of the fluidic system (Copyright: cellasys GmbH)

5.2   Numerical Models for Reaction and Diffusion

In sensor-based microphysiometry, a coupling of metabolic reactions – associated with sinks or sources of species such as glucose, protons, or oxygen – and diffusion of these different species between the cell culture and the sensor is analyzed. The situation is complicated by additional chemical reactions such as the binding of protons to buffering substances. To complete the physical model, response characteristics of the sensors have to be considered to account for the difference of "true" and "measured" values. Ultimately, the analysis provides an estimate of the assay outputs, e.g., the time course of extracellular acidification (i.e., dpH/dt) or the OUR (dO$_2$/dt) by fitting experimental sensor data with a transfer function of the sensor response synthesized in the physical model with appropriate input parameters such as diffusion constants, spatial distribution of the active cell mass, and sensor response time [79]. Today, powerful software tools are available coupling multiple physical principles expressed in form of differential equations and yielding three-dimensional and dynamic parameter distributions on the basis of finite element calculus.

While the situation of reaction–diffusion processes is quite clearly understood and controlled in the case of homogeneous cell monolayers, the level of complexity is much higher in the case of 3D specimens such as slices of viable tissues [80]. Here, the local metabolic activities and the diffusion constants of the different species may vary significantly. In addition, the diffusion of metabolic acids (H$^+$ ions formed, e.g., from released CO$_2$ or lactic acid) is superposed by local variations of buffer capacity resulting in free and bound H$^+$ ions and an effective diffusion constant that depends on the molecular masses of the buffering species. Taking into account all these constraints, the model will yield spatially and temporally resolved estimates of the pH value and the concentration of dissolved oxygen in the interior of such 3D multicellular specimen (Fig. 6). This kind of modeling helps considerably to keep the respective levels in a range of physiological tolerance.

In principle, any metabolic activity itself is potentially a function of extracellular pH, dissolved oxygen, or other variables of the extracellular microenvironment, and must be considered as feedback factors (cf. Fig. 1). Therefore, it may not be appropriate to assume constant rates of extracellular acidification and oxygen consumption in each measurement interval with declining pH and O$_2$. In case of O$_2$, it appears that OUR are quite independent of ambient O$_2$ up to rather low O$_2$ concentrations of about 10–20 μM [36, 81] which are, however, regularly observed under hypoxic conditions [82, 83]. The sensitivity of oxygen uptake and acid release rates to some central substrates of core energy metabolism have been recently studied in human mamma carcinoma models [19, 84]. To account for these sensitivities, it is recommended to keep the measurement intervals short so that any deviation of the ambient conditions of the cells remains negligible. This, in turn, requires a high sampling rate, fast sensor response, and high signal to noise ratio of the sensor data. Another factor that has to be carefully included – particularly in measurement geometries with high surface to volume ratios – is the influence of gas-absorbing

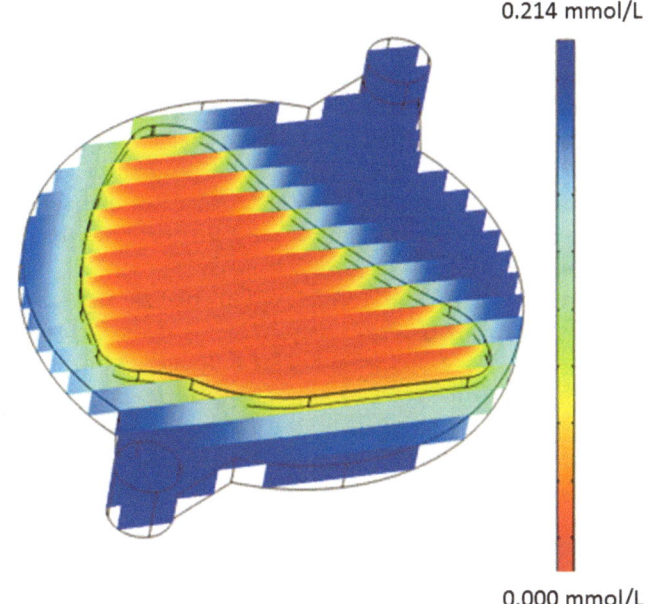

0.214 mmol/L

0.000 mmol/L

Fig. 6 Calculated distribution of the dissolved oxygen concentration in a microreaction chamber with a chicken spleen tissue slice at the end of a 20 min rest interval. Micrographs have been used to define the position and the geometry of the slice for the simulation. The rate of oxygen consumption during the rest cycle has been determined from parameter estimation, i.e., by fitting of an experimental, time-dependent sensor response $CO_2(t)$ with the model's transfer function. The respiratory activity and oxygen diffusion constant within the slice was assumed to be homogeneous, resulting in the spatial distribution of CO_2 as shown in the image (Copyright: TUM)

polymers confining the measurement chambers since any absorption and desorption of CO_2 and O_2 will attenuate the measurable EAR and OUR [85].

Certainly, the precision of the model depends on a precise knowledge of the input parameters and the sensitivity of the model suffers from any inaccuracy of these inputs. To evaluate the predictive value of a model, an experimental validation with a narrow margin of errors both, in the model inputs and the experimental design is necessary [79].

6 Selected Applications

6.1 Toxicology

Until now, the most advanced applications of microphysiometry are in the field of toxicology. Here, the *Cytosensor Microphysiometer* was used for assessment of eye irritation in various studies [86–88]. In 2009, this technique was retrospectively validated by the European Centre for the Validation of Methods [89] to correctly identify compounds from two out of three European Union and GHS (Globally Harmonized System of Classification and Labeling of Chemicals) classification categories in ocular irritation. The method investigates the EAR of L929 mouse fibroblast when exposed to seven dilutions of a given chemical under test and returns the "metabolic rate decrement by 50%" (MRD50) value. This value is the concentration of the chemical that causes a reduction of EAR by 50% compared to the basal rate. In a prediction model, the MRD50 is used to determine the classification category.

In a recent study, HepG2 cells were employed to compare monitoring EAR and OUR with conventional methods of cell analysis. In this study, the usefulness of the XF analyzer for a mechanistic understanding of drug-induced mitochondrial dysfunction was presented [90]. In another series of experiments on HepG2 cells, it was shown that an automated microphysiometry is capable of monitoring living cells for periods longer than 1 month (Fig. 7). The EAR was continuously monitored and once per day the drug Trovafloxacin (an antibiotic which was withdrawn from the market in 1999 due to hepatotoxicity) has been added to the culture medium for 1 h.

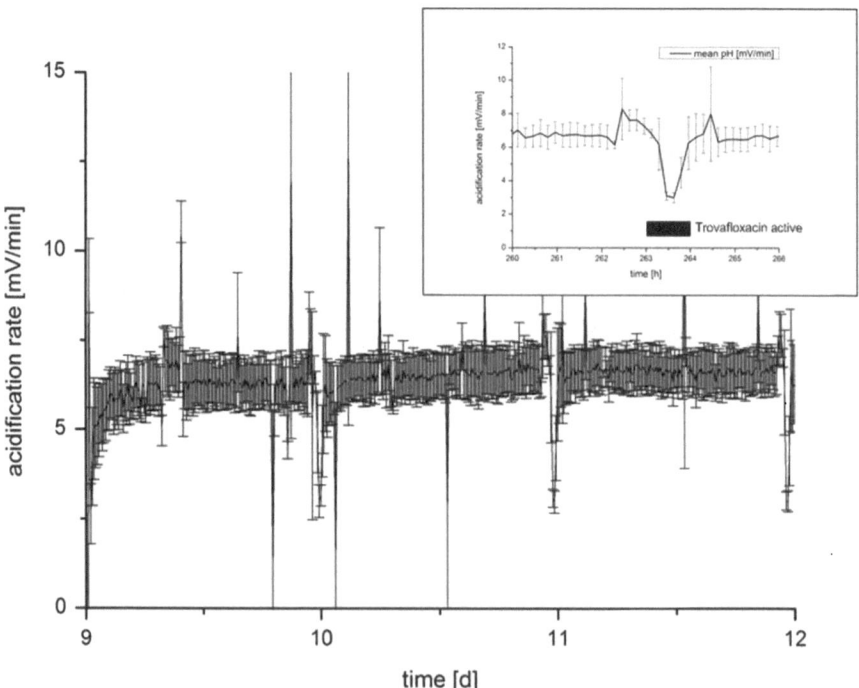

Fig. 7 Section of a long-term monitoring experiment ($n = 2$) of the extracellular acidification rate (EAR) of HepG2 cells in a repeated dose toxicology assay with repeated applications of Trovafloxacin. The inset shows a zoom into the reduced EAR due to the addition of Trovafloxacin at the end of day 11. The culture medium was RPMI 1640 medium without sodium bicarbonate, with Gentamicin, and 10% fetal bovine serum (Copyright: cellasys GmbH)

After 28 days, 0.2% sodium dodecyl sulfate (SDS) was added as a positive control. The results show reversible reductions of EAR under the repeated application of Trovafloxacin.

The retrospective validation of the *Cytosensor microphysiometer* method for eye irritation and ongoing research with more complex cellular models are proofs for the usefulness of microphysiometric systems in the field of toxicology.

The transition into a validated assay was achieved with the *Cytosensor Microphysiometer* assay for eye irritation. The ongoing development of Organ-on-Chip and Human-on-Chip technology is an interesting challenge to the field [91, 92]. In summary, the incorporation of microphysiometry into human-on-chip models can be a significant improvement of drug development and toxicology [93].

6.2 Clinical Cancer Research

A personalized approach to anticancer chemotherapy requires predictive testing with biopsy material to avoid ineffective therapies on the patient. In this context, there is a still unmet demand in cancer medicine for physiologically relevant, organotypic, ex-vivo test models. Although urgently needed and despite the availability of clinical data supporting the performance of such tests [94], they are not widely practiced. Until recently, only few predictive assays have been introduced into clinical routine and these assays are based on selected molecular markers that interact with distinct drugs ("companion diagnostics"). However, the benefit of molecular tests for treatments using conventional chemotherapeutic drugs remains doubtful [95]. An open question is, whether a multigene profiling of a patient's tumors underpinned by next generation sequencing methods is able to reveal genes that are addressable by tailored treatment [96].

The starting point for any predictive test is a cellular sample derived from the cancer patient. This may be a surgical or a biopsied piece of tissue or a liquid sample from peripheral blood, potentially containing circulating cancer cells. Thus, the amount of cellular material available for a predictive test may be very small. Typically, a 10 mg core biopsy taken from a solid tumor is equivalent to a cell number of roughly $2–4 \times 10^6$ [97], while the number of circulating tumor cells detected in peripheral blood is typically only at 10^0 per mL [98]. In addition, any sampling strategy has to account for the intratumoral heterogeneity on a phenotypic and genotypic level.

The general perspectives of label-free cell monitoring technologies for the prediction of anticancer drug efficacy have been reviewed some years ago [99]. For all these approaches, it is important that the cellular sample is kept viable after explantation, during preparation (i.e., dissection into the number of subsamples required for the test array) and throughout the testing. The microphysiometric approach allows for an ex-vivo model therapy using patient-derived, organotypic tissue samples in combination with a metabolic monitoring before, during, and eventually after drug application [100, 101]. This kind of referencing of a biological response to the condition of each individual tissue section prior to the treatment is unique to the sensor-based, dynamic readout. In data analysis, it produces a valuable reduction of statistical variance.

As outlined above, the preparation of viable sections from the original patient sample should maintain the histologic structure of the tumor. In a recent proof-of-principle study on human mamma carcinoma, slicing of surgical samples with a vibratome technique was followed by monitoring of extracellular acidification and cellular oxygen consumption of these specimens [68]. In this study, doxorubicin and an alkylating agent (chloracetaldehyde) have been applied in the ex-vivo therapy model and both, rapid cytotoxic effects as well as delayed responses becoming manifest after several days of incubation were revealed (Fig. 8).

A noticeable finding in the results presented in Fig. 8 is the steady decay of metabolic rates even in the untreated control samples. In that particular case, the slices have been perfused only along one of the opposing flat surfaces. Since OUR obviously declines much faster than EAR, the ratio EAR/OUR increases, a cue pointing out a progressively increasing anaerobic metabolism under conditions of oxygen deficiency. On the one hand, such findings support the significance of multiple parameters recorded in parallel, on the other hand they document how crucial it is to provide a uniform supply of oxygen (and nutrients).

In principle, the microphysiometric sensor readout is applicable to all tumor entities provided that a minimum number and density of viable cells can be collected. Obviously, this minimum cell density depends on the size and the quality of the sensor and the specific cell activity. From our own experience, a total number of at least 10^4–10^5 viable cells are required to come up with a feasible assay and a sufficient dynamic range. Moreover, a crucial aspect for the quality of the test is an adequate and uniform maintenance of the sections. Since diffusion is the only mechanism of transporting nutrients and oxygen into the tissue, the thickness of the sections is limited to a few hundreds of micrometers. The fluidic maintenance

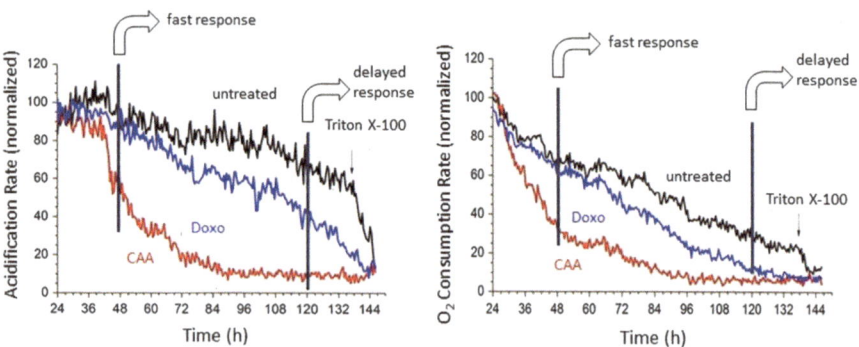

Fig. 8 Result of a 6-day test on a surgical human mamma carcinoma sample. Left: EAR and right: OUR. For EAR and OUR, a number of $n = 4$ and $n = 5$ (respectively) of equally treated slices have been analyzed. *Doxo* doxorubicin (5 μM), *CAA* chloracetaldehyde (50 μM). Time of drug addition was at $t = 24$ h. At the end of the test, 0.2% of the detergent Triton X-100 has been added as positive control (Copyright: TUM)

system has to provide steep concentration gradients from the tissue surfaces to the core by convective transport.

Whether microphysiometry can provide an additional building block for an improved pharmaceutical cancer therapy depends on the outcome of the upcoming clinical trials. The ultimate expectation is that patients treated on the basis of a preceding ex-vivo drug test have a significantly longer survival or improved quality of life. Evidently, it will take clinical trials to underpin that hypothesis.

However, there is yet another role for organotypic ex-vivo test models in cancer research aside from a direct clinical application. The complex and dynamic relationship between tumor metabolism and the microenvironment emphasizes the importance of studying metabolic regulation using appropriate in vitro model systems, as well as the need for more sophisticated measurements of cell metabolism and the associated microenvironmental conditions in human tumors [102]. In that context, microphysiometric platforms may become very valuable tools in cancer research, for instance, for a pre-validation of molecular markers or signatures with a postulated predictive meaning or for a preselection of the vast number of potential target-specific single or combination therapies in the run-up to expensive clinical trials [103]. The profiling ability with respect to tumor metabolism renders microphysiometry particularly interesting for those novel agents that are directed against the core energy metabolism in solid tumors, which is recognized as one of the hallmarks of cancer [104].

Acknowledgement The work presented here was in part funded by the German *Bundesministerium für Bildung und Forschung* during the BIOGRAPHY project (BMBF No. 02PN2241). The authors want to thank the *Deutscher Tierschutzbund – Akademie für Tierschutz* (German Animal Welfare Federation – Animal Welfare Academy, Neubiberg, Germany) and the colleagues at *Heinz Nixdorf-Lehrstuhl für Medizinische Elektronik of Technische Universität München*, Munich, Germany. Special thanks to Dr. Regina Kleinhans for the measurements shown in Fig. 8 and to Dr. Frank Alexander Jr. for proofreading.

References

1. Alajoki ML, Bayter GT, Bemiss WR, Blau D, Bousse LJ, Chan SDH, Dawes TD, Hahnenberger KM, Hamilton JM, Lam P, McReynolds RJ, Modlin DN, Owicki C, Parce JW, Redington D, Stevenson K, Wada HG, Williams J (1997) High-performance microphysiometry in drug discovery. In: Devlin JP (ed) High throughput screening – the discovery of bioactive substances. Marcel Dekker, New York, pp 427–442
2. Roberts WS, Lonsdale DJ, Griffiths J, Higson SPJ (2007) Advances in the application of scanning electrochemical microscopy to bioanalytical systems. Biosens Bioelectron 23:301–318
3. Ges IA, Baudenbacher F (2010) Enzyme electrodes to monitor glucose consumption of single cardiac myocytes in sub-nanoliter volumes. Biosens Bioelectron 25:1019–1024
4. Lehmann M, Baumann W, Brischwein M, Gahle H, Freund I, Ehret R, Drechsler S, Palzer H, Kleintges M, Sieben U, Wolf B (2001) Simultaneous measurement of cellular resporation and acidification with a single CMOS ISFET. Biosens Bioelectron 16:195–203
5. Yotter RA, Wilson M (2004) Sensor technologies for monitoring metabolic activity in single cells – part II: nonoptical methods and applications. IEEE Sensors J 4:412–429

6. Wolf B, Brischwein M, Grothe H, Stepper C, Ressler J, Weyh T (2006) Lab-on-a-chip systems for cellular assays. In: Urban G (ed) BioMEMS series: microsystems, vol 16. Springer, Dordrecht, pp 269–308

7. Burgess RM, Ho KT, Brack W, Lamoreex M (2013) Effects-directed analysis (EDA) and toxicity identification evaluation (TIE): complementary but different approaches for diagnosing causes of environmental toxicity. Environ Toxicol Chem 32:1935–1945

8. Diers AR, Vayalil PK, Oliva CR, Griguer CE, Darley-Usmar V, Hurst DR, Welch DR, Landar A (2013) Mitochondrial bioenergetics of metastatic breast cancer cells in response to dynamic changes in oxygen tension: effects of HIF-1a. PLoS One 8:e68348

9. Hafeman DG, Parce JW, McConnell H (1988) Light-addressable potentiometric sensor for biochemical systems. Science 240:1182–1185

10. Owicki JC, Parce JW (1990) Bioassays with a microphysiometer. Nature 344:271–272

11. Liu Q, Wu C, Cai H, Hu N, Zhou J, Wang P (2014) Cell based biosensors and their application in biomedicine. Chem Rev 114:6423–6461

12. Thomas Jr CA, Springer PA, Loeb GE, Berwald-Netter Y, Okun LM (1972) A miniature microelectrode array to monitor the bioelectric activity of cultured cells. Exp Cell Res 74:61–66

13. Giaever I, Keese CR (1984) Monitoring fibroblast behavior in tissue culture with an applied electric filed. PNAS 81:3761–3764

14. Parce JW, Owicki JC, Kercso KM, Sigal GB, Wada HG, Muir VC, Bousse LJ, Ross KL, Sikic BI, McConnell HM (1989) Detection of cell-affecting agents with a silicon biosensor. Science 246:243–246

15. McConnel HM, Owicki JC, Parce JW, Miller DL, Baxter GT, Wada HG, Pitchford S (1992) The cytosensor microphysiometer: biological applications of silicon technology. Science 257:1906–1912

16. Wolf B, Brischwein M, Baumann W, Ehret R, Kraus M (1998) Monitoring of cellular signaling and metabolism with modular sensor-technique: the PhysioControl-Microsystem (PCM®). Biosens Bioelectron 13:501–509

17. Eklund SE, Taylor D, Kozlov E, Prokop A, Cliffel DE (2004) A microphysiometer for simultaneous measurement of changes in extracellular glucose, lactate, oxygen, and acidificaion rate. Anal Chem 76:519–527

18. Weltin A, Slotwinski K, Kieninger J, Moser I, Jobst G, Wego M, Ehret R, Urban G (2014) Cell culture monitoring for drug screening and cancer research: a transparent, microfluidic, multi-sensor microsystem. Lab Chip 14:138–146

19. Demmel F, Brischwein M, Wolf P, Huber F, Pfister C, Wolf B (2015) Nutrient depletion and metabolic profiles in breast carcinoma cell lines measured with a label-free platform. Physiol Meas 36:1367–1381

20. Owicki JC, Parce W (1992) Biosensors based on the energy metabolism of living cells. The physical chemistry and cell biology of extracellular acidification. Biosens Bioelectron 7:255–272

21. Bergveld P (1970) Development of an ion-sensitive solid-state device for neurophysiological measurements. IEEE Trans Biomed Eng 17(1):70–71

22. Bergveld P (2003) Thirty years of isfetology – what happened in the past 30 years and what may happen in the next 30 years. Sens Actuators B Chem 88(1):1–20

23. Owicki JC, Bousse LJ, Hafeman DG, Kirk GL, Olson JD, Wada HG, Parce JW (1994) The light-addressable potentiometric sensor: principles and biological applications. Annu Rev Biophys Biomol Struct 23:87–113

24. Glab S, Hulanicki A, Edwall G, Ingman F (1989) Metal–metal oxide and metal oxide electrodes as pH sensors. Anal Chem 21(1):29–47

25. Qin Y, Kwon HJ, Howlader MMR, Deen MJ (2015) Microfabricated electrochemical pH and free chlorine sensors for water quality monitoring: recent advances and research challenges. RSC Adv 5:69086–69109

26. Nakao M, Inoue S, Yoshinobu T, Iwasaki H (1996) High-resolution pH imaging sensor for microscopic observation of microorganisms. Sens Actuators B Chem 34:234–239

27. Ito Y (1998) High-spatial resolution LAPS. Sens Actuators B Chem 52:107–111

28. Martinoia S, Rosso N, Grattarola M, Lorenzelli L, Margesin B, Zen M (2001) Development of ISFET array-based microsystems for bioelectrochemical measurements of cell populations. Biosens Bioelectron 16:1043–1050
29. Nemeth B, Piechocinski MS, Cumming DRS (2012) High-resolution real-time ion-camera system using a CMOS-based chemical sensor array for proton imaging. Sens Actuators B 171–172:747–752
30. Caroll S, Baldwin RP (2010) Self-calibrating microfabricated iridium-oxide pH electrode array for remote monitoring. Anal Chem 82:878–885
31. Wu C-C, Lin W-C, Fu S-Y (2011) The open container-used microfluidic chip using IrOx ultramicroelectrodes for the in situ measurement of extracellular acidification. Biosens Bioelectron 26:4191–4197
32. Simonis A, Lüth H, Wang J, Schöning MJ (2004) New concepts of miniaturised reference electrodes in silicon technology for potentiometric sensor systems. Sens Actuators B 103:429–435
33. Yang H, Kang SK, Choi CA, Kim H, Shin D-H, Kim YS, Kim YT (2004) An iridium oxide reference electrode for use in microfabricated biosensors and biochips. Lab Chip 4:42–46
34. Liebsch G, Klimant I, Krause C, Wolfbeis OS (2001) Fluorescent imaging of pH with optical sensors using time domain dual lifetime referencing. Anal Chem 73:4354–4363
35. Herst PM, Berridge MV (2007) Cell surface oxygen consumption: a major contributor to cellular oxygen consumption in glycolytic cancer cell lines. Biochim Biophys Acta 1767:170–177
36. Wagner BA, Venkataraman S, Buettner GR (2011) The rate of oxygen utilization by cells. Free Radic Biol Med 51:700–712
37. Clark JR, Leland C (1959) Electrochemical device for chemical analysis. Patent US2913386A
38. McDonagh C, Kolle C, McEvoy AK (2001) Phase fluorometric dissolved oxygen sensor. Sens Actuators B 74:124–130
39. Pinsent BRW, Pearson L, Roughton FJW (1956) The kinetics of combination of carbon dioxide with hydroxide ions. Trans Faraday Soc 52:1512–1521
40. Burke CS, Markey A, Nooney RI, Byrne P, McDonagh C (2006) Development of an optical sensor probe for the detection of dissolved carbon dioxide. Sens Actuators B 119:288–294
41. Zilberman Y, Ameri SK, Sonkusale SR (2014) Microfluidic optoelectronic sensor based on a composite halochromic material for dissolved carbon dioxide detection. Sens Actuators B 194:404–409
42. Toghill KE, Compton RG (2010) Electrochemical non-enzymatic glucose sensors: a perspective and an evaluation. Int J Electrochem Sci 5:1246–1301
43. Mross S, Zimmermann T, Winkin N, Kraft M, Vogt H (2015) Integrated multi-sensor system for parallel in-situ monitoring of cell nutrients, metabolites and cell mass in biotechnological processes. Proc Eng 120:372–375
44. Eklund SE, Thompson RG, Snider RM, Carney CK, Wright DW, Wikswo JP, Cliffel DE (2009) Metabolic discrimination of select list agents by monitoring cellular responses in a multianalyte microphysiometer. Sensors 9:2117–2133
45. Moser I, Jobst G, Urban GA (2002) Biosensor arrays for simultaneous measurement of glucose, lactate, glutamate, and glutamine. Biosens Bioelectron 17:297–302
46. Donner JS, Thompson SA, Kreuzer MP, Baffou G, Quidant R (2012) Mapping intracellular temperature using green fluorescent temperature. Nano Lett 12:2107–2111
47. Baffou G (2014) A critique of methods for temperature imaging in single cells. Nat Methods 11:899–901
48. Lee W, Fon W, Axelrod BW, Roukes ML (2009) High sensitivity microfluidic calorimeters for biological and chemical applications. PNAS 106:15225–15230
49. Chancellor EP, Wikswo JP, Baudenbacher F, Radparvar M, Osterman D (2004) Heat conduction calorimeter for massively parallel high throughput measurements with picoliter sample volumes. Appl Phys Lett 85:2408–2410

50. Lerchner J, Wolf A, Buchholz F, Mertens F, Neu TR, Harms H, Maskow T (2008) Miniaturized calorimetry – a new method for real-time biofilm activity analysis. J Microbiol Methods 74:74–81
51. Inomata N, Toda M, Sato M, Ishijima A, Ono T (2012) Pico calorimeter for detection of heat produced in an individual brown fat cell. Appl Phys Lett 100:154104
52. Pettersen EO, Ebbesen P, Gieling RG, Williams KJ, Dubois L, Lambin P, Ward C, Meehan J, Kunkler IH, Langdon SP, Ree AH, Flatmark K, Lyng H, Calzada MJ, Peso LD, Landazuri MO, Görlach A, Flamm H, Kieninger J, Urban G, Weltin A, Singleton DC, Haider S, Buffa FM, Harris AL, Scozzafava A, Supuran CT, Moser I, Jobst G, Busk M, Toustrup K, Overgaard J, Alsner J, Pouyssegur J, Chiche J, Mazure N, Marchiq I, Parks S, Ahmed A, Ashcroft M, Pastorekova S, Cao Y, Rouschop KM, Wouters BG, Koritzinsky M, Mujcic H, Cojocari D (2015) Targeting tumour hypoxia to prevent cancer metastasis. From biology, biosensing and technology to drug development: the METOXIA consortium. J Enzyme Inhib Med Chem 30:689–721
53. Parrinello S, Samper E, Krtolica A, Goldstein J, Melov S, Campisi J (2003) Oxygen sensitivity severely limits the replicative lifespan of murine fibroblasts. Nat Cell Biol 5:741–747
54. Wolf P, Brischwein M, Kleinhans R, Demmel F, Schwarzenberger T, Pfister C, Wolf B (2013) Automated platform for sensor-based monitoring and controlled assays of living cells and tissues. Biosens Bioelectron 50:111–117
55. Thedinga E, Ullrich A, Drechsler S, Niendorf R, Kob A, Runge D, Keuer A, Freund I, Lehmann M, Ehret R (2007) In vitro system for the prediction of hepatoxic effects in primary hepatocytes. ALTEX 24:22–34
56. Ferrick DA, Neilson A, Beeson C (2008) Advances in measuring cellular bioenergetics using extracellular flux. Drug Discov Today 13:269–274
57. Pfister C, Bozsak C, Wolf P, Demmel F, Brischwein M (2015) Cell shape-dependent shear stress on adherent cells in a micro-physiologic system as revealed by FEM. Physiol Meas 36:955–966
58. Hoch E, Hirth T, Tovar GEM, Borchers K (2013) Chemical tailoring of gelatin to adjust its chemical and physical properties for functional bioprinting. J Mater Chem B 1:5675–5685
59. Wiest J, Stadthagen T, Schmidhuber M, Brischwein M, Ressler J, Raeder U, Grothe H, Melzer A, Wolf B (2006) Intelligent mobile lab for metabolics in environmental monitoring. Anal Lett 39(8):1759–1771
60. Hakanson M, Cukierman E, Charnley M (2014) Miniaturized pre-clinical cancer models as research and diagnostic tools. Adv Drug Deliv Rev 69–70:52–66
61. Baker M (2009) Stem cells: fast and furious. Nature 458:962–965
62. Abbott A (2003) Cell culture: biology's new dimension. Nature 424:870–872
63. Griffith LG, Swartz MA (2006) Capturing complex 3D tissue physiology in vitro. Nat Rev Mol Cell Biol 7:211–224
64. Meads M, Gatenby R, Dalton W (2009) Environment-mediated drug resistance: a major contribution to minimal residual disease. Nat Rev Cancer 9:665–674
65. van der Valk J, Bieback K, Buta C, Cochrane B, Dirks WG, Fu J, Hickman JJ, Hohensee C, Kolar R, Liebsch M, Pistolla F, Schulz M, Thieme D, Weber T, Wiest J, Winkler S, Gstraunthaler G (2017) Fetal bovine serum (FBS): past – present – future. ALTEX. https://doi.org/10.14573/altex.1705101
66. Huh D, Matthews BD, Mammoto A, Montoya-Zavala M, Yuan Hsin H, Ingber DE (2010) Reconstituting organ-level function on a chip. Science 328:1662–1668
67. Ravi M, Paramesh V, Kaviya SR, Anuradha E, Paul Solomon FD (2015) 3D cell culture systems: advantages and applications. J Cell Physiol 230:16–26
68. Kleinhans R, Brischwein M, Wang P, Becker B, Demmel F, Schwarzenberger T, Zottmann M, Wolf P, Niendorf A, Wolf B (2012) Sensor-based cell and tissue screening for personalized cancer chemotherapy. Med Biol Eng Comput 50:117–126
69. Bugge A, Dib L, Collin S (2014) Measuring respiratory activity of adipocytes and adipose tissues in real time. Methods Enzymol 538:233–247

70. Alexander FA, Eggert S, Wiest J (2017) A novel lab-on-a-chip platform for spheroid metabolism monitoring. Cytotechnology. https://doi.org/10.1007/s10616-017-0152-x (Epub ahead of print)

71. Hirschhaeuser F, Menne H, Dittfeld C, West J, Mueller-Klieser W, Kunz-Schughart LA (2010) Multicellular tumor spheroids: an underestimated tool is catching up again. J Biotechnol 148:3–15

72. Wenzel C, Riefke B, Gründemann S, Krebs A, Christian S, Prinz F, Osterland M, Golfier S, Räse S, Ansari N, Esner M, Bickle M, Pampaloni F, Mattheyer C, Stelzer EH, Parczyk K, Prechtl S, Steigemann S (2014) 3D high-content screening for the identification of compounds that target cells in dormant tumor spheroid regions. Exp Cell Res 323:131–143

73. Kondo J, Endo H, Okuyama H, Ishikawa O, Iishi H, Tsujii M, Ohue M, Inoue M (2011) Retaining cell–cell contact enables preparation and culture of spheroids composed of pure primary cancer cells from colorectal cancer. PNAS 108:6234–6240

74. Vaira V, Fedele G, Pyne S, Fasoli E, Zadra G, Bailey D, Snyder E, Faversani A, Coggi G, Flavin R, Bosari S, Loda M (2010) Preclinical model of organotypic culture for pharmacodynamic profiling of human tumors. PNAS 107(18):8352–8356

75. Metzger R, Deglmann CJ, Hoerrlein S, Zapf S, Hilfrich J (2001) Towards in-vitro prediction of an in-vivo cytostatic response of human tumor cells with a fast chemosensitivity assay. Toxicology 166:97–108

76. Voiculescu I, Li F, Liu F, Zhang X, Cancel LM, Tarbell JM, Khademhosseini A (2013) Study of long-term viability of endothelial cells for lab-on-a-chip devices. Sens Actuators B 182:696–705

77. Pancrazio JJ, Gray SA, Shubin YS, Kulagina N, Cuttino DS, Shaffer KM, Eisemann K, Curran A, Zim B, Gross GW, O'Shaughnessy TJ (2003) A portable microelectrode array recording system incorporating cultured neuronal networks for neurotoxin detection. Biosens Bioelectron 18:1339–1347

78. Wiest J, Namias A, Pfister C, Wolf P, Demmel F, Brischwein M (2016) Data processing in cellular microphysiometry. IEEE Trans Biomed Eng 63(11):2368–2375. https://doi.org/10.1109/TBME.2016.2533868

79. Grundl D, Zhang X, Messaoud S, Pfister C, Demmel F, Mommer MS, Wolf B, Brischwein M (2013) Reaction-diffusion modelling for microphysiometry on cellular specimens. Med Biol Eng Comput 51:387–395

80. Pfister C, Forstmeier C, Biedermann J, Schermuly J, Demmel F, Wolf P, Kaspers B, Brischwein M (2016) Estimation of dynamic metabolic activity in micro-tissue cultures from sensor recordings with an FEM model. Med Biol Eng Comput 54:763–772

81. Wilson DF, Erecifiska M, Drown C, Silvers IA (1979) The oxygen dependence of cellular energy metabolism. Arch Biochem Biophys 195:485–493

82. Koh MY, Powis G (2012) Passing the baton: the HIF-switch. Trends Biochem Sci 37:364–372

83. Vaupel P, Mayer A (2012) Availability, not respiratory capacity governs oxygen consumption of solid tumors. Int J Biochem Cell Biol 44:1477–1481

84. Otto AM, Hintermair J, Janzon C (2015) NADH-linked metabolic plasticity of MCF-7 breast cancer cells surviving in a nutrient-deprived microenvironment. J Cell Biochem 116:822–835

85. Gerencser AA, Neilson A, Choi SW, Edman U, Yadava N, Oh RJ, Ferrick DA, Nicholls DG, Brand MD (2009) Quantitative microplate-based respirometry with correction for oxygen diffusion. Anal Chem 81:6868–6878

86. Brantom PG, Bruner LH, Chamberlain M, De Silva O, Dupuis J, Earl LK, Lovell DP, Pape WJ, Uttley M, Bagley DM, Baker FW, Bracher M, Courtellemont P, Declercq L, Freeman S, Steiling W, Walker AP, Carr GJ, Dami N, Thomas G, Harbell J, Jones PA, Pfannenbecker U, Southee JA, Tcheng M, Argembeaux H, Castelli D, Clothier R, Esdaile DJ, Itigaki H, Jung K, Kasai Y, Kojima H, Kristen U, Larnicol M, Lewis RW, Marenus K, Moreno O, Peterson A, Rasmussen ES, Robles C, Stern M (1997) A summary report of the COLIPA international validation study on alternatives to the draize rabbit eye irritation test. Toxicol In Vitro 11:141–179

87. Balls M, Botham PA, Bruner LH, Spielmann H (1995) The EC/HO international validation study on alternatives to the draize eye irritation test. Toxicol In Vitro 9:871–929

88. Gettings SD, Lordo RA, Hintze KL, Bagley DM, Casterton PL, Chudkowski M, Curren RD, Demetrulias JL, Dipasquale LC, Earl LK, Feder PI, Galli CL, Glaza SM, Gordon VC, Janus J, Kurtz PJ, Marenus KD, Moral J, Pape WJ, Renskers KJ, Rheins LA, Roddy MT, Rozen MG, Tedeschi JP, Zyracki J (1996) The CFTA evaluation of alternatives program: an evaluation of in vitro alternatives to the draize primary eye irritation test. (Phase III) Surfactant-based formulations. Food Chem Toxicol 34:79–117

89. Hartung T, Bruner L, Curren R, Eskes C, Goldberg A, McNamee P, Scott L, Zuang V (2010) First alternative method validated by a retrospective weight-of-evidence approach to replace the draize eye test for the identification of non-irritant substances for a defined applicability domain. ALTEX 27:43–51

90. Kamalian L, Chadwick AE, Bayliss M, French NS, Monshouwer M, Snoeys J, Park BK (2015) The utility of HepG2 cells to identify direct mitochondrial dysfunction in the absence of cell death. Toxicol In Vitro 29:732–740

91. Maschmeyer I, Lorenz AK, Schimek K, Hasenberg T, Ramme AP, Hübner J, Lindner M, Drewell C, Bauer S, Thomas A, Sambo NS, Sonntag F, Lauster R, Marx U (2015) A four-organ-chip for interconnected long-term co-culture of human intestine, liver, skin and kidney equivalents. Lab Chip 15:2688–2699

92. Wikswo JP, Curtis EL, Eagleton ZE, Evans BC, Kole A, Hofmeister LH, Matloff WJ (2013) Scaling and systems biology for integrating multiple organs-on-a-chip. Lab Chip 13:3496–3511

93. Marx U, Andersson TB, Bahinski A, Beilmann M, Beken S, Cassee FR, Cirit M, Daneshian M, Fitzpatrick S, Frey O, Gaertner C, Giese C, Griffith L, Hartung T, Heringa MB, Hoeng J, de Jong WH, Kojima H, Duehnl J, Luch A, Maschmeyer I, Sakharov D, Sips AJAM, Steger-Hartmann T, Tagle DA, Tonevitsky A, Tralau T, Tsyb S, van de Stolpe A, Vandebriel R, Vulto P, Wang J, Wiest J, Rodenburg M, Roth A (2016) Biology-inspired microphysiological system approaches to solve the prediction dilemma of substance testing. ALTEX 33(3):272–321. https://doi.org/10.14573/altex.1603161

94. Halfter K, Ditsch N, Kolberg HC, Fischer H, Hauzenberger T, Edler von Koch F, Bauerfeind I, von Minckwitz G, Funke I, Crispin A (2015) Prospective cohort study using the breast cancer spheroid model as a predictor for response to neoadjuvant therapy – the SpheroNEO study. BMC Cancer 15:519

95. Majewski IJ, Bernards R (2011) Taming the dragon: genomic biomarkers to individualize the treatment of cancer. Nat Med 17:304–312

96. O'Brien CP, Taylor SE, O'Leary JJ, Finn SP (2014) Molecular testing in oncology: problems, pitfalls and progress. Lung Cancer 83:309–315

97. Lyng H, Haraldseth O, Rofstad EK (2000) Measurement of cell density and necrotic fraction in human melanoma xenografts by diffusion weighted magnetic resonance imaging. Magn Reson Med 43:828–836

98. Alunni-Fabbroni M, Sandri MT (2010) Circulating tumour cells in clinical practice: methods of detection and possible characterization. Methods 50:289–297

99. Ona T, Shibata J (2010) Advanced dynamic monitoring of cellular status using label-free and non-invasive cell-based sensing technology for the prediction of anticancer drug efficacy. Anal Bioanal Chem 398:2505–2533

100. Henning T, Brischwein M, Baumann W, Ehret R, Freund I, Kammerer R, Lehmann M, Schwinde A, Wolf B (2001) Approach to a multiparametric sensor-chip-based tumor chemosensitivity assay. Anti-Cancer Drugs 12:21–32

101. Mestres P, Morguet A, Schmidt W, Kob A, Thedinga E (2006) A new method to assess drug sensitivity on breast tumor acute slices preparation. Ann N Y Acad Sci 1091:460–469

102. Cairns RA, Harris IS, Mak T (2011) Regulation of cancer cell metabolism. Nat Rev 11:85–95

103. Weigelt B (2008) Unraveling the microenvironmental influences on the normal mammary gland and breast cancer. Semin Cancer Biol 18:311–321

104. Hanahan D, Weinberg RA (2011) Hallmarks of cancer: the next generation. Cell 144:646–674

BIOREV (2019) 2: 189–218
https://doi.org/10.1007/11663_2018_3
© Springer International Publishing AG, part of Springer Nature 2018
Published online: 28 April 2018

Optical Waveguide-Based Cellular Assays

Y. Fang

Contents

Abstract Optical waveguides have been widely used to develop a range of biosensing platforms. Among them, resonant waveguide grating (RWG) has found broad applications in monitoring cell phenotypic responses in native cells mostly due to its adoption to microtiter plates, the de facto footprint for drug discovery. This chapter first reviews RWG biosensor configurations and reader systems, followed by a discussion about how to apply them to facilitate drug discovery, elucidate receptor biology, and perform single-cell analysis.

Y. Fang (✉)
Biochemical Technologies, Science and Technology Division, Corning Incorporated, Corning, NY, USA
e-mail: fangy2@corning.com

Keywords Biosensor · Cell phenotype · Drug discovery · Drug residence time · High-throughput screening · Ligand-directed functional selectivity · Optical waveguide · Polypharmacology · Receptor pharmacology · Resonant waveguide grating · Single-cell analysis

1 Introduction

Optical waveguides use an inhomogeneous structure to guide light so that it propagates within the high refractive index waveguide by total internal reflection, which, in turn, generates an electromagnetic field (i.e., evanescent wave) extending into the surrounding low refractive index substrate or medium for biosensing. The evanescent wave is surface bound since its amplitude exponentially decays over the distance from the surface. In the past decades, a wide range of biosensing platforms have been developed based on optical waveguides, such as fiber optic [1, 2], optical grating coupler or optical waveguide lightmode spectroscopy (OWLS) [3, 4], resonant mirror [5], diffractive grating [6, 7], reverse symmetry waveguide [8], and metal clad waveguide [9]. Most of these biosensors generally have a short penetration depth (~200 nm) for biosensing, except for metal clad and reverse symmetry waveguides [10]. Similar to surface plasmon resonance (SPR), almost all waveguide-based biosensors were originally developed for and have found applications in biomolecular interaction analysis [10].

Cell-based assays are the workhorse for drug discovery. The past decade has witnessed increasing applications of waveguide-based biosensors to perform such assays. These biosensors were initially used to study several cellular processes associated with cell-substrate interactions including cell adhesion [11–14], proliferation and metabolic activity [15], and cell death [16]. Since the development of resonant waveguide grating (RWG) a decade ago, it has come to the realization that these biosensors also offer an unprecedented approach for studying receptor pharmacology [17–21]. In the recent years, RWG has become a leading label-free optical platform in drug discovery, mostly due to its ability for drug screening with high throughput based on both direct binding and cell phenotypic responses [22–30].

This paper first provides an overview of typical RWG biosensor configurations and reader systems, followed by a discussion about cell phenotypes examined and applications found in early drug discovery, as well as receptor biology and single-cell analysis.

2 RWG Biosensors

RWG belongs to a family of optical biosensors that uses a diffractive nanograting structure to couple light into its high refractive index waveguide and to provide a leaky mode such that an evanescent wave extending ~200 nm into the medium or cells [20]. RWG is the first label-free optical biosensor to be adopted in microtiter

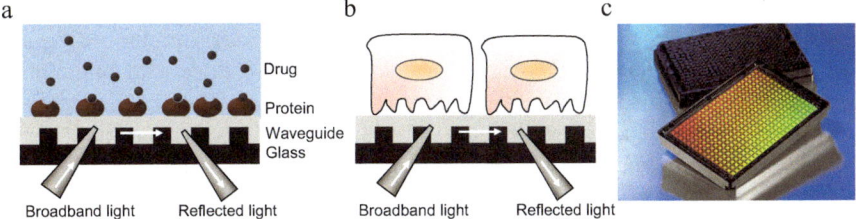

Fig. 1 Resonant waveguide grating (RWG) for biosensing. (**a**) Affinity-based biochemical assay. (**b**) Functional cell-based assays. For biochemical assays, RWG measures the wavelength shift in the reflected light arising from the binding of small molecules to the proteins immobilized on the sensor surface. For cell-based assay, RWG measures the wavelength shift in the reflected light arising from the drug-induced dynamic mass redistribution (DMR) within the cells. (**c**) A colored light image of a typical 384-well Epic® plate from Corning, which was obtained by illuminating the plate with visible light under an angle

plates (e.g., 96-, 384-, and 1536-well), enabling high-throughput screening (HTS) [31–34]. Since microplates are the de facto footprint for drug profiling and screening, RWG can be easily integrated into various drug discovery processes.

Common to Epic® biosensor microplates from Corning is that there is an RWG biosensor located within the bottom of each well. The biosensor typically has a three-layer structure, that is, a glass substrate with a refractive index of 1.50, a polymeric nanograting structure with a period of 500 nm, and a waveguide thin film with a high refractive index (e.g., Nb_2O_5 with a thickness of 150 nm and a refractive index of 2.36) (Fig. 1) [20]. For biochemical assays, each well contains two sensors, one coated with a polymer hydrogel designed to facilitate immobilization of the protein target with minimal non-specific background binding for the sample and another coated with another hydrogel to prevent non-specific binding for intra-well referencing. For cell-based assays, there is a single biosensor per well, and the biosensor microplates are cell culture treated or coated with an extracellular matrix (ECM) protein.

Commercial RWG sensors generally have a low quality factor (~500) [35]. This is due to practical considerations for screening, where sensitivity and reproducibility are equally important. Of note, many new-generation optical biosensors such as whispering gallery resonators have much higher quality factor ($>10^8$), thus permitting single-molecule sensitivity [36, 37].

3 RWG Readers

Critical to the adoption of RWG in basic research and drug discovery is to develop reader systems tailored toward specific applications. In the past years, several RWG readers have been developed and commercialized. These readers differ greatly in operational details, dimensions, temporal and spatial resolution, and applications, but all share the basic principle. That is, the resonance light is coupled into, propagates within, and eventually leaks out and reflects back from the waveguide (Fig. 1). All commercial RWG readers can achieve the resonance with a nominally normal incident angle (i.e., illumination perpendicular to senor surface) and monitor it with a reflection interrogation, both of which are critical to screening using biosensor microplates [38, 39]. In comparison, SPR requires a large incident angle and has high angular sensitivity [40], while OWLS measures the angle at which incident light is guided with constructive interference to the photodiode detectors at the end of the waveguide [15, 41].

3.1 Angular Interrogation-Based Imaging Systems

One of the early research RWG instruments was based on angular interrogation scheme, where the resonance was observed at a critical angle of incidence at the sensor [19, 20]. This system splits a light source into an array of 7×7 light beams to illuminate simultaneously 7×7 biosensors within a 96-well microplate (Fig. 2a). Each incident light has a dimension of $200 \times 3,000$ μm when illuminating the

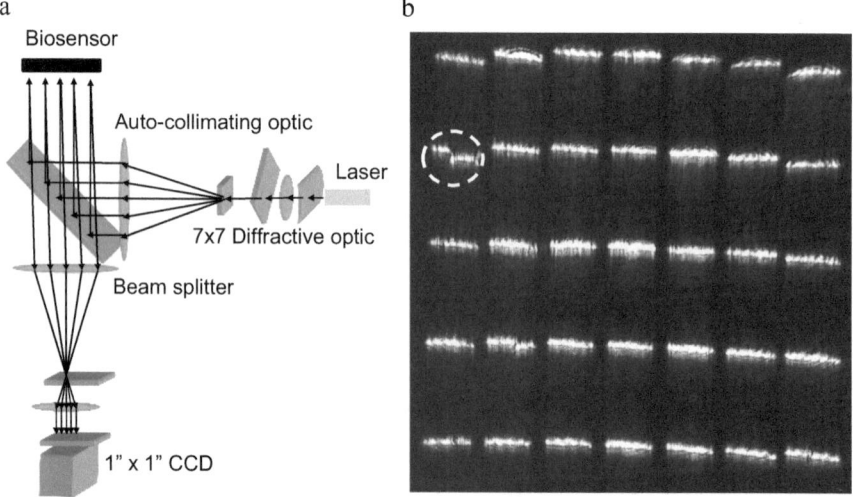

Fig. 2 Angular interrogation-based RWG imaging system. (**a**) Schematic drawing showing the optical setup, where a laser light is split into an array of 7×7 light beams, such that each of them illuminates a biosensor located within one well of a 96-well Epic® microplate. The reflected lights are received with a CCD camera as an image of 7×7 resonant bands. (**b**) An actual resonant band image of 5×7 sensors in a 96-well Epic® microplate having A431 cells. The readout parameter is a change in pixel of the central position of the resonant band as imaged by the CCD camera that scales with a shift in wavelength of the reflected light. The broken circle indicated that the cell density was not even across this specific sensor, which was confirmed by light microscopic imaging. Figure was reproduced with permission from Fang et al. [19]

sensor, and the resonance image is collected using a charge-coupled device (CCD) camera every 3 s (Fig. 2b). However, this system is difficult to scale up to the whole plate with different densities.

3.2 Wavelength Interrogation-Based Scanning Systems

The two first-generation commercial RWG readers, Epic® from Corning and BIND® from SRU Biosystems, employ a wavelength interrogation scheme to achieve resonance. Both use an array of integrated fiber optics to illuminate all sensors within a single column or row of the microplate and a lateral motion stage to scan the whole microplate across the fiber optics (Fig. 3). Both systems record a real-time, averaged shift in resonance wavelength in picometer (pm) for each sensor. There are, however, clear differences between the two systems. Epic® is a stand-alone and HTS-compatible system consisting of a reader with a temperature-controlled unit, an internal liquid handling unit, and robotics for processing plates [31]. Furthermore, Epic® uses 16 fiber optic heads for illumination with incident and collection of reflected light, scans a plate every 6 s, and reports an averaged signal every two scans. Epic® is scalable to 96-, 384-, and 1536-well microplates. On the other hand, BIND® is a batch top reader without robotics and temperature control and requires an external liquid handling system for sample delivery [42]. In addition, BIND® uses a linear array of eight illumination/detection fiber optic heads to scan the plate with a temporal resolution of ~15 s.

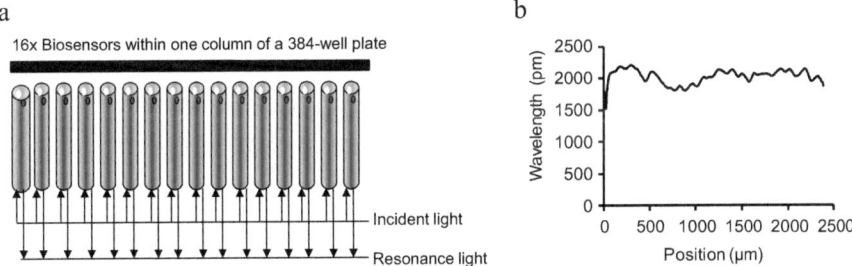

Fig. 3 Wavelength interrogation-based scanning RWG system. (**a**) Schematic drawing showing the main component of the system – a linear array of 16 illumination/detection heads is used to scan a 384-well microplate, one column at a time. In each of the 16 heads, there is an optical fiber to send a laser light to illuminate a biosensor located in each well and a second optical fiber to receive the reflected light. (**b**) The resonance wavelengths along a single line (100 μm in width) across a single sensor, scanning in steps of 4 μm

3.3 Whole-Plate Imaging Systems

There are several RWG readers for whole-plate imaging. The BIND® scanner developed by SRU Biosystems uses a narrow entrance slit to direct a broadband light-emitting diode (LED) light for free-space illumination of the biosensor plate and a CCD camera to collect the reflected light. The CCD rapidly acquires a spatial resonance map with a temporal resolution up to 30 s and a spatial resolution up to 150×150 μm by scanning the plate line by line [43]. To construct a two-dimensional (2D) resonance image, a scanning stage translates the biosensor across the illumination line in small spatial increments. Epic® can also be operated in a scanning mode to collect an endpoint image of the microplate after binding or cell responses [44]. However, both approaches measure endpoint results [44]. In addition, both systems have drawbacks associated with line scanning in principle, which results in time domain difference among different areas of a sensor or different sensors, thus introducing artifacts when reading the kinetics of rapid cellular responses.

To improve temporal and spatial resolution, a swept wavelength interrogation-based whole-plate imager was developed recently [45]. This imager uses a light beam, together with a high-resolution filter, to sweep wavelengths from 825 to 840 nm in a stepwise fashion (100 pm every 20 ms) for illuminating all biosensors of a microplate. A high-speed complementary metal-oxide semiconductor (CMOS) digital camera is used to record the reflected light (Fig. 4a). For each sweeping cycle, 150 images (one per wavelength) are acquired and processed to generate a resonance image of the whole plate (Fig. 4b). This system has a spatial resolution of 90×90 μm and a temporal resolution of 3 s. This imager can be placed inside typical cell culture incubator owing to its small footprint, thus permitting cell assays under physiological conditions.

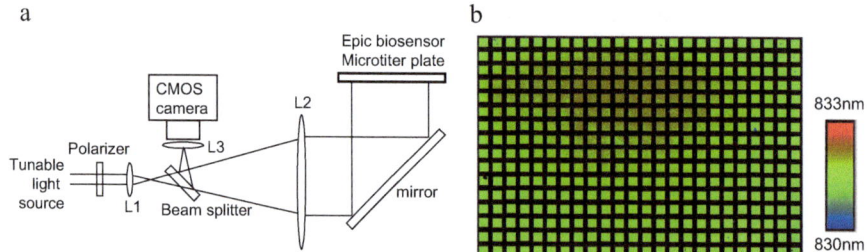

Fig. 4 A swept wavelength interrogation-based whole-plate RWG imager. (**a**) Schematic drawing of the imager, which consists of three components: a tunable light source to sweep the wavelength range, an array of optical components to guide and expand the light paths illuminating at a normal incident angle all biosensors within a plate, and a CMOS camera to record the resonant wavelength image. (**b**) A false-colored resonant wavelength image of a fibronectin-coated 384-well Epic® biosensor plate having confluent HEK293 cells. The color scale bar indicates the resonance wavelength

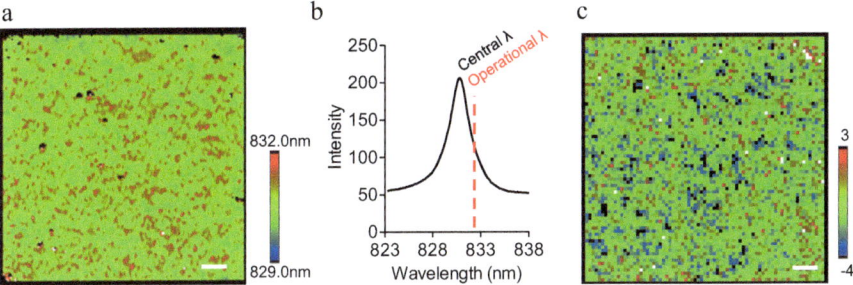

Fig. 5 High-frequency RWG imager for monitoring rapid cell responses. (**a**) A resonance wavelength image of a biosensor having human iPS-derived cardiomyocytes after 11 days of culture at 37°C. (**b**) The operational light is set to a wavelength at the right waist of the resonance wavelength distribution of all pixels of the biosensor. (**c**) The intensity modulation image of the same biosensor with a colored scale showing the net change of the intensity (in percent) at a specific time after the initial intensity of each pixel is normalized to 100%. Scale bar: 200 μm (**a**, **c**). Color scale bars indicate the resonance wavelength (**a**) or intensity change after normalization (**b**). Figure was reproduced with permission from Ferrie et al. [51]

3.4   High-Resolution Imaging Systems

Three different techniques have been used to achieve high-resolution label-free imaging using RWG. The first approach uses a transmission acquisition mode based on upright microscopy [46]. Here, an angle-tunable laser is used for wide-field illumination with a fixed wavelength, and transmission efficiency is measured as a function of incident angle using an electron-multiplying CCD camera. Besides label-free imaging, this setup also enables evanescent wave-enhanced fluorescence imaging after sample drying [47, 48].

The second approach uses reflection acquisition mode with an inverted microscope [49]. Here, the light from an LED source is first converted from a circular spot to a more concentrated linear beam, which is then used to illuminate the biosensor. The resonance light, once reflected, is collected using the microscope objective. This imager enables assays with live cells.

The third approach uses the swept wavelength interrogation-based imager (cf. Sect. 3.3) to focus on a 3 × 4 biosensor array within a 384-well microplate, enabling label-free imaging with a spatial solution of 12 μm and a temporal resolution of 3 s [50]. Although the resonance light has a relatively long propagation length within the waveguide, this imager enables single-cell imaging (Fig. 5a).

3.5   High-Frequency Intensity-Based Imaging System

A high-frequency RWG imager was recently developed to measure ultrafast cell responses [51]. Here, a swept wavelength interrogation-based high-resolution RWG imager is first used to determine the resonance wavelength at each location in the wavelength-imaging mode (Fig. 5a). A central wavelength value is then determined by averaging resonance peaks of all pixels, so an operating light source can be set to the right waist point of the resonance spectrum, typically 400 pm higher than the central wavelength (Fig. 5b). The intensity modulated image is finally monitored with high frequency (up to 60 Hz) and high spatial resolution (Fig. 5c). This imager allows automatic switching between wavelength- and intensity-imaging modes. The wavelength-imaging mode is useful for measuring typical dynamic mass redistribution (DMR) signals of cells upon stimulation, whereas the intensity-imaging mode enables measuring rapid cellular responses.

3.6 Microfluidic Biosensor Systems

Microfluidics offer a robust means for precise control of cell phenotype, compound concentration gradients, and drug-target interactions. Integration of RWG with microfluidics can provide additional dimensions for investigating the cellular and molecular mechanisms of receptor signaling and drug action [52, 53]. Several microfluidic RWG devices have been reported. The first one uses a central *common well* in each row of a 96-well plate as an access point for introduction or withdrawal of reagents via flow channels [54] that make individual fluidic contact to the other *11 analyte* wells in the same row (5 to the left and 6 to the right of the central *common well* in a given row). The flow channels describe serpentine patterns, providing an equal length of the flow path from each *analyte well* to the *common well* in the same row. A pneumatic fluid driving force is used to deliver test solutions via the *common well* or to remove solutions from the *analyte wells*. All 11 flow paths run in parallel close to *common well* such that they are simultaneously monitored by RWG line imaging. This setup has been used for analysis of the interactions of up to 8×11 receptor-ligand interactions. However, this device has not been used for cell-based assays.

The second device uses a polydimethylsiloxane (PDMS)-based microfluidic chamber [55, 56]. PDMS is a widely used material for fabricating microfluidic devices in general laboratories. Here, a sensor array having a footprint of a 384-well microplate is microfabricated onto a glass substrate, and each sensor has dimensions of 2×2 mm. The microfluidic device is made of PDMS using soft lithography and contains a 4×4 array of chambers, each having three inlets and one outlet (Fig. 6a). The PDMS microfluidic structure is attached to the sensor array substrate, so each chamber covers one sensor in the bottom. The flow is generated using syringe pumps connected to the inlets.

The third device has a footprint of conventional 96-well microplates and consists of 32 (Fig. 6b) or 96 individually addressable microchambers, each containing two inlets and one outlet [57]. The two inlets are located at one side of a sensor, opposite to the outlet. Such a T-shaped design ensures that no mixing occurs when two solutions flow side by side, enabling intra-sensor referencing. The flow is obtained

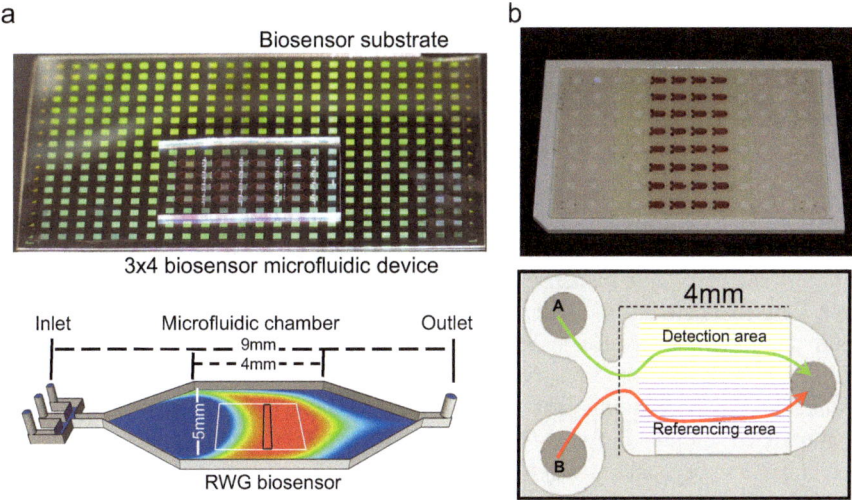

Fig. 6 Two microfluidic RWG devices. (**a**) A PDMS replica-based RWG device, where a PDMS replica containing 4×4 microchambers is attached to a biosensor substrate, so each microchamber has one sensor within its bottom substrate, three inlets, and one outlet. (**a**, bottom) Schematic of the biosensor microchamber, wherein an agonist solution, 2 μl in total, is perfused at a flow rate of 1 μl/min between two perfusion periods with the assay buffer, thus creating 2 min pulse stimulation to the cells located within the detection area (black box) of a biosensor (white box). Reproduced from Goral et al. [56] through the Creative Commons Attribution License. (**b**) A bottom view photo of a microinjection molded RWG microfluidic device with a biosensor plate containing a 4×8 functional biosensor array. A colored buffer solution was introduced to illustrate the microchamber geometry. (**b**, bottom) A schematic of side-by-side perfusion scheme using this T-shaped microchamber. Reproduced from Zaytseva et al. [57] with permission

using a fluidic head consisting of 96 pins connected to electronically controlled pumps. This system enables simultaneous measurements of cellular responses from 32 sensors using the whole-plate imager.

4 Optical Waveguide-Based Cell Assays

Optical waveguide-based cellular assays permit studying a wide range of cell phenotypes, thereby enabling investigation of drug action on different cell phenotypes. These phenotypes include cell adhesion, proliferation, cell death, differentiation, migration, invasion, pathogen and viral infection, receptor signaling, cell-to-cell communication, and cardiomyocyte beating (reviewed in [28, 29, 58]). Furthermore, these biosensors offer an unprecedented flexibility in assay formats to investigate the biochemical, molecular, and cellular mechanisms of action (MOA) of drug molecules in almost all cell types including immortalized, stem, and primary cells, regardless of being anchorage-dependent or not [29].

4.1 Cell Phenotype profiling

Cell adhesion to a surface is a dynamic multistep process, leading to robust biosensor signals that are sensitive to cell types, density, surface chemistry, and drug molecules [11–13, 20]. Orgovan et al. used RWG to determine the dependence of HeLa cell spreading kinetics on the average surface density of pre-adsorbed Arg-Gly-Asp (RGD) motifs that integrins bind to [13]. Results showed that the spreading rate constant was independent of the density of surface-immobilized RGD peptides. In contrast, the maximum biosensor response, a direct function of the effective contact area between cell and surface, increased with increasing RGD surface density until saturation at high densities. A two-dimensional dissociation constant of $(1,753 \pm 243)$ μm^{-2} (corresponding to a dissociation constant of ~ 30 μM) was estimated for the binding between cell surface integrins and the RGD peptides immobilized on the biosensor surface [13]. Besides monitoring cell adhesion, RWG can also be used to study the remodeling of cell adhesion arising from receptor signaling [19, 59] and the impact of surface coating on receptor signaling [60]. For instance, the activation of epidermal growth factor (EGF) receptor was found to induce cell de-adhesion [19]; in contrast, the activation of β_2-adrenergic receptor (β_2-AR) increased cell adhesion [59]. The ECM coating was found to have significant impact on the biosensor signature, potency, and efficacy of seven purinergic P2Y receptor agonists in human embryonic kidney HEK293 cells [60].

Cell proliferation is known to increase cell confluency, generally resulting in an increased biosensor signal. Hug et al. used OWLS to monitor the proliferation of two fibroblast and one hepatoma cell lines [15]. Results showed that OWLS detected multiple phases within the entire life cycle of cells ranging from adhesion to spreading to proliferation with fine resolution. Notably, they found that the exponentially growing hepatoma cells resulted in a linear increase of the sensor signal, probably due to the exponentially decreased cell contact area.

Cell death is often accompanied with a loss of biomass and/or cell detachment, leading to a decreased biosensor signal. RWG not only can assess compound-induced cell toxicity but also detect apoptotic cells. Apoptotic recognition is linked to the innate apoptotic immunity involving anti-inflammatory and immunosuppressive responses. Pattabiraman et al. applied a photonic crystal biosensor to detect the recognition of apoptotic cells by viable adherent responder cells [61]. Results showed that the biosensor detected early recognition-specific events in responder cells, and the innate apoptotic cell recognition was independent on species and involved cytoskeletal remodeling but not phagocytosis.

Cell differentiation rewires signaling networks and circuits, resulting in distinct changes of cell phenotype and functions. RWG not only can track the entire differentiation process but also quantitatively measure the functions and differentiated products of stem cells at the receptor level. Pai et al. used RWG to profile functional G-protein-coupled receptors (GPCRs) expressed in human neural progenitor stem cells before and after differentiation [62]. Results showed that dopamine D_1 and D_4 receptors underwent marked alterations in expression and signaling during the differentiation process.

Cancer cell invasion through surrounding extracellular matrices is the first critical step to metastasis. This process can be studied at the single-cell level using the abovementioned high-resolution RWG imager [63, 64]. Here, the biosensor was coated with Matrigel, a complex mixture of ECM proteins and growth factors. A single spheroid of cancer cells was then placed on the top of the Matrigel, and the adhesion event after the cells dissociate from the spheroid and invade through the Matrigel was monitored in real time at single-cell level using the imager. Results showed that EGF accelerated the invasion of the colon cancer cell line HT29, while vandetanib, a multi-target kinase inhibitor, inhibited its invasion [63]; furthermore, the deletion of phosphatase and tensin homologues (PTEN) also accelerated cancer cell invasion, leading to a superior sensitivity to phosphoinositide 3-kinase (PI3K) inhibitors [64]. PTEN is a negative regulator of PI3K.

Pathogen and viral infection often leads to cytopathic effect in cells, generally resulting in a robust negative biosensor signal. RWG can investigate the entire life cycle of infection and screen molecules that inhibit different steps of viral infection [65, 66]. Owens et al. employed RWG to investigate the entire infection process of HeLa cells with two different human rhinovirus strains, HRV14 and HRV16, and found that it was possible to screen inhibitors that modulated distinct stages of viral infection [66].

Receptor signaling often results in marked protein trafficking and remodeling of cytoskeletal structure, cell adhesion, and even cell morphology, all of which can lead to detectable biosensor signals. RWG has the sensitivity to detect tiny changes in local refractive index, a direct function of local cellular mass. In this regard, RWG-enabled cell assays are often referred to as *dynamic mass redistribution* (DMR) assays [20]. RWG simply acts as a recorder to monitor in real time the receptor activated signaling waves, leading to a characteristic DMR signal that mirrors the innate complexity of receptor biology and drug pharmacology (reviewed in [21, 25]).

Cardiomyocytes under optimal culture conditions can result in synchronized beating, leading to a characteristic, pulsatile biosensor signal. Ferrie et al. applied a high-frequency RWG imager to detect the dynamic beating of induced pluripotent stem cell (iPS)-derived cardiomyocytes [51]. Results showed that these cells gave rise to a resonance intensity modulation pattern consisting of a series of peaks and valleys, which was sensitive to β_2-adrenoceptor agonists (e.g., adrenalin) and cardiotoxic drug molecules such as cisapride (Fig. 7).

4.2   Assay Formats

Unlike label-based techniques that often are limited to examine a specific molecular MOA, RWG permits a wide range of assay formats to probe distinct aspects of receptor pharmacology. The choice of assay formats and conditions depends on the cell phenotype examined [67, 68]. Figure 8 summarizes common assay formats for receptor pharmacology assessment (Fig. 8) [69]. In these assays the cells generally

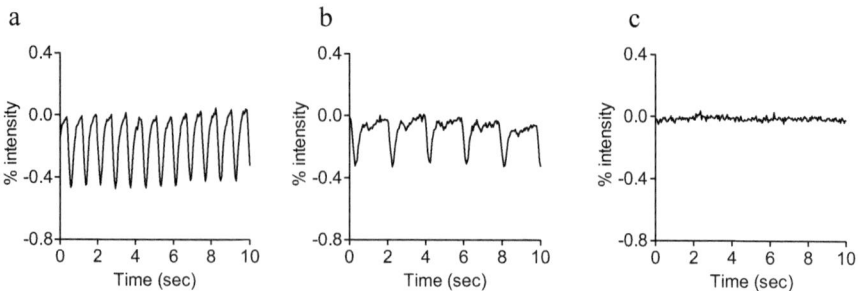

Fig. 7 Data recorded with a high-frequency RWG imager for assessing drug-induced cardiotoxicity. The resonance intensity modulation pattern of human iPS-derived cardiomyocytes at 37°C before (**a**), 5 min (**b**), and 15 min (**c**) after treatment with 10 μM cisapride. The cells were cultured within the biosensor plate for 11 days

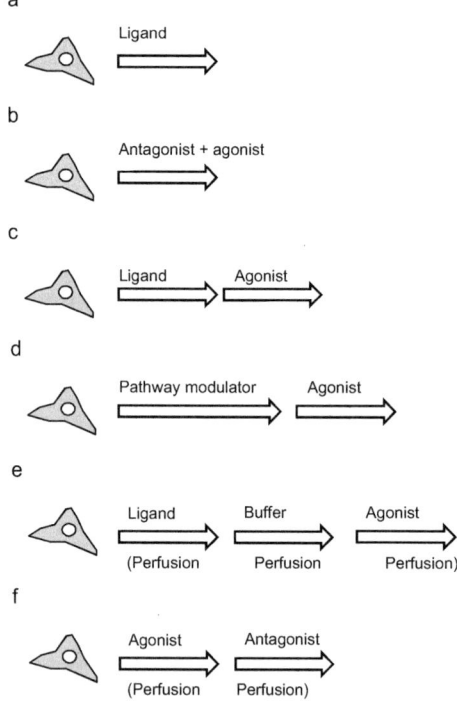

Fig. 8 RWG provides great flexibility in DMR assay formats for receptor pharmacology. (**a**) Agonism assay. (**b**) Competitive antagonism assay. (**c**) Antagonism/desensitization assay. (**d**) Pathway deconvolution assay. (**e**) Ligand washout assay. (**f**) Antagonist reverse assay. Except for perfusion which is performed using active pump-equipped microfluidic biosensor devices, all other assays are done in biosensor microplates. Figure was reproduced with permission from Ferrie et al. [69]

form a confluent monolayer to achieve robust results with minimal background signal [25, 70].

Agonism assays directly measure the cell response upon stimulation with an agonistic compound and are often used for cell phenotypic profiling.

Competitive antagonism assay determines the ability of an antagonistic compound to compete with an agonist for the binding and activation of the target receptor. Here, both ligands are mixed together to co-stimulate the cells.

Antagonism/desensitization assays investigate signal changes upon stimulation of the cells with a known agonist that is specific to the target receptor when the cells have been pretreated with a *test ligand* prior to agonist addition. The *test ligand* is an agonist for the receptor when it triggers a signal similar to the known agonist and at the same time desensitizes the cells responding to the subsequent agonist stimulation. In contrast, the ligand is an antagonist for the receptor when it does not trigger any signal but blocks the agonist response. Combining this type of experiments with competitive antagonist assays is useful to determine whether a ligand is an allosteric modulator or not.

Pathway deconvolution assays study the signaling pathway(s) of the target receptor. This is performed by pretreating the cells with known pathway modulators (e.g., G-protein toxins, interference RNA, kinase inhibitors), followed by agonist stimulation.

Ligand washout assays examine whether the effect of the ligand on the receptor is long- or short-lasting. This is performed by treating the cells with the ligand, followed by the removal of the ligand through washing or perfusion with a microfluidic device. A second stimulation step with a well-known agonist of the receptor can also be included to further ascertain the effect of the ligand examined in the first stimulation.

Antagonist reverse assays examine the ability of an antagonist to reverse the biosensor signal of an agonist for the target receptor. Here, the cells are first stimulated with an agonist for the receptor, followed by the treatment with an antagonist for the same receptor. A washing step can also be introduced between the two steps. Critical to this assay is that the agonist is known to trigger a sustained biosensor signal [69].

5 Applications in Drug Discovery

Drug discovery is a long and costly process, often starting with a hypothetical target for a disease or a disease-relevant phenotypic effect of a chemical (e.g., a natural product). RWG has found applications in various stages of the drug discovery process including hit identification, lead optimization, target engagement determination, and drug safety assessment (Fig. 9) [71, 72]. This section reviews the primary applications of waveguide-based biosensors, in particular RWG, in drug discovery.

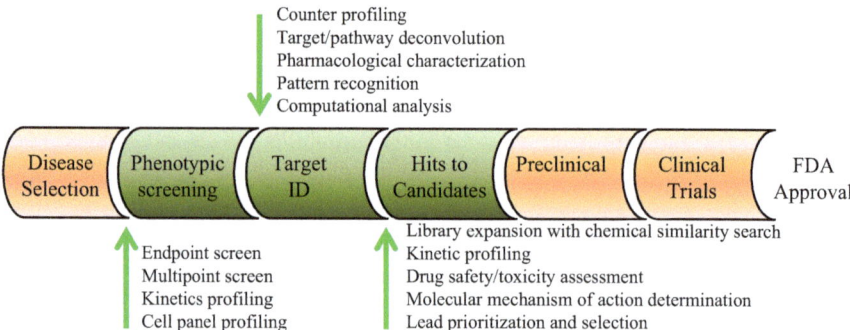

Counter profiling
Target/pathway deconvolution
Pharmacological characterization
Pattern recognition
Computational analysis

| Disease Selection | Phenotypic screening | Target ID | Hits to Candidates | Preclinical | Clinical Trials | FDA Approval |

Endpoint screen
Multipoint screen
Kinetics profiling
Cell panel profiling

Library expansion with chemical similarity search
Kinetic profiling
Drug safety/toxicity assessment
Molecular mechanism of action determination
Lead prioritization and selection

Fig. 9 RWG has broad applicability in early drug discovery. Combining computational approaches with label-free cell phenotypic profiling and screening techniques can be used for high-throughput screening, target engagement determination, compound library expansion, lead optimization, molecular mechanism of action determination, drug safety/toxicity assessment, and lead prioritization and selection. Reproduced from Fang [29] through the Creative Commons Attribution License

5.1 Screening

Hits are mostly identified using HTS. Owing to its integration within microplates, in particular high-density microplates (e.g., 1536-well), RWG has found applications in (ultra)HTS for hit identification using both biochemical [73, 74] and cell-based assays [32–34, 75, 76]. In particular, DMR screens can be performed in different formats, including endpoint, multipoint, kinetic profiling, and cell panel profiling (reviewed in detail in [29]). Unlike conventional binding assays, RWG enables unbiased detection of all types of ligands to the target receptor. Wells et al. performed an affinity-based screen using RWG and identified inhibitors and enhancers for the $G_{\beta\gamma}$-SNARE interaction [73]. SNARE refers to *soluble N-ethylmaleimide-sensitive factor attachment protein receptor*, which is involved in a downstream event of voltage-dependent calcium channels and signaling mediated through $G_{i/o}$-coupled presynaptic GPCRs.

RWG-enabled cell-based screens (also termed DMR screens) can discover different classes of ligands for the receptor of interest, all within a single campaign. Gitschier et al. developed a triple-addition screen using RWG to identify agonists, allosteric potentiators, and antagonists for the muscarinic acetylcholine receptor M1 [34]. This screen was performed in three sequential steps, that is, addition of compound (agonist mode), acetylcholine at its EC20 dose (potentiator mode), and acetylcholine at its EC80 dose (antagonist mode), respectively. Acetylcholine is the nonselective, orthosteric muscarinic receptor agonist. The DMR at a specific time point for each step was recorded. Screening a library of approximately 180,000 compounds identified 0.6% hits in agonist mode, 0.05% hits in potentiator mode, and 3% hits in antagonist mode. Follow-up studies confirmed 675 to be agonists and 45 to be positive allosteric modulators.

DMR screens can identify ligands for multiple targets, all within a single campaign. Zhang et al. [77] used RWG to screen 320 fractions extracted from two traditional Chinese medicinal (TCM) plants using three cell lines A431, A549, and HT29. 160 fractions were obtained from high-performance liquid chromatography separation of extracts from each plant. Results showed that there were multiple fractions containing compounds active at GPR35, and both plants were rich in niacin, an agonist known to activate GPR109A in A431 cells.

DMR screens typically result in much lower false positives and negatives, compared to conventional label-based screens [78, 79]. Dodgson et al. used RWG to screen 10,000 compounds for muscarinic M_3 receptor antagonism and compared results with a FLIPR-based approach (fluorescent imaging plate reader reading intracellular calcium mobilization) [33]. They found that the RWG screen had a hit rate of 0.5%, whereas the FLIPR screen had a hit rate of 0.8% using the same inhibition threshold. Only 72 compounds were active in both assays, with 857 compounds found to be active in RWG-based screens and 785 compounds active in the FLIPR assay. Almost all FLIPR hits were confirmed to be false positives, except for three compounds that displayed low potency. In contrast, 392 compounds that were active in the DMR screen were further examined, 149 of which were found to have a pIC_{50} of 4.7 to >6.6.

5.2 Lead Optimization

Once hits are identified and confirmed, the next critical step is to expand hits and to generate and optimize lead candidate compounds. Hit expansion can be done by synthesis of analogs or by searching for compounds based on chemical similarity [79]. Chemical optimization of lead candidates aims to improve potency, reduce off-target activities, and achieve desired physicochemical/metabolic properties.

RWG can be used to screen promiscuous or pan-assay interference compounds, thus quickly eliminating false positives from a screen and facilitating optimization by medicinal chemistry. Self-aggregation of small molecules is one of the common mechanisms resulting in promiscuous inhibition of target proteins, leading to false positives [80, 81]. Many natural products are prone to aggregation and difficult to optimize [82]. RWG and photonic crystal biosensors, in general, offer a simple, quantitative, direct measurement of small-molecule aggregation by measuring increased mass density arising from adsorption of aggregators to the surface [83].

RWG-enabled cell phenotypic profiling (also named DMR profiling) allows for quantitative structure-activity relationship (SAR) analysis with respect to efficacy, potency, mode of action, and off-target effects of lead-like compounds, all within a single campaign. Schmidt et al. screened a targeted library of small carboxylic acids using DMR assays and identified ligands selective to *free fatty acid receptor 2* (FFA2; GPR43) [84]. FFA2 is a GPCR for short-chain fatty acids and implicated in inflammatory and metabolic disorders. SAR analysis revealed a general rule to predict selectivity of small carboxylic acids for the orthosteric-binding site: ligands with sp2- or sp-hybridized α-carbons prefer FFA2, but ligands with substituted sp3-hybridized α-carbons preferentially activate FFA3, a closely related receptor.

Combining DMR profiling with ligand similarity search is powerful to quickly expand a hit library and perform effective SAR analysis, leading to the identification of lead-like compounds [79]. In recent years, we had combined DMR profiling with chemical similarity search to discover different classes of ligands for GPR35 (Fig. 10). DMR screening of an internal chemical library of 660 compounds led us to identify several 2-(4-methylfuran-2(5H)-ylidene)malononitrile compounds to be GPR35 agonists [85]. Based on their chemical similarity to tyrphostins, we hypothesized and confirmed that a group of tyrphostins such as tyrphostin-25 [86] and multiple tyrosine metabolites including 5,6-dihydroxyindole-2-carboxylic acid (DHICA) and 3-nitro-L-tyrosine are GPR35 agonists [87, 88]. These findings led us to further discover that several natural phenols [89], the fragment compound catechol and pyrogallol [90], and a group of nitrophenol-containing drugs such as

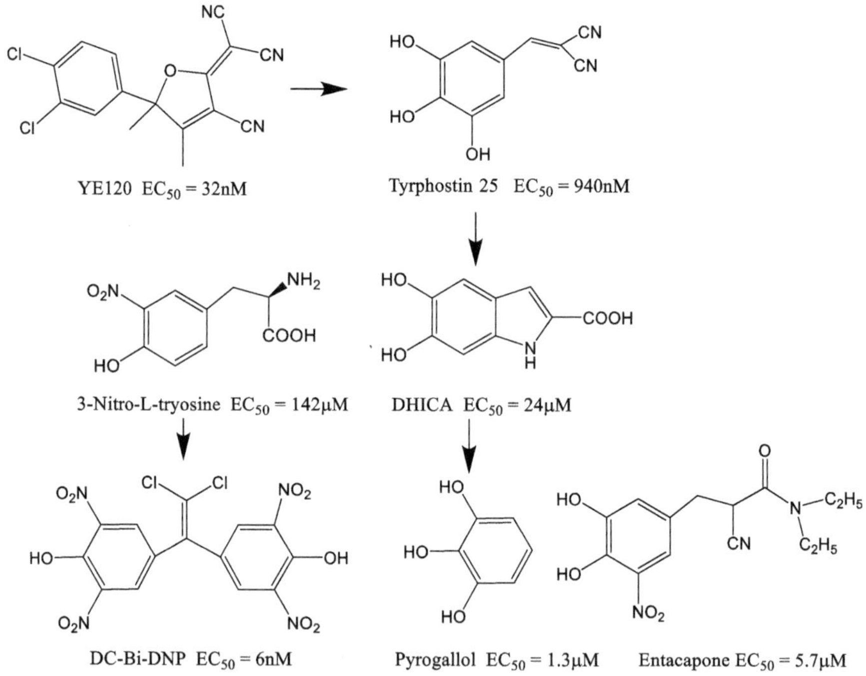

Fig. 10 Combining label-free cell-based screening with chemical similarity search quickly identifies novel ligands. YE120 was first identified as a full agonist for GPR35 in HT29 from our DMR screen. SAR results led us to search for and profile known malononitrile-containing compounds from public databases, yielding a family of tyrphostins (e.g., tyrphostin-25) and multiple tyrosine metabolites (e.g., DHICA and 3-nitro-L-tyrosine) as GPR35 agonists. Substructure search of active hits identified from public databases led us to discover DC-Bi-DNP as one of the most potent agonists known for GPR35 and nitrophenol-containing drugs such as entacapone as GPR35 agonists. EC_{50} values reported were obtained using dynamic mass redistribution assays in HT29 cells

entacapone, tolcapone, and nitecapone [91] were GPR35 agonists, among which 4,4'-(2,2-dichloroethene-1,1-diyl)bis(2,6-dinitrophenol) (DC-Bi-DNP) displays high potency with an EC_{50} of 6 nM [91].

5.3   Lead Selection

Effective lead prioritization and selection for preclinical testing are essential to reduce the cost for drug discovery. Current approaches for lead selection are mostly idiosyncratic and unpredictable. DMR assays offer an attractive approach to prioritize and select leads, based on analysis of ligand-directed functional selectivity, on-target pharmacology, biochemical mechanism of action, off-target effects, and drug toxicity. Ligand-directed functional selectivity refers to the ligand-dependent selectivity for a specific signal transduction pathway over another downstream of the same receptor [67, 92]. On-target pharmacology refers to the functional activity of a drug after binding to a specific target [67]. Biochemical MOA refers to the interaction between a drug and its target that creates a binding kinetics-dependent response [93]. Drug residence time (reciprocal of the off rate) is the amount of time that a drug spends on its target and has been found to play important roles in drug pharmacology [94]. Polypharmacology describes the interaction of a single drug with multiple targets, resulting in either beneficial effects via a synergistic mechanism or detrimental effects via off-targets [67].

DMR profiling is useful to detect ligand-directed functional selectivity or *biased agonism*. Selective activation of signaling pathways downstream of a GPCR may lead to safer and more effective drug therapies. DMR assays provide a real-time, unbiased means to measure receptor signaling, permitting multi-parametric analysis of ligand pharmacology [95, 96]. Brust et al. combined DMR assays with other conventional molecular assays (e.g., extracellular signal-regulated kinase phosphorylation, heterologous sensitization) to examine several dopamine D_2 receptor ligands [97, 98]. Results showed that aripiprazole, a third-generation antipsychotic drug used to treat schizophrenia, was found to display a unique functional profile at this receptor – a partial agonist for $G_{\alpha i/o}$, a robust antagonist for $G_{\beta\gamma}$ signaling, and a weak partial agonist for both heterologous sensitization and DMR [98].

DMR profiling of a family of ligands for a receptor in a panel of cellular backgrounds offers a high-resolution self-referencing pharmacology heat map to differentiate these ligands with fine details. This technique is referred to *RWG-based label-free integrative pharmacology on-target approach* [99–102]. We had applied this approach to analyze a family of ligands for β_2-AR [99, 100] and another for μ-, δ-, and κ-opioid receptors [101, 102]. This approach leverages the power of RWG to detect the functional receptome and biased agonism of a given ligand. The cell panel may consist of the parental cell after pretreatment with different probes or manipulating procedures (e.g., pathway modulators, gene expression, deletion, or RNAi knockdown) or a panel of different recombinant or disease-relevant cell lines. This approach enables the classification of lead-like molecules into distinct clusters, each of which may share a common molecular MOA, so representative lead-like molecules from each cluster can be selected for in vivo testing.

DMR profiling provides insights about the biochemical MOA. First, DMR profiling, together with the determination of binding kinetics, can relate the cell phenotypic response to the binding kinetics for a specific ligand-receptor interaction [103]. DMR profiling of six agonists for the endogenous muscarinic M3 receptor in six different native cell lines revealed that there was a positive correlation between the efficacy and receptor residence time of these ligands [103]. Similar results were found for a family of agonists acting on the adenosine A2A receptor using impedance-based cell assay [104]. Second, DMR profiling under microfluidics can directly be used to elucidate the biochemical MOA of lead molecules [55, 56, 105, 106]. By controlling the duration of the receptor's exposure to the ligands, we found that the sustainability of the positive DMR arising from the activation of the β_2-AR in A431 was dependent on the residence time of the agonist examined [56]. Furthermore, this approach also assisted to elucidate the MOA of different kinase inhibitors. The whole-cell efficacy of three inhibitors known to block EGF receptor signaling in A431 and HT29 was found to be related mostly to cellular uptake and efflux via efflux transporters instead of receptor residence time [106].

DMR profiling can provide insights about the safety of lead-like molecules. Drug-induced toxicities in the liver, heart, kidney, and brain currently account for more than 70% of drug attrition and withdrawal from the drug discovery process [107]. First, DMR profiling using primary cells that are prone to drug-induced toxicity is useful to assess potential adverse reactions of lead-like molecules. Second, the high-frequency RWG imager can be used to examine compound-induced cardiotoxicity using primary or stem cell-derived cardiomyocytes. These cells represent highly physiologically relevant and robust systems to assess accurately compound-induced cardiotoxicity. We recently found using the imager that after culture for 11 days, iPS-derived cardiomyocytes form a monolayer on the biosensor surface and beat in a synchronized fashion, resulting in a regular beating pattern consisting of a series of minimal valleys [51]. The treatment of cardiomyocytes with cisapride, a known toxic compound, for 15 min completely stopped their beating (Fig. 7).

DMR screening and profiling is also useful for assessing off-target effects owing to its unbiased nature and wide target/pathway coverage [67, 84]. Almost all drugs display clinically relevant polypharmacology, although drugs are traditionally optimized against a single protein [108, 109]. For instance, DMR profiling revealed that six opioid ligands including BNTX, ICI 199441, β-funaltrexamine, etonitazenyl isothiocyanate, dynorphin A 2–13, and nociceptin 1–13 activated opioid-like receptor-1 and other non-opioid target(s) in HEK293 cells [102].

5.4 Target Identification

Conventional phenotypic approaches mostly rely on discrete, descriptive, empirical, endpoint results obtained from cells, tissues, or animals, which, by themselves and individually, may offer little insight about the MOA of drugs [110]. DMR assays have advantages in identifying target engagement, since they provide high information content and are easily combined with chemical biology, computational approaches, or gene manipulation tools (e.g., gene overexpression, knockdown with RNAi, or clustered regularly interspaced short palindromic repeats (CRISPR)-mediated gene deletion or editing) [67, 79]. DMR agonist profiling of a library of 72 ion channel ligands showed that pinacidil, a known K_{ATP} channel opener, triggered a robust negative DMR in A431, A549, HT29, and HepG2C3A, but not HepG2 cells (Fig. 11) [111]. Given that the expression and function of K_{ATP} channel in the liver cell line HepG2C3A is unknown, RNAi-mediated knockdown of known K_{ATP} channel components and pathway proteins was used to decode the pinacidil activated channel to be a SUR2-containing K_{ATP} channel. The K_{ATP} channel activation mediated signaling through Rho kinase, Janus kinase-2, and Janus kinase-3.

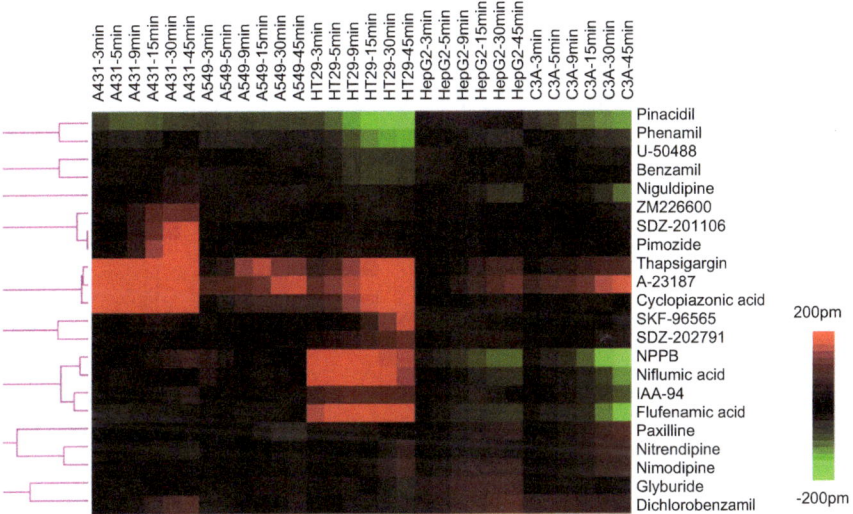

Fig. 11 Label-free cell phenotypic profiling for identifying mechanism of action of drugs in cells. DMR heat map of ion channel ligands that are active in five cell lines: A431, A549, HT29, HepG2, and C3A. This heat map was obtained using clustering analysis of the DMR profiles of the ion channel ligands in these five cell lines. For each ligand, the real responses at six discrete time points post stimulation (3, 5, 9, 15, 30, 45 min) in each cell line were used for the cluster analysis. Only ligands that led to a DMR greater than 50 pm or less than −50 pm in at least one cell line were included in this analysis. Reproduced from Sun et al. [111] through the Creative Commons Attribution License

5.5 Discovery of Novel Receptor Biology

DMR profiling enables dissection of complex receptor signaling patterns even in primary human cells with unprecedented accuracy [18, 19, 112–117]. Recently, Hennen et al. applied DMR assays to assist the discovery of a small-molecule agonist MDL29,951 for GPR17 [116]. MDL29,951 was found to selectively activate GPR17 in primary oligodendrocytes, leading to the activation of the entire set of intracellular adaptor proteins for GPCRs: G-proteins of the $G_{\alpha i}$, $G_{\alpha s}$, and $G_{\alpha q}$ subfamily, as well as β-arrestins.

Combining RWG with microfluidics provides an additional lever to deconvolute receptor signaling. DMR profiling using a panel of assays including microfluidics-based assays suggests that the activation of the β2-AR in A431 cells triggered multiple signaling waves (Fig. 12) [69]. Epinephrine at low doses (e.g., 10 nM) only activated rapid $G_{\alpha s}$-dependent and independent waves initiated at the cell surface, which were reversed by both the cell membrane non-permeable antagonist sotalol and the permeable antagonist propranolol. In contrast, epinephrine at high

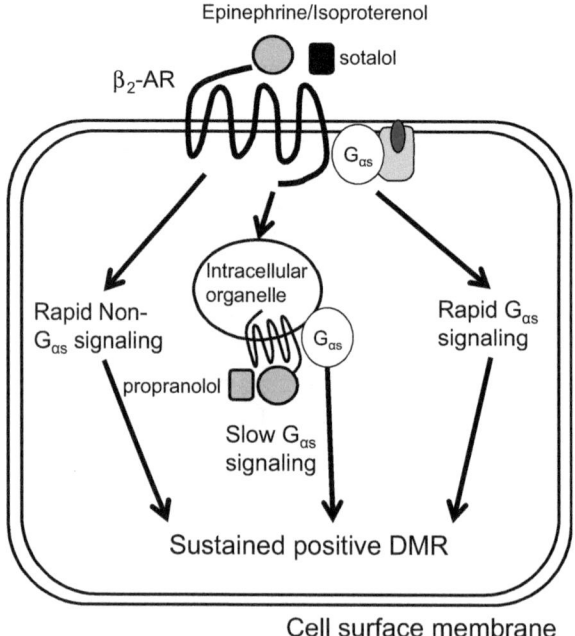

Fig. 12 The activation of the β2-ARs in A431 triggers multiple signaling waves. The rapid Gαs-dependent wave initiated at the cell surface is activated by epinephrine at low doses (e.g., 10 nM) and reversed by the hydrophilic antagonist sotalol. The rapid Gαs-independent wave initiated at the cell surface is activated by epinephrine at high doses (e.g., 10 μM), while the slow $G_{\alpha s}$-dependent intracellular wave is also activated by epinephrine at high doses but cannot be reversed by sotalol. All three signaling waves contribute to the long-term sustainability of the epinephrine DMR response. Reproduced from Ferrie et al. [69] with permission

doses (e.g., 10 µM) activated an additional slow $G_{\alpha s}$-dependent intracellular wave, which was selectively reversed by propranolol but not sotalol. Furthermore, pulse stimulation as short as <1 min was found to impair the slow signaling wave [56, 69, 118]. These results are consistent with a novel GPCR signaling model in which the activation of the β_2-AR triggers a slow signaling originating from the endosome and not from the cell plasma membrane [119].

The DMR signal of receptor activation is believed to arise from remodeling of cell-matrix adhesion, cytoskeletal structure, changes in cell morphology, as well as protein trafficking such as receptor endocytosis and multi-protein assembly processes such as *purinosome* dynamics. *Purinosomes* are putative, dynamic cellular bodies formed by cellular enzymes involved in purine biosynthesis [120]. *Purinosomes* were first observed in HeLa cells using fluorescence microscopy in response to purine depletion [121]. Later they were found to be regulated by casein kinase-2 [122]. Two known casein kinase-2 inhibitors, DMAT and TBB, behaved differently in regulating purinosomes: DMAT promoted the purinosome complexes, whereas TBB disassembled them [122]. DMR screening using a library of GPCR agonists against the DMR arising from subsequent stimulation with DMAT and TBB led to the discovery that the activation of several $G\alpha i$-coupled receptors such as α_{2A}-AR in HeLa promoted purinosome formation via $G\alpha i$-mediated signaling [123]. Recently, we further discovered that purinosomes were spatially colocalized and functionally linked with mitochondria [124]. A short hairpin RNA (shRNA) kinome screen using DMR assays suggested a putative kinase network linking the activation of α_{2A}-AR to purinosome assembly which is additionally influenced by the *mechanistic target of rapamycin* (mTOR). These results suggest that the purinosome assembly and disassembly processes are under the control of GPCR signaling, and there is an mTOR-mediated link between purinosomes and mitochondria, opening a possibility to discover novel anticancer drugs targeting GPCRs [125].

5.6   Single-Cell Analysis

High-resolution RWG imagers permit single-cell analysis, which can reveal hetero-geneity in cellular processes and offer new insights into cell biology [126]. Hetero-geneity or cell-to-cell variability is intrinsic to many cell signaling pathways and processes such as differentiation. Recently, we applied a high-resolution RWG imager to study cell-to-cell variability of EGF receptor signaling in A431 [50]. Results showed that under physiological conditions EGF stimulated robust DMR signals that share similar characteristics but with variable amplitudes and kinetics at the single-cell level (Fig. 13a). The positive EGF response at 2 min post stimulation gave rise to a distribution having two peaks centered at 100 pm and 600 pm, respectively (Fig. 13b). This is possibly due to the presence of two distinct quiescent states achieved through contact inhibition and serum withdrawal, respectively, which, in turn, has different basal activity of the protein kinase C pathway

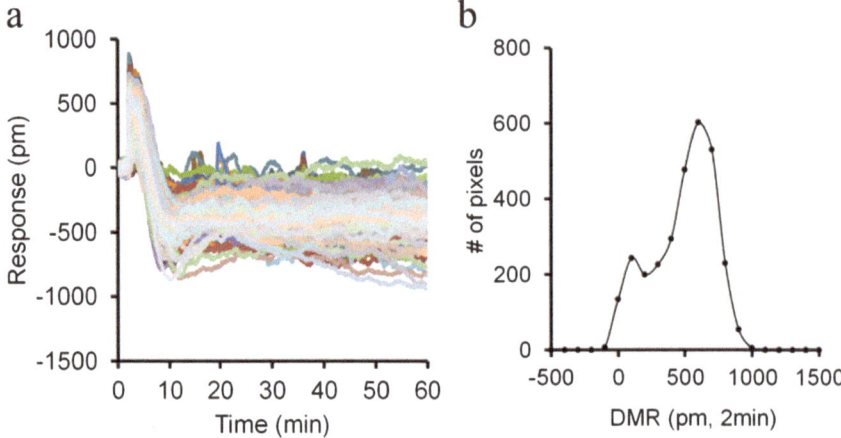

Fig. 13 Cell-to-cell variability of the DMR signal arising from the activation of epidermal growth factor receptor in quiescent A431 cells. (**a**) The real-time DMR of 166 pixels across the biosensor. (**b**) Histogram of the EGF responses at 2 min with a bin size of 100 pm

[127]. Single-cell monitoring with RWG also enables the detection of cancer cell invasion through Matrigel coating in a 3D spheroid model [63, 64].

6 Summary and Perspective

Waveguide-based biosensors, in particular RWG, have found increasing applications in cell analysis. The whole-plate RWG readers are useful for label-free cell phenotypic profiling and screening, as well as receptor and cell biology studies under physiological conditions. High-resolution RWG imagers can be used to perform single-cell analysis, while combining it with microfluidics enables elucidating the biochemical MOA of drugs and discovering novel receptor biology. In particular, a high-frequency RWG imager is valuable to assess drug-induced cardiotoxicity.

However, since RWG-enabled cell assays are label-free, there are disadvantages associated with the lack of molecular details, the relatively low power to resolve the full innate complexity of drug pharmacology, and the limited applicability to some important classes of drug targets (e.g., nuclear receptors). Therefore, combining label-free with multiple other techniques, in particular label-dependent approaches (e.g., total internal reflection fluorescence imaging), would be beneficial to fully exploit the potential of these biosensors for cell biology [128–130]. Furthermore, there are still unmet needs for label-free cell phenotypic profiling as, for instance, 3D cell models and subcellular imaging. Linking in vitro results with in vivo effects may require the development of new or the refinement of existing systems and methodologies.

References

1. Bosch ME, Sánchez AJR, Rojas FS, Ojeda CB (2007) Recent development in optical fiber biosensors. Sensors 7(6):797–859
2. Pospíšilová M, Kuncová G, Trögl J (2015) Fiber-optic chemical sensors and fiber-optic biosensors. Sensors 15(10):25208–25259
3. Tiefenthaler K, Lukosz W (1989) Sensitivity of grating couplers as integrated-optical chemical sensors. J Opt Soc Am B 6:209–220
4. Vörös J, Ramsden JJ, Csúcs G, Szendrő I, De Paul SM, Textor M, Spencer ND (2002) Optical grating coupler biosensors. Biomaterials 23:3699–3710
5. Cush R, Cronin JM, Stewart WJ, Maule CH, Molloy J, Goddard NJ (1993) The resonant mirror: a novel optical biosensor for direct sensing of biomolecular interactions Part I: principle of operation and associated instrumentation. Biosens Bioelectron 8(7–8):347–354
6. Cunningham B, Li P, Lin B, Pepper J (2002) Colorimetric resonant reflection as a direct biochemical assay technique. Sens Actuators B 81:316–328
7. Nirsch M, Reuter F, Vörös J (2011) Review of transducer principles for label-free biomolecular interaction analysis. Biosensors 1(3):70–92
8. Horvath R, Pedersen HC, Larsen NB (2002) Demonstration of reverse symmetry waveguide sensing in aqueous solutions. Appl Phys Lett 81(12):2166–2168
9. Zourob M, Mohr S, Brown BJT, Fielden PR, McDonnell M, Goddard NJ (2003) The development of a metal clad leaky waveguide sensor for the detection of particles. Sens Actuators B 90(1–3):296–307
10. Fan X, White IM, Shopova SI, Zhu H, Suter JD, Sun Y (2008) Sensitive optical biosensors for unlabeled targets: a review. Anal Chim Acta 620(1):8–26
11. Li SY, Ramsden JJ, Prenosil JE, Heinzle E (1994) Measurement of adhesion and spreading kinetics of baby hamster kidney and hybridoma cells using an integrated optical method. Biotechnol Prog 10(5):520–524
12. Ramsden JJ, Horvath R (2009) Optical biosensors for cell adhesion. J Recept Signal Transduct Res 29(1–2):211–223
13. Orgovan N, Peter B, Bősze S, Ramsden JJ, Szabó B, Horvath R (2014) Dependence of cancer cell adhesion kinetics on integrin ligand surface density measured by a high-throughput label-free resonant waveguide grating biosensor. Sci Rep 4:4034
14. Fang Y (2010) Label-free and non-invasive biosensor cellular assays for cell adhesion. J Adhes Sci Technol 24:1011–1021
15. Hug TS, Prenosil JE, Maier P, Morbidelli M (2002) Optical waveguide lightmode spectroscopy (OWLS) to monitor cell proliferation quantitatively. Biotechnol Bioeng 80(2):213–221
16. Hug TS (2004) Biophysical methods for monitoring cell-substrate interactions in drug discovery. Assay Drug Dev Technol 1(3):479–488
17. Fang Y, Ferrie AM, Fontaine NH, Yuen PK (2005) Optical biosensors for monitoring dynamic mass redistribution in living cells mediated by epidermal growth factor receptor activation. Conf Proc IEEE Eng Med Biol Soc 1:666–669
18. Fang Y, Li G, Peng J (2005) Optical biosensor provides insights for bradykinin B2 receptor signaling in A431 cells. FEBS Lett 579:6365–6374
19. Fang Y, Ferrie AM, Fontaine NH, Yuen PK (2005) Characteristics of dynamic mass redistribution of EGF receptor signaling in living cells measured with label free optical biosensors. Anal Chem 77:5720–5725
20. Fang Y, Ferrie AM, Fontaine NH, Mauro J, Balakrishnan J (2006) Resonant waveguide grating biosensor for living cell sensing. Biophys J 91:1925–1940
21. Fang Y (2010) Label-free receptor assays. Drug Discov Today Technol 7:e5–e11
22. Fang Y (2006) Label-free cell-based assays with optical biosensors in drug discovery. Assay Drug Dev Technol 4:583–595
23. Fang Y, Frutos AG, Verklereen R (2008) Label-free cell assays for GPCR screening. Comb Chem High Throughput Screen 11:357–369

24. Scott CW, Peters MF (2010) Label-free whole-cell assays: expanding the scope of GPCR screening. Drug Discov Today 15(17–18):704–716
25. Fang Y (2011) The development of label-free cellular assays for drug discovery. Expert Opin Drug Discovery 6:1285–1298
26. Fang Y (2012) Ligand-receptor interaction platforms and their applications for drug discovery. Expert Opin Drug Discovery 7:969–988
27. Rocheville M, Martin J, Jerman J, Kostenis E (2013) Mining the potential of label-free biosensors for seven-transmembrane receptor drug discovery. Prog Mol Biol Transl Sci 115:123–142
28. Fang Y (2014) Label-free cell phenotypic drug discovery. Comb Chem High Throughput Screen 17(7):566–578
29. Fang Y (2014) Label-free drug discovery. Front Pharmacol 5:52
30. Grundmann M, Kostenis E (2015) Label-free biosensor assays in GPCR screening. Methods Mol Biol 1272:199–213
31. Li G, Ferrie AM, Fang Y (2006) Label-free profiling of endogenous G-protein-coupled receptors using a cell-based high throughput screening technology. J Assoc Lab Autom 11:181–187
32. Tran E, Fang Y (2008) Duplexed label-free G-protein-coupled receptor assays for high throughput screening. J Biomol Screen 13:975–985
33. Dodgson K, Gedge L, Murray DC, Coldwell M (2009) A 100 K well screen for a muscarinic receptor using the Epic® label-free system – a reflection on the benefits of the label-free approach to screening seven-transmembrane receptors. J Recept Signal Transduct 29 (3–4):163–172
34. Gitschier HJ, Bergerone AB, Randle DH, Bacon CE, Baez M, Yang P, Broad LM, Goldsmith PJ, Felder CC, Schober DA (2015) Triple-addition label-free assays for high-throughput screening of muscarinic M1 receptor agonists, antagonists, and allosteric modulators. Methods Pharmacol Toxicol 53:197–214
35. Orgovana N, Kovacs B, Farkas E, Szabó B, Zaytseva N, Fang Y, Horvath R (2014) Bulk and surface sensitivity of a resonant waveguide grating imager. Appl Phys Lett 104:083506
36. Vollmer F, Arnold S (2008) Whispering-gallery-mode biosensing: label free detection down to single molecules. Nat Methods 5:591–596
37. Pal S, Fauchet PM, Miller BL (2012) 1-D and 2-D photonic crystals as optical methods for amplifying biomolecular recognition. Anal Chem 84(21):8900–8908
38. Fang Y (2007) Non-invasive optical biosensor for probing cell signaling. Sensors 7:2316–2329
39. Fang Y (2015) Resonant waveguide grating imagers for single cell analysis and high throughput screening. Proc SPIE 9550:95500P
40. Sun X, Shu X, Chen C (2015) Grating surface plasmon resonance sensor: angular sensitivity, metal oxidization effect of Al-based device in optimal structure. Appl Optics 54(6):1548–1554
41. Konradi R, Textor M, Reimhult E (2012) Using complementary acoustic and optical techniques for quantitative monitoring of biomolecular adsorption at interfaces. Biosensors 2(4):341–376
42. Cunningham BT, Li P, Schulz S, Lin B, Baird C, Gerstenmaier J, Genick C, Wang F, Fine E, Laing L (2004) Label-free assays on the BIND system. J Biomol Screen 9(6):481–490
43. Li P, Lin B, Gerstenmaier J, Cunningham BT (2004) A new method for label-free imaging of biomolecular interactions. Sens Actuators B 99(1):6–13
44. Fang Y, Fang J, Tran E, Xie X, Hallstrom M, Frutos AG (2009) High-throughput analysis of biomolecular interactions and cellular responses with resonant waveguide grating biosensors. In: Cooper MA (ed) Label-free biosensors: techniques and applications. Cambridge University Press, New York, pp 206–222
45. Ferrie AM, Wu Q, Fang Y (2010) Resonant waveguide grating imager for live cell sensing. Appl Phys Lett 97:223704

46. Block ID, Mathias PC, Ganesh N, Jones ID, Dorvel BR, Chaudhery V, Vodkin L, Bashir R, Cunningham BT (2009) A detection instrument for enhanced fluorescence and label-free imaging on photonic crystal surfaces. Opt Express 17:13222–13235

47. Block ID, Mathias PC, Jones SI, Vodkin LO, Cunningham BT (2009) Optimizing the spatial resolution of photonic crystal label-free imaging. Appl Optics 48:6567–6574

48. Choi CJ, Belobraydich AR, Chan LL, Mathias PC, Cunningham BT (2010) Comparison of label-free biosensing in microplate, microfluidic, and spot-based affinity capture assays. Anal Biochem 405(1):1–10

49. Chen WL, Long KD, Lu M, Chaudhery V, Yu H, Choi JS, Polans J, Zhuo Y, Harley BAC, Cunningham BT (2013) Photonic crystal enhanced microscopy for imaging of live cell adhesion. Analyst 138:5886–5894

50. Ferrie AM, Deichmann OD, Wu Q, Fang Y (2012) High resolution resonant waveguide grating imager for cell cluster analysis under physiological condition. Appl Phys Lett 100:223701

51. Ferrie AM, Wu Q, Deichmann O, Fang Y (2014) High frequency resonant waveguide grating imager for assessing drug-induced cardiotoxicity. Appl Phys Lett 104:183702

52. Orgovan N, Patko D, Hos C, Kurunczi S, Szabó B, Ramsden JJ, Horvath R (2014) Sample handling in surface sensitive chemical and biological sensing: a practical review of basic fluidics and analyte transport. Adv Colloid Interface Sci 211:1–16

53. Fang Y (2013) Microfluidic biosensor systems for cell biology and drug discovery. In: Panzarella S, Maroni W (eds) Microfluidics: control, manipulation and behavioral applications. Nova, New York, pp 51–78

54. Choi CJ, Cunningham BT (2007) A 96-well microplate incorporating a replica molded microfluidic network integrated with photonic crystal biosensors for high throughput kinetic biomolecular interaction analysis. Lab Chip 7:550–556

55. Goral V, Wu Q, Sun H, Fang Y (2011) Label-free optical biosensor with microfluidics for sensing ligand-directed functional selectivity on trafficking of thrombin receptor. FEBS Lett 585:1054–1060

56. Goral V, Jin Y, Sun H, Ferrie AM, Wu Q, Fang Y (2011) Agonist-directed desensitization of the β_2-adrenergic receptor. PLoS One 6:e19282

57. Zaytseva N, Miller W, Goral V, Hepburn J, Fang Y (2011) Microfluidic resonant waveguide grating biosensor system for whole cell sensing. Appl Phys Lett 96:163703

58. Fang Y (2011) Label-free biosensors for cell biology. Intl J Electrochem 2011:e460850

59. Zaytseva Z, Lynn JG, Wu Q, Mudaliar DJ, Kuang PQ, Fang Y (2013) Resonant waveguide grating biosensor-enabled label-free and fluorescence detection of cell adhesion. Sens Actuators B 188:1064–1072

60. Tran E, Sun H, Fang Y (2012) Dynamic mass redistribution assays decodes surface influence on signaling of endogenous purinergic receptors. Assay Drug Dev Technol 10:37–45

61. Pattabiraman G, Lidstone EA, Palasiewicz K, Cunningham BT, Ucker DS (2014) Recognition of apoptotic cells by viable cells is specific, ubiquitous, and species independent: analysis using photonic crystal biosensors. Mol Biol Cell 25(11):1704–1714

62. Pai S, Verrier F, Sun H, Hu H, Ferrie AM, Eshraghi A, Fang Y (2012) Dynamic mass redistribution assay decodes differentiation of a neural progenitor stem cell. J Biomol Screen 17:1180–1191

63. Febles NK, Ferrie AM, Fang Y (2014) Label-free single cell quantification of the invasion of spheroidal colon cancer cells through 3D Matrigel. Anal Chem 86(17):8842–8849

64. Chandrasekaran S, Deng H, Fang Y (2015) PTEN deletion potentiates invasion of colorectal cancer spheroidal cells through 3D Matrigel. Integr Biol 7(3):324–334

65. McCoy MH, Wang E (2005) Use of electric cell-substrate impedance sensing as a tool for quantifying cytopathic effect in influenza A virus infected MDCK cells in real-time. J Virol Methods 130(1–2):157–161

66. Owens RM, Wang CQ, You JA, Jiambutr J, Xu AS, Marala RB, Jin MM (2009) Real-time quantitation of viral replication and inhibitor potency using a label-free optical biosensor. J Recept Signal Transduct 29(3–4):195–201
67. Fang Y (2013) Troubleshooting and deconvoluting label-free cell phenotypic assays in drug discovery. J Pharmacol Toxicol Methods 67(1):69–81
68. Fang Y (2016) Compound annotation with real time cellular activity profiles to improve drug discovery. Expert Opin Drug Discovery 11(3):269–280
69. Ferrie AM, Wang C, Deng H, Fang Y (2013) Label-free optical biosensor with microfluidics identifies an intracellular signalling wave mediated through the β_2-adrerengic receptor. Integr Biol 5(10):1253–1261
70. Fang Y (2015) Label-free cell phenotypic profiling and screening: techniques, experimental design and data assessment. Methods Pharmacol Toxicol 53:233–252
71. Fang Y (2015) Are label-free investigations the best approach to drug discovery? Future Med Chem 7(12):1561–1564
72. Folmer RHA (2016) Integrating biophysics with HTS-driven drug discovery projects. Drug Discov Today 21(3):491–498
73. Wells CA, Betke KM, Lindsley CW, Hamm HE (2012) Label-free detection of G-protein-SNARE interactions and screening for small molecule modulators. ACS Chem Nerosci 3 (1):69–78
74. Geschwindner S, Carlsson JF, Knecht W (2012) Application of optical biosensors in small-molecule screening activities. Sensors 12(4):4311–4323
75. Larsson N, Sundström L, Ryberg E, Frostne L (2013) Cell based label-free assays in GPCR drug discovery. Eur Pharm Rev 18(4):13–16
76. Lee MY, Mun J, Lee JH, Lee S, Lee BH, Oh K-S (2014) A comparison of assay performance between the calcium mobilization and the dynamic mass redistribution technologies for the human urotensin receptor. Assay Drug Dev Technol 12(6):361–368
77. Zhang X, Deng H, Xiao Y, Xue X, Ferrie AM, Tran E, Liang X, Fang Y (2014) Label-free cell phenotypic profiling identifies pharmacologically active compounds in two traditional Chinese medicinal plants. RSC Adv 4(50):26368–26377
78. Malo N, Hanley JA, Cerquozzi S, Pelletier J, Nadon R (2006) Statistical practice in high-throughput screening data analysis. Nat Biotechnol 24(2):167–175
79. Fang Y (2015) Combining label-free cell phenotypic profiling with computational approaches for novel drug discovery. Expert Opin Drug Discovery 10(4):331–343
80. McGovern SL, Caselli E, Grigorieff N, Shoichet BK (2002) A common mechanism underlying promiscuous inhibitors from virtual and high-throughput screening. J Med Chem 45 (8):1712–1722
81. Baell J, Walters MA (2014) Chemistry: chemical con artists foil drug discovery. Nature 513 (7519):481–483
82. Cragg GM, Newman DJ (2013) Natural products: a continuing source of novel drug leads. Biochim Biophys Acta 1830(6):3670–3695
83. Chan LL, Lidstone EA, Finch KE, Heeres JT, Hergenrother PJ, Cunningham BT (2009) A method for identifying small-molecule aggregators using photonic crystal biosensor microplates. JALA Charlottesv Va 14(6):348–359
84. Schmidt J, Smith NJ, Christiansen E, Tikhonova IG, Grundmann M, Hudson BD, Ward RJ, Drewke C, Milligan G, Kostenis E, Ulven T (2011) Selective orthosteric free fatty acid receptor 2 (FFA2) agonists: Identification of the structural and chemical requirements for selective activation of FFA2 versus FFA3. J Biol Chem 286(12):10628–10640
85. Deng H, Hu H, He M, Hu J, Niu W, Ferrie AM, Fang Y (2011) Discovery of 2-(4-methylfuran-2(5H)-ylidene)malononitrile and thieno[3,2-b]thiophene-2-carboxylic acid derivatives as G-protein-coupled receptor-35 (GPR35) agonists. J Med Chem 54:7385–7396
86. Deng H, Hu H, Fang Y (2011) Tyrphostin analogs are GPR35 agonists. FEBS Lett 585:1957–1962

87. Deng H, Hu H, Fang Y (2012) Multiple tyrosine metabolites are GPR35 agonists. Sci Rep 2:373
88. Deng H, Fang Y (2012) Synthesis and agonistic activity at the GPR35 of 5,6-dihydroxyindole-2-carboxylic acid analogs. ACS Med Chem Lett 3:550–554
89. Deng H, Hu H, Ling S, Ferrie AM, Fang Y (2012) Discovery of natural phenols as G-protein-coupled receptor-35 (GPR35) agonists. ACS Med Chem Lett 3:165–169
90. Deng H, Fang Y (2013) The three catecholics benserazide, catechol and pyrogallol are GPR35 agonists. Pharmaceuticals 6:500–509
91. Deng H, Fang Y (2012) Discovery of nitrophenols as GPR35 agonists. Med Chem Commun 3:1270–1274
92. Kenakin T, Miller LJ (2010) Seven transmembrane receptors as shapeshifting proteins: the impact of allosteric modulation and functional selectivity on new drug discovery. Pharmacol Rev 62:265–304
93. Swinney DC (2004) Biochemical mechanisms of drug action: what does it take for success? Nat Rev Drug Discov 3:801–808
94. Copeland RA, Pompliano DL, Meek TD (2006) Drug-target residence time and its implications for lead optimization. Nat Rev Drug Discov 5:730–739
95. Fang Y, Ferrie AM (2008) Label-free optical biosensor for ligand-directed functional selectivity acting on β_2-adrenoceptor in living cells. FEBS Lett 582:558–564
96. Ferrie AM, Goral V, Wang C, Fang Y (2015) Label-free functional selectivity assays. Methods Mol Biol 1272:227–246
97. Brust TF, Hayes MP, Roman DL, Burris KD, Watts VJ (2015) Bias analyses of preclinical and clinical D2 dopamine ligands: studies with immediate and complex signaling pathways. J Pharmacol Exp Ther 352(3):480–493
98. Brust TF, Hayes MP, Roman DL, Watts VJ (2015) New functional activity of aripiprazole revealed: Robust antagonism of D2 dopamine receptor-stimulated G$\beta\gamma$ signaling. Biochem Pharmacol 93(1):85–91
99. Ferrie AM, Sun H, Fang Y (2011) Label-free integrative pharmacology on-target of drugs at the 2-adrenergic receptor. Sci Rep 1:33
100. Ferrie AM, Sun H, Zaytseva N, Fang Y (2014) Divergent label-free cell phenotypic pharmacology of ligands at the overexpressed β_2-adrenergic receptors. Sci Rep 4:3828
101. Morse M, Tran E, Levension RL, Fang Y (2011) Ligand-directed functional selectivity at the mu opioid receptor revealed by label-free on-target pharmacology. PLoS One 6:e25643
102. Morse M, Sun H, Tran E, Levenson R, Fang Y (2013) Label-free integrative pharmacology on-target of opioid ligands at the opioid receptor family. BMC Pharmacol Toxicol 14:17
103. Deng H, Sun H, Fang Y (2013) Label-free cell phenotypic assessment of the biased agonism and efficacy of agonists at the endogenous muscarinic M3 receptors. J Pharmacol Toxicol Methods 68(3):323–333
104. Guo D, Mulder-Krieger T, IJzerman AP, Heitman LH (2012) Functional efficacy of adenosine A_2A receptor agonists is positively correlated to their receptor residence time. Br J Pharmacol 166(6):1846–1859
105. Deng H, Wang C, Su M, Fang Y (2012) Probing biochemical mechanisms of action of muscarinic M3 receptor antagonists with label-free whole-cell assays. Anal Chem 84:8232–8239
106. Deng H, Wang C, Fang Y (2013) Label-free cell phenotypic assessment of the molecular mechanism of action of epidermal growth factor receptor inhibitors. RSC Adv 3:10370–10378
107. Wilke RA, Lin DW, Roden DM, Watkins PB, Flockhart D, Zineh I, Giacomini KM, Krauss RM (2007) Identifying genetic risk factors for serious adverse drug reactions: current progress and challenges. Nat Rev Drug Discov 6(11):904–916
108. Paolini GV, Shapland RH, van Hoorn WP, Mason JS, Hopkins AL (2006) Global mapping of pharmacological space. Nat Biotechnol 24(7):805–815
109. Yildirim MA, Goh KI, Cusick ME, Barabási AL, Vidal M (2007) Drug-target network. Nat Biotechnol 25(10):1119–1126

110. Feng Y, Mitchison TJ, Bender A, Young DW, Tallarico JA (2009) Multi-parameter pheno-typic profiling: using cellular effects to characterize small-molecule compounds. Nat Rev Drug Discov 8(7):567–578
111. Sun H, Wei Y, Xiong Q, Li M, Lahiri J, Fang Y (2014) Label-free cell phenotypic profiling decodes the composition and signaling of an endogenous ATP-sensitive potassium channel. Sci Rep 4:4934
112. Schröder R, Janssen N, Schmidt J, Kebig A, Merten N, Hennen S, Müller A, Blättermann S, Mohr-Andrä M, Zahn S, Wenzel J, Smith NJ, Gomeza J, Drewke C, Milligan G, Mohr K, Kostenis E (2010) Deconvolution of complex G-protein-coupled receptor signaling in live cells using dynamic mass redistribution measurements. Nat Biotechnol 28(9):943–949
113. Schmidt J, Liebscher K, Merten N, Grundmann M, Mielenz M, Sauerwein H, Christiansen E, Due-Hansen ME, Ulven T, Ullrich S, Gomeza J, Drewke C, Kostenis E (2011) Conjugated linoleic acids mediate insulin release through islet G-protein-coupled receptor FFA1/GPR40. J Biol Chem 286(14):11890–11894
114. Schröder R, Schmidt J, Blättermann S, Peters L, Janssen N, Grundmann M, Seemann W, Kaufel D, Merten N, Drewke C, Gomeza J, Milligan G, Mohr K, Kostenis E (2011) Applying label-free dynamic mass redistribution technology to frame signaling of G-protein-coupled receptors noninvasively in living cells. Nat Protoc 6(11):1748–1760
115. Ahmedat AS, Warnken M, Seemann WK, Mohr K, Kostenis E, Juergens UR, Racké K (2013) Pro-fibrotic processes in human lung fibroblasts are driven by an autocrine/paracrine endothelinergic system. Br J Pharmacol 168(2):471–487
116. Hennen S, Wang H, Peters L, Merten N, Simon K, Spinrath A, Blättermann S, Akkari R, Schrage R, Schröder R, Schulz D, Vermeiren C, Zimmermann K, Kehraus S, Drewke C, Pfeifer A, König GM, Mohr K, Gillard M, Müller CE, Lu QR, Gomeza J, Kostenis E (2013) Decoding signaling and function of the orphan g protein-coupled receptor GPR17 with a small-molecule agonist. Sci Signal 6(298):ra93
117. Tilley DG, Repas AA, Carter RL (2015) Label-free profiling of endogenous receptor responses in primary isolated cardiac cells. Methods Pharmacol Toxicol 53:169–182
118. Ahmed D, Muddana H, Lu M, French J, Ozcelik A, Fang Y, Butler P, Benkovic S, Manz A, Huang TJ (2014) Acoustofluidic chemical waveform generator and switch. Anal Chem 86 (23):11803–11810
119. Irannejad R, Tomshine JC, Tomshine JR, Chevalier M, Mahoney JP, Steyaert J, Rasmussen SG, Sunahara RK, El-Samad H, Huang B, von Zastrow M (2013) Conformational biosensors reveal GPCR signalling from endosomes. Nature 495(7442):534–538
120. Zhao H, French JB, Fang Y, Benkovic SJ (2013) The purinosome, a multi-protein complex involved in the de novo biosynthesis of purines in humans. Chem Commun 49:4444–4452
121. An S, Kumar R, Sheets ED, Benkovic SJ (2008) Reversible compartmentalization of de novo purine biosynthetic complexes in living cells. Science 320:103–106
122. An S, Kyoung M, Allen JJ, Shokat KM, Benkovic SJ (2010) Dynamic regulation of a metabolic multi-enzyme complex by protein kinase CK2. J Biol Chem 285:11093–11099
123. Verrier F, An S, Ferrie AM, Sun H, Kyoung M, Fang Y, Benkovic SJ (2011) G-protein-coupled receptor signaling regulates the dynamics of a metabolic multienzyme complex. Nat Chem Biol 7:909–915
124. French JB, Jones SA, Deng H, Hu H, Pedley AM, Kim D, Chan CY, Hu H, Pugh RJ, Zhao H, Zhang Y, Huang TJ, Fang Y, Zhuang X, Benkovic SJ (2016) Spatial colocalization and functional link of purinosomes with mitochondria. Science 351(6274):733–737
125. Fang Y, French J, Zhao H, Benkovic S (2013) G-protein-coupled receptor regulation of de novo purine biosynthesis: a novel druggable mechanism. Biotechnol Genet Eng Rev 29 (1):31–48
126. Fang Y (2015) Label-free chemical and phenotypic profiling of living cells. Sci Lett 4:156
127. Fang Y (2010) Probing cancer signaling with resonant waveguide grating biosensors. Expert Opin Drug Discovery 5:1237–1248

128. Chen M, Zaytseva NV, Wu Q, Li M, Fang Y (2013) Microplate-compatible total internal reflection fluorescence microscopy for receptor pharmacology. Appl Phys Lett 102:193702
129. Fang Y (2015) Total internal reflection fluorescence quantification of receptor pharmacology. Biosensors 5:223–240
130. Michaelis S, Wegener J, Robelek R (2013) Label-free monitoring of cell-based assays: combining impedance analysis with SPR for multiparametric cell profiling. Biosens Bioelectron 49:63–70

BIOREV (2019) 2: 219–272
https://doi.org/10.1007/11663_2019_6
© Springer Nature Switzerland AG 2019
Published online: 15 May 2019

Label-Free Quantitative In Vitro Live Cell Imaging with Digital Holographic Microscopy

B. Kemper, A. Bauwens, D. Bettenworth, M. Götte, B. Greve, L. Kastl, S. Ketelhut, P. Lenz, S. Mues, J. Schnekenburger, and A. Vollmer

Contents

B. Kemper (✉), L. Kastl, S. Ketelhut, S. Mues, and J. Schnekenburger
Biomedical Technology Center, University of Muenster, Münster, Germany
e-mail: bkemper@uni-muenster.de

A. Bauwens
Institute of Hygiene, University Hospital Muenster, Münster, Germany

D. Bettenworth
Department of Medicine B, University of Muenster, Münster, Germany

M. Götte
Department of Gynecology and Obstetrics, University Hospital Muenster, Münster, Germany

B. Greve
Department of Radiotherapy – Radiooncology, University Hospital Muenster, Münster, Germany

P. Lenz
Department of Medicine B, University Hospital Muenster, Münster, Germany

Institute of Palliative Care, University Hospital Muenster, Münster, Germany

A. Vollmer
Center for Biomedical Optics and Photonics, University of Muenster, Münster, Germany

Abstract   Label-free quantitative in vitro imaging of living cell cultures with light microscopy is an important tool for various research fields in the life sciences. Digital holographic microscopy (DHM) provides contactless, minimally invasive quantitative phase contrast imaging and can be integrated as a module in common research microscopes. Due to the numerical reconstruction of quantitative phase images, multi-focus imaging is achieved from a single digital hologram. The evaluation of the recorded quantitative phase contrast images allows the extraction of data for simplified object tracking and image segmentation. The special DHM feature of numerical autofocusing avoids mechanical focus realignment. As quantitative DHM phase imaging is based on the detection of optical path length changes in transmission, the method only requires low light intensities for object illumination which minimizes the interaction with the sample. Thus, minimally invasive long-term time-lapse investigations for quantitative monitoring of dynamic changes of cell morphology, motility, and proliferation are accessible. In addition, the integral cellular refractive index, which is related to intracellular solute concentrations as well as cellular volume and dry mass, is available. The chapter starts with an introduction to DHM for live cell observation and procedures for the extraction of biophysical parameters from quantitative DHM phase contrast images. After the physical basis has been laid out, several selected applications of in vitro live cell analysis are described. This includes the characterization of suspended cells and spherical intracellular organelles as well as the quantification of the cellular response to osmotic stimulation, drugs, toxins, nanomaterials, and genetic modifications. Subsequent paragraphs illustrate how DHM can be applied to quantify cell motility, migration, and the morphology of adherent cell cultures. Finally, phenotyping based on cell thickness determination, dynamic multimodal imaging of cellular growth, proliferation, and wound healing in vitro as well as applications in toxicity testing of pathogens and the characterization of cell nanomaterial interactions are demonstrated.

Keywords Biophysical cell analysis · Digital holographic microscopy · Label-free in vitro imaging · Quantitative microscopy · Quantitative phase imaging

1   Introduction

In vitro imaging of living single cells with light microscopy is essential for many research areas in biology and medicine as live cell imaging provides important insights into various cellular properties such as motility or biomechanics which are related to cell structure changes. Furthermore, there is a strong need to quantify the dynamic response of cells to drugs, toxins, pathogens, and nanomaterials. Ideal imaging approaches for the investigation of life cell activities should be minimally invasive to affect the living specimen under investigation as little as possible and provide quantitative measurement data. Fluorescence microscopy has been extensively applied for live cell imaging [1, 2] and allows high-resolution 3D imaging of cellular structures and organelles with a localization accuracy down to a few nanometers (see [3] and references therein). However, while the specificity of fluorescence signals is high due to a large number of available fluorescence-labeled antibodies or otherwise specific fluorescence labeling [4], specific autofluorescence in biological samples [5], fluorescence lifetime imaging (FLIM) [6], and long-term experience with these techniques, there still remain challenges in the application of these methods for live cell imaging. For example, long-term in vitro investigations are problematic as many fluorescence dyes are (photo)toxic, require high intensities of the excitation light, or may interact with the substances to be tested [7]. Endogenous fluorophores like the green fluorescent protein (GFP) [8] allow extended live cell investigations but commonly require genetic engineering, e.g., by chemical [9] or optical [10] transfection. Thus, in the past years also, many label-free methods such as optical coherence tomography (OCT) [11, 12], mid-IR imaging [13, 14], and Raman spectroscopy [15] were explored for its usefulness in the life sciences. In addition, in order to make use of the high accuracy of light diffraction and interferometry-based metrology, various quantitative phase imaging techniques were developed and continuously improved with the aim of high-resolution label-free quantitative imaging of living cells and tissues in various applications [16–38]. Quantitative phase imaging is based on the detection of optical path length differences that are induced by a mainly transparent specimen against the surrounding environment. In contrast to, e.g., OCT and fluorescence microscopy, the illumination is performed in transmission. Thus, only low light intensities are required which minimizes the interaction with the sample and makes the technology in particular suitable for label-free and minimally invasive long-term live cell imaging in vitro.

This chapter focuses on quantitative label-free in vitro live cell imaging with digital holographic microscopy (DHM), a variant of quantitative phase microscopy. Holographic interferometric metrology is a well-established tool in industrial nondestructive testing and quality control [39–41]. In combination with microscopy, digital holography provides label-free, quantitative phase contrast imaging [17, 19, 20, 25, 42]. The reconstruction of digitally captured holograms is performed numerically. Thus, in comparison with Zernike's phase contrast microscopy [43] and differential interference contrast (DIC) [44] or OCT, DHM offers quantitative phase contrast including the option for subsequent numerical focus correction (multi-focus imaging) from single holograms. In difference to fluorescence microscopy and spectroscopic techniques that provide molecular specificity, DHM returns complementary global or *wholistic* information about the morphology and the content of living cellular specimens. The evaluation of quantitative DHM phase images permits the extraction of integral biophysical parameters globally quantifying the intracellular content, dynamic morphology, and volume changes as well as cell motility.

Aiming to illustrate the method and to demonstrate the various application fields of DHM, several variants of DHM that have been found suitable for efficient label-free in vitro analysis of living cell cultures and single cells are presented in a short review. The first paragraph describes different experimental setups that allow the modular integration of DHM into common research microscopes and the use of DHM in an interdisciplinary biomedical laboratory environment. Then, the numerical reconstruction of quantitative phase images from digitally captured holograms and holographic autofocusing is illustrated. Afterward, we explain the evaluation of quantitative DHM phase images by image segmentation, the quantification of cell motility, and the extraction of global biophysical parameters such as refractive index, volume, and dry mass. The focus of the subsequent paragraphs is to demonstrate the application of DHM in different assay formats. First, quantitative DHM phase imaging is used for the characterization of suspended single cells and cellular organelles as well as for monitoring the intracellular water content and for the quantification of the response of cells to genetic modifications and the influence of cytokines. Then, we describe the application of DHM for the time-resolved monitoring of transfection-induced motility changes, label-free imaging of cytokinesis, and cell phenotyping by measuring the cell thickness. Afterward, multimodal analysis of mixed cell cultures and in vitro wound healing assays as well as the characterization of toxicity effects from pathogens and cell-nanomaterial interactions is demonstrated. Finally, future prospects and challenges are discussed.

2 Digital Holographic Microscopy for Live Cell Imaging

Digital holographic microscopy is based on the classic holographic principles [45–47] with the difference that hologram recording is performed with a digital sensor and that the reconstruction is performed numerically with a computer [48]. However, during the past decade, various different approaches for the implementation of digital holographic microscopy (DHM) and quantitative phase imaging have been developed (for an overview see, e.g., [25, 49–51] and references therein). For label-free imaging of living cells in a biomedical laboratory environment, the combination

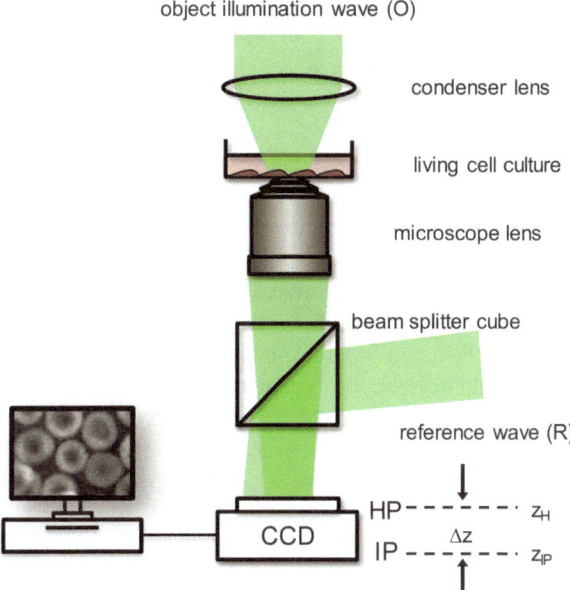

Fig. 1 Schematic of an inverted off-axis digital holographic microscope with transmitting light illumination for label-free quantitative imaging of living cell cultures. A laser beam is divided by a beam splitter into an *object wave* that illuminates the specimen via a condenser lens and an undisturbed *reference wave* not passing through the sample. The *object wave* interferes with the slightly tilted *reference wave* (off-axis geometry) on a digital sensor, e.g., a charge-coupled device (CCD). Morphological changes within the biological specimen induce changes of the optical path length of the object wave, which are then encoded in the resulting interference pattern (digital off-axis hologram). *HP* hologram plane located at $z = z_H$, *IP* image plane located at $z = z_{IP}$

of robust and flexible DHM setups with common research microscopes is very important for easier acceptance in the community and to simplify its combination with other optical and nonoptical imaging methods [52–57]. Thus, after a general introduction to DHM, two principles for quantitative phase imaging with DHM are described that are suitable for the modular integration into various inverted research microscopes. In order to ensure a vibration-insensitive data acquisition, off-axis arrangements in Mach-Zehnder and Michelson interferometer configurations are described as these allow a fast and robust single-shot acquisition of digital off-axis holograms.

2.1 Principles of Off-Axis Digital Holographic Microscopy

Figure 1 schematically depicts the setup of an off-axis digital holographic microscope. An inverted microscope arrangement, in combination with an illumination of the sample in transmission, enables the investigation of mainly transparent samples in a liquid, for example, living cell cultures in a Petri dish filled with cell culture medium. The coherent light of a laser is divided into an *object illumination wave O* and a *reference wave R*. In analogy to bright-field microscopy with white light illumination, a condenser lens is used to achieve homogeneous illumination of the sample. The reference wave is guided directly via a beam splitter cube to a digital image recording device, for example, a charge-coupled device (CCD) or a complementary metal oxide semiconductor (CMOS) [48], that digitizes the holograms in the hologram plane HP at $z = z_H$. Holographic off-axis geometry is achieved by a slight tilt of the *reference wave* against the wave front of the *object wave*. The system's lateral resolution is mainly restricted by the pixel pitch of the applied image recording sensor. The transmitted *object wave* that includes the information about the sample is magnified by a microscope lens. The magnification is selected such that the smallest structures that can be resolved within the diffraction limit of the optical imaging system are oversampled with the hologram recording device [22] so that the lateral resolution is not decreased by numerical reconstruction algorithms as, for example, described in Sects. 2.3 and 2.4.

The upper panel of Fig. 2 shows an optical fiber-based modular Mach-Zehnder interferometer concept for digital holographic microscopy. The system is designed for the integration into common inverted research microscopes and thus allows investigating living cells in a physiological environment [52].

The light of a laser, here a frequency-doubled neodymium-doped yttrium aluminum garnet (Nd:YAG) laser, $\lambda = 532$ nm with a large coherence length >1 m, is divided into an object illumination wave (object wave) and a reference wave. Variable light guidance is achieved by coupling of the light into single-mode optical fibers. The illumination of the sample with coherent laser light is performed in transmission by introduction of the object wave into the microscope's condenser. This permits an adjustable optimized illumination of the sample. The setup relies on common microscope objectives. The reference wave is guided via an optical fiber directly to an interferometric unit that is adapted to one of the microscope's camera ports. In difference to the setup that is shown in Fig. 1, the beam splitter tilts the reference wave front against the wave front of the object wave to provide off-axis holography. The superimposed object and reference waves are recorded with a CCD camera and transferred to an image processing unit for reconstruction and evaluation of the digitized holograms.

The lower panel of Fig. 2 illustrates the modular integration of DHM into an inverted research microscope (iMIC, TiLL Photonics GmbH, Munich, Germany) that is equipped with a temperature-controlled incubation chamber (Solent Scientific Ltd., Segensworth, United Kingdom) for live cell imaging at physiological temperature.

A drawback of Mach-Zehnder interferometer-based quantitative phase imaging arrangements, as sketched in Fig. 2, is the requirement for a separate reference wave. This induces a reduction of phase stability and requires a precise adjustment of the intensity ratio between object and reference wave. To overcome these problems,

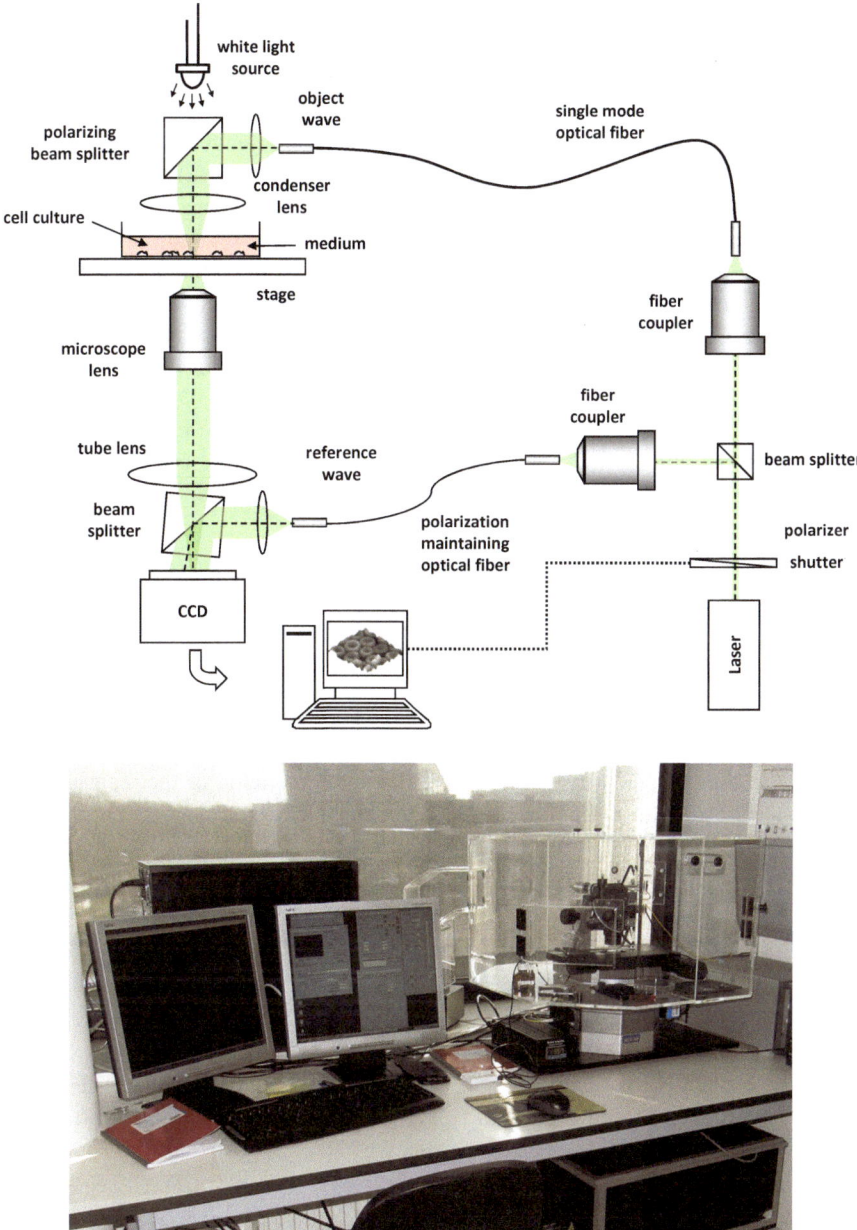

Fig. 2 Upper panel: schematic of the modular integration of digital holography into an inverted microscope (adapted from [52]). Lower panel: photo of a research microscope (iMIC, TiLL Photonics GmbH, Munich, Germany, modified for DHM) including a heating chamber (Solent Scientific Ltd., Segensworth, United Kingdom) for live cell imaging at physiological temperatures (adapted from [58])

various approaches were reported (see [59] and included references). For instance, a Michelson interferometer approach for DHM avoids a separate reference wave [59]. The advantage of this setup is that different from other self-interference or common path-based approaches no additional components like temporal phase shifting devices or customized reflective surfaces are required. Moreover, no diffractive optical elements (DOEs) or spatial light modulators (SLMs) are needed, which are expensive, complicated to align, and may affect the object wave by spatial filtering which influences the lateral resolution. Figure 3 shows on the left a sketch of the self-interference DHM principle and on the right a photo of a Michelson interferometer-based DHM setup attached to a modified inverted microscope (OptikaXDS-3, OPTIKA SRL, Italy) as routinely used in cell culture experiments. The sample is illuminated by light from a laser diode ($\lambda = 532$ nm) with a low coherence length <1 mm and a single-mode optical fiber via the illumination path of the microscope's white light source. A tunable lens allows the modulation of the object illumination for enhanced quantitative phase imaging (see Sect. 3.12 and [60]). After passing the microscope objective, the light is coupled into a Michelson interferometer that consists of a beam splitter cube and two mirrors. One of the mirrors within the Michelson interferometer is tilted by a slight angle α such that the image of an area of the sample that contains no biological objects was superposed on the image of the specimen to create a suitable spatial carrier fringe pattern for off-axis holography (for details see [59]). Due to the Michelson interferometer arrangement, two wave fronts with almost identical curvatures are superimposed in areas without specimen. This is even fulfilled for an imaging geometry with two slightly divergent waves that differ from a collimated arrangement [60], as it is sketched in the left panel of Fig. 3 to simplify the illustration of the measurement principle. The digital holograms are recorded by a CCD sensor. The numerical calculation of the quantitative DHM phase contrast images from the resulting self-interference digital holograms is performed as described in Sects. 2.3 and 2.4.

2.2   Evaluation of Digital Off-Axis Holograms

The intensity distribution I_H of the interferogram in the hologram plane HP that is created with the experimental configurations in Figs. 1, 2, or 3 by interference of the object wave with the slightly tilted reference wave is:

$$
\begin{aligned}
I_H(x, y, z_H) &= O(x, y, z_H)O^*(x, y, z_H) + R(x, y, z_H)R^*(x, y, z_H) \\
&\quad + O(x, y, z_H)R^*(x, y, z_H) + R(x, y, z_H)O^*(x, y, z_H) \\
&= I_O(x, y, z_H) + I_R(x, y, z_H) + 2\sqrt{I_O(x, y, z_H)I_R(x, y, z_H)} \cos \Delta\phi_{HP}(x, y, z_H)
\end{aligned}
$$

$$(1)$$

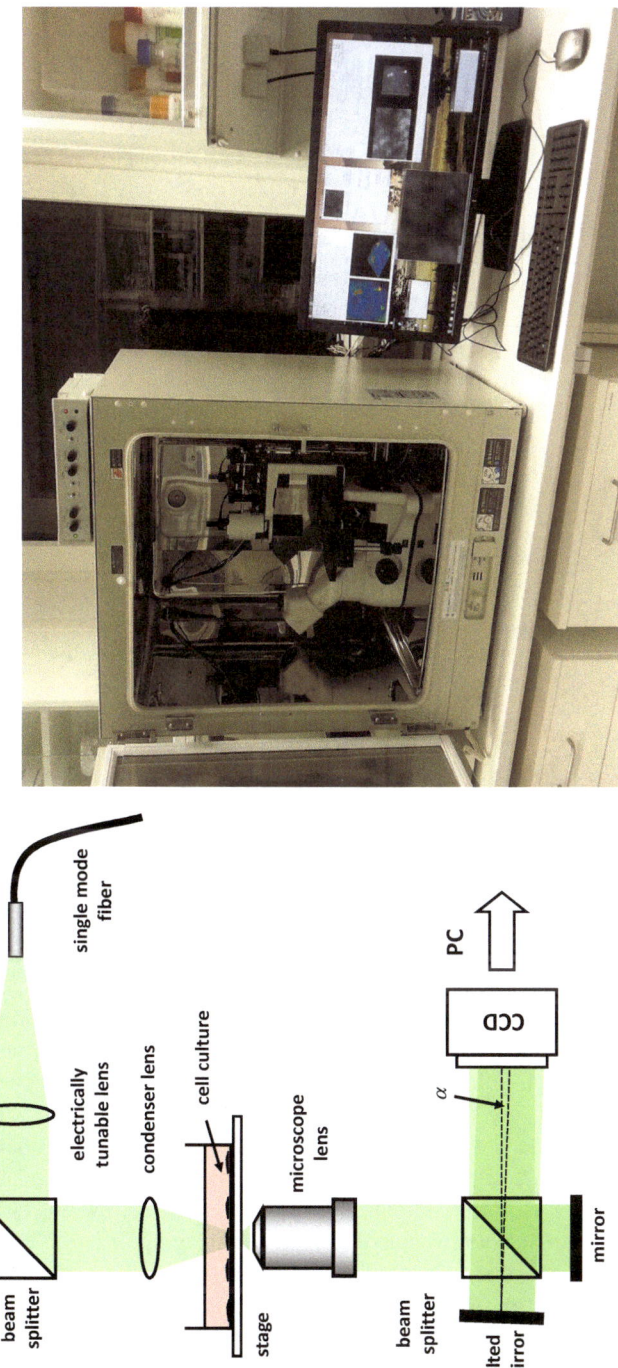

Fig. 3 Left panel: schematic of a setup for Michelson interferometer-based self-interference DHM for observation of transparent samples in transmission (adapted from [59]). Right panel: self-interference DHM adapted to a cell culture microscope. The system is operated inside a culture incubator which is utilized to achieve physiological temperature and CO_2 atmosphere.

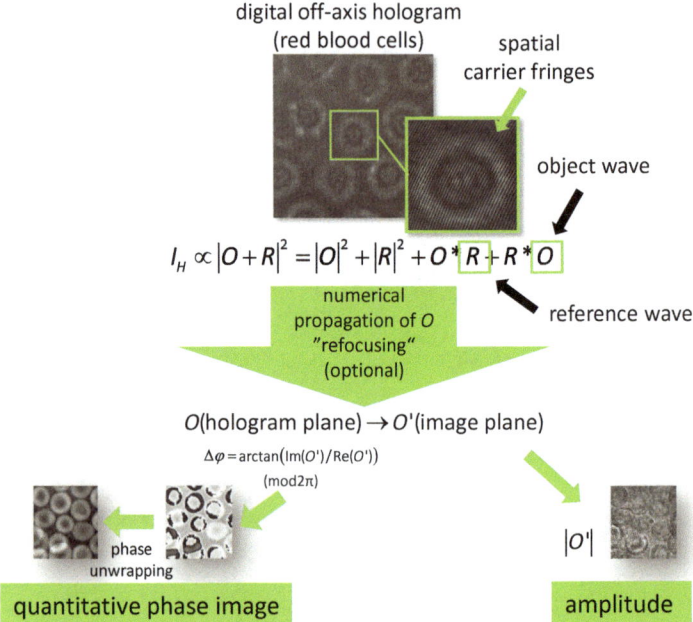

digital off-axis hologram
(red blood cells)

spatial
carrier fringes

object wave

$$I_H \propto |O+R|^2 = |O|^2 + |R|^2 + O^*R + R^*O$$

numerical
propagation of O
"refocusing"
(optional)

reference wave

O(hologram plane) $\rightarrow O'$(image plane)

$\Delta\varphi = \arctan(\mathrm{Im}(O')/\mathrm{Re}(O'))$
$(\mathrm{mod}\,2\pi)$

phase
unwrapping

$|O'|$

quantitative phase image

amplitude

Fig. 4 Evaluation of digital off-axis holograms for multi-focus quantitative phase imaging

with $I_O = OO^* = |O|^2$ and $I_R = RR^* = |R|^2$ (*represents the conjugate complex terms). The parameter $\Delta\phi_H(x, y, z_H) = \phi_R(x, y, z_H) - \phi_O(x, y, z_H)$ denotes the phase difference between O and R in the hologram plane $z = z_H$. In the presence of a sample in the optical path of O, the phase distribution represents the sum $\phi_O(x, y, z_H) = \phi_{O_H}$ $(x, y, z_H) + \Delta\varphi_S(x, y, z_H)$ with the pure object wave phase $\phi_{O_H}(x, y, z_H)$ (i.e., without sample) and the phase change $\Delta\varphi_s(x, y, z_H)$ that is caused by the sample. In the following paragraphs, it is explained how the sample-induced phase change $\Delta\varphi_s(x, y, z_{IP})$ in the image plane $z = z_{IP}$ is retrieved from the intensity distribution in Eq. (1).

During the last decade, many methods for the numerical reconstruction of digital off-axis holograms have been developed and adapted to various applications (for an overview, see [25, 48, 61–64] and included references). Here, the numerical reconstruction of digital holograms is illustrated by a concept that is performed in two subsequent steps (see Fig. 4 for illustration). First the complex object wave O is reconstructed in the hologram plane. This can be performed by spatial phase shifting or by Fourier transformation-based spatial filtering in spatial frequency domain (see Sects. 2.3 and 2.4). Both methods provide the retrieval of the object wave and were found highly applicable for quantitative phase imaging of living cells. For focused imaging of objects that have been recorded out-of-focus, for example, during long-term microscopic observations, numerical refocusing is required. Therefore, in an optional subsequent step, O is numerically propagated to the image plane. This is typically achieved by the numerical implementation of the Fresnel-Huygens principle [48, 65]. Here the numerical wave propagation is explained for a variant of the

convolution method in which the *convolution theorem* and the Fresnel approximation [22, 66] are applied. The propagation of $O(x, y, z_H)$ from the hologram plane to
the image plane z_{IP} that is located at $z_{IP} = z_H + \Delta z$ in the distance Δz to the hologram
plane (see Fig. 1) is performed by using the equation:

$$O(x, y, z_{IP} = z_H + \Delta z) = F^{-1}\{F\{O(x, y, z_H)\}\exp(i\pi\lambda\Delta z(\nu^2 + \mu^2))\} \quad (2)$$

In Eq. (2), λ is the illuminating laser light wave length, ν and μ are the coordinates
in frequency domain, and F denotes a Fourier transformation. The advantage of this
approach is that the size of the propagated wave field is preserved during the
refocusing process. This is a particular advantage for numerical autofocusing as
described in Sect. 2.5 because it simplifies the comparison of the images from
different focal planes. In contrast to propagation by digital Fresnel transformation
[16, 17], Eq. (2) also allows for refocusing of only slightly defocused images of the
sample close to the hologram plane [22]. It is noteworthy that numerical refocusing
may be performed with other common numerical propagation methods as well,
including in particular more general approaches of the *convolution method* [48]
and the *angular spectrum method* [64, 67]. During the propagation process, the
parameter Δz in Eq. (2) is set such that the holographic amplitude image $|O|$ appears
focused like a microscopic bright-field image under white light illumination. In the
special case that the image of the sample is focused in the hologram plane with
$\Delta z = 0$ and thus $z_{IP} = z_H$, the reconstruction process is faster because no propagation
of O by Eq. (2) is required.

In addition to the absolute amplitude $|O(x, y, z_{IP})|$ that represents the image of the
sample, the numerically reconstructed and propagated complex object wave $O(x, y, z_{IP})$ provides the phase information $\Delta\varphi_S(x, y, z_{IP})$ of the sample according to Eq. (3):

$$\Delta\varphi_S(x, y, z_{IP}) = \arctan\frac{\text{Im}\{O(x, y, z_{IP})\}}{\text{Re}\{O(x, y, z_{IP})\}} \ (\text{mod } 2\pi) \quad (3)$$

After elimination of the 2π ambiguity, caused by the periodic properties of the
arcustangent function, by phase unwrapping [39], the data obtained by Eq. (3) can be
used for quantitative phase imaging (cf. Fig. 4).

In *transmission mode* as sketched in Figs. 1, 2, and 3, the induced phase change

$$\Delta\varphi_S(x, y, z_{IP}) = \frac{2\pi}{\lambda}(n_S - n_{medium})d_S(x, y) \quad (4)$$

of a semitransparent sample in the image plane, located at $z = z_{IP}$, is influenced by
the sample thickness d_S, the refractive index n_S of the sample, and the refractive
index of the medium n_{medium} that surrounds the specimen under study. Thus, for
quantitative evaluation of phase images, information about the cellular refractive
index is required. Several interferometric and holographic methods to determine the
cellular refractive index have been developed (see, e.g., reference collections in

Fig. 5 Principle of label-free quantitative phase contrast imaging. The object-induced phase shift $\Delta\varphi_{cell}$ depends on the cell thickness d_{cell}, the cellular refractive index n_{cell}, and the refractive index n_{medium} of the surrounding medium

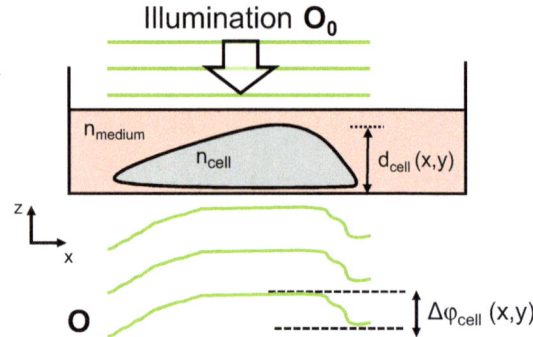

[25, 68, 69] and Sect. 2.8). However, it is not possible to unequivocally decouple sample *thickness* and its *refractive index* for every measurement scenario.

For cells in cell culture medium with the refractive index n_{medium} and under the assumption of a homogeneously distributed, integral cellular refractive index $n_S := n_{cell}$ (cf. Fig. 5), the cell thickness $d_s := d_{cell}(x, y)$ can be determined by measuring the optical path length change $\Delta\varphi_S := \Delta\varphi_{cell}$ of the cells compared to the surrounding medium [19, 22]:

$$d_{cell}(x, y, z_{IP}) = \frac{\lambda \Delta\varphi_{cell}(x, y, z_{IP})}{2\pi} \cdot \frac{1}{n_{cell} - n_{medium}} \qquad (5)$$

The parameter λ in Eq. (5) represents the wavelength of the applied laser light. For adherently growing cells, the parameter d_{cell} is an integral indicator for cell shape (Fig. 5). But the interpretations based on Eq. (5) have to be handled with care as cells that are exposed to toxins or non-isotonic media (see [22, 70] and Sects. 3.2, 3.3, 3.10, and 3.11) may not only change their shape but also alter their integral cellular refractive index.

2.3 Spatial Phase Shifting-Based Reconstruction of Off-Axis Holograms

This section discusses the numerical reconstruction of digital off-axis holograms by spatial phase shifting. In spatial phase shifting holography, neighboring pixels of single off-axis holograms are evaluated by common interferometric phase shifting algorithms [71, 72]. The required spatial phase shift between *object wave* and *reference wave* is generated by an adequate off-axis tilt of one of the waves (cf. Figs. 1, 2, 3). However, the interferogram equation in Eq. (1) can be also solved pixel-wise within a squared area of pixels, in practice, for example, 5 × 5 pixels around the hologram pixel under analysis. As described in detail in [17] and [73],

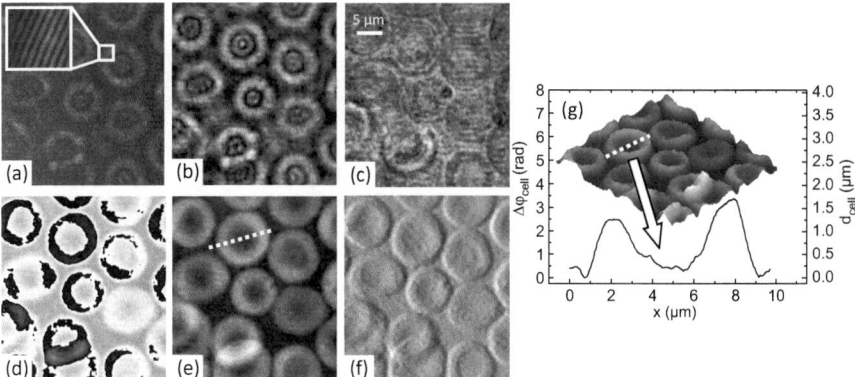

Fig. 6 Evaluation of digital off-axis holograms by spatial phase shifting. (**a**) Slightly defocused digital hologram of living human red blood cells in phosphate-buffered saline (PBS), (**b**) reconstructed amplitude image in the hologram plane (not in focus), (**c**) numerically refocused amplitude distribution, (**d**) reconstructed phase distribution modulo 2π corresponding to (**c**) coded to 256 gray levels; (**e**) unwrapped phase distribution coded to 256 gray levels; (**f**) first derivative of (**e**) in x-direction, (**g**) pseudo-3D representation of the phase distribution in (**e**) with cross section through a cell (adapted from [76])

the resulting nonlinear problem is transformed to a linear one by defining appropriate substitutions. The resulting robust reconstruction method for the complex object wave O was found particularly suitable for the application in DHM and has been applied successfully for the analysis of living cells [17, 22]. To minimize phase noise, the optimum spatial phase shift per pixel should be close to 90° or $\pi/4$ [74, 75]. If the sample is not imaged sharply onto the hologram acquisition device during the hologram recording, a subsequent propagation of the reconstructed object wave to the image plane can be performed with Eq. (2) as described in Sect. 2.2.

Figure 6 illustrates the reconstruction process of a digital off-axis hologram by spatial phase shifting-based reconstruction. Figure 6a shows an off-axis hologram of human red blood cells (RBCs) in phosphate-buffered saline (PBS). The cells were recorded slightly defocused with a 63× microscope objective (NA = 0.75) using a setup as sketched in Fig. 2 including a frequency-doubled Nd:YAG laser ($\lambda = 532$ nm) as light source. The enlarged area in Fig. 6a shows part of the carrier fringe pattern that was used for holographic encoding of the object wave. Figure 6b depicts the reconstructed out-of-focus amplitude image of the RBCs in the hologram plane (CCD sensor plane) and after numerical propagation to the in-focus image plane (Fig. 6c) using Eq. (2). This image corresponds to a microscopic bright-field image under illumination with coherent laser light. Figure 6d shows the phase distribution modulo 2π. The quantitative phase image after elimination of the 2π ambiguity is added as Fig. 6e. Figure 6g illustrates the outcome of a cell thickness profile along a cross section through the phase data. The thickness values d_{cell} of the RBC were calculated from the phase contrast $\Delta\varphi_{cell}$ using Eq. (5) (cf. Sect. 2.2) assuming for the integral cellular refractive index $n_{RBC} = 1.400$ [77]. The refractive index of the buffer solution $n_{medium} = 1.337$ was obtained by an Abbe refractometer.

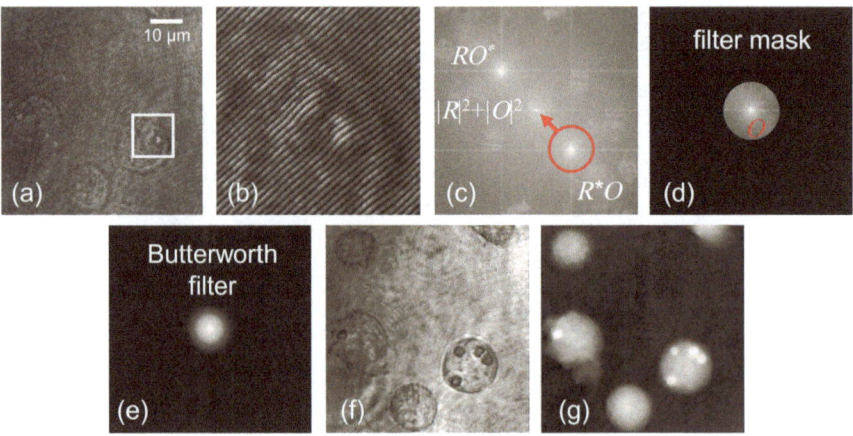

Fig. 7 Evaluation of digital off-axis holograms by spatial filtering. (**a**) Digital off-axis hologram of suspended human fibrosarcoma cells (HT1080) with internalized silica microspheres (diameter ≈3.44 μm), (**b**) enlarged part of the hologram (see box in (**a**)), (**c**) 2D frequency spectrum of (**a**) with the spatially separated "zero-order intensity" ($|R|^2 + |O|^2$), image (R^*O), and twin image (RO^*), (**d**) spectrum of the object wave after application of a spatial filter and elimination of the phase shift due to the carrier fringe pattern, (**e**) data in (**d**) after Butterworth filtering, (**f**) reconstructed amplitude $|O|$, (**g**) quantitative phase image (adapted from [81])

Figure 6g visualizes the cell morphology by a gray level-encoded pseudo-3D representation of the cell thickness retrieved from the data in Fig. 6e.

2.4 Hologram Evaluation by Spatial Filtering

Beside spatial phase shifting techniques that are described in Sect. 2.3, spatial filtering based on Fourier transformation represents another efficient common approach for the retrieval of the *object wave* in digital holography. Fourier transformation methods were first applied in optical holography for fringe pattern analysis and phase retrieval from double exposure interferograms [78, 79]. In order to reconstruct the *object wave* in the hologram plane from digitally recorded off-axis holograms by spatial filtering, the intensity distribution $I_H(x, y, z_H)$ given in Eq. (1) is evaluated by a two-dimensional Fourier transform following [20, 78]. Figure 7 illustrates the analysis of digital off-axis holograms by spatial filtering using suspended human fibrosarcoma cells (HT1080, [80]) with internalized silica microspheres (diameter ≈3.44 μm) in cell culture medium as example. Figure 7a shows a digital off-axis hologram of the cells that was recorded with an experimental setup as shown in Fig. 3 using a 532 nm laser illumination. Figure 7b shows an enlarged part of the carrier fringe pattern of the off-axis hologram (box in Fig. 7a) used for holographic encoding of the object wave.

Figure 7c depicts the 2D frequency spectrum that is retrieved numerically by a fast Fourier transformation (FFT) from Fig. 7a for an adequate off-axis angle between *object* and *reference wave* for which the components "zero-order intensity" ($|R|^2 + |O|^2$), image (R^*O), and twin image (RO^*) appear spatially separated. In the next evaluation step, the components $|R|^2 + |O|^2$ and RO^* are eliminated by application of a suitable spatial filter mask. Then, the contribution of the phase-conjugated reference wave is removed by a shift of the term R^*O toward the center of the spectrum. The remaining parts of the spectrum (Fig. 7d) are filtered with a Butterworth filter to reduce disturbing spatial frequencies and numerical artifacts like the ringing effect (Fig. 7e). Afterward, the *object wave O* in spatial domain is calculated by an inverse FFT. Figure 7f shows the resulting amplitude distribution $|O|$. In Fig. 7g the unwrapped phase contrast image is depicted after background correction. If the sample is not imaged in focus of the image sensor during the hologram recording, a numerical propagation of O to the image plane can be performed by using Eq. (2) in analogy to spatial phase shifting (cf. Sect. 2.2).

2.5 Automated Holographic Refocusing

As described in Sect. 2.2, digital holography allows for adjustable refocusing of the sample by variation of the propagation distance Δz in the numerical recipe summarized in Eq. (2). The combination of this feature with image sharpness quantification algorithms provides the option for autofocusing (for an overview see [76] and references therein). Pure phase objects with negligible absorption such as technical reflective specimens or biological cells are sharply focused when the amplitude distributions show the least contours [82, 83]. In contrast to bright-field microscopy using white light, this setting is of particular interest in digital holography as the amplitude and phase distributions are accessible simultaneously. Those focal settings producing the least contrast in amplitude images provide the best spatial resolution in the quantitative phase contrast distribution.

Various methods for holographic autofocusing of transparent and absorbing specimens have been reported (see [76]). In [83] different numerical methods that calculate a scalar focus value to quantify the image sharpness were compared using common criteria from bright-field microscopy to evaluate image sharpness [84, 85]. In agreement with previously reported results for microscopic imaging with white light illumination [86, 87], the evaluation of weighted and band-pass-filtered power spectra was identified as a robust determination of image sharpness in DHM [83, 88] in combination with the *convolution method* described in Sect. 2.2. The principle of holographic autofocusing is illustrated in Fig. 8 applying the band-pass-filtered power spectra for image sharpness quantification.

Figure 8 shows results recorded from formalin-fixed pancreatic adenocarcinoma cells (PaTu 8988 S) on a glass carrier slide with a setup as depicted in Fig. 2 ($\lambda = 532$ nm, $63\times$ microscope lens, NA $= 0.75$). An off-axis hologram of the slightly defocused cells was recorded and evaluated by spatial phase shifting as

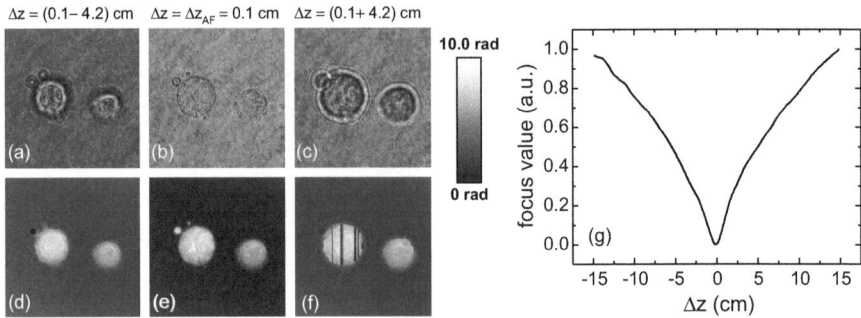

Fig. 8 Principle of holographic autofocusing. Comparison of defocused (**a**), (**c**), (**d**), (**f**) and focused (**b**), (**e**) reconstructed amplitude (**a**)–(**c**) and unwrapped phase distributions (**d**)–(**f**) obtained from fixed pancreatic tumor cells (PaTu8988 S) on an object carrier glass slide in transmission. (**g**) Focus values calculated by spectral weighted analysis in dependence of the propagation distance Δz (adapted from [76])

described in Sect. 2.3. Figure 8 shows the impact of (de-)focusing on the reconstructed amplitudes (a–c) and the unwrapped phase distributions (d–f). In the case of focused imaging, the phase specimens almost disappear in the reconstructed amplitude distribution (Fig. 8b), while the object appears focused in the unwrapped phase distribution in Fig. 8e with maximum contrast. In both defocused cases, the amplitude distributions (Fig. 8a, c) show diffraction patterns and thus are visible with enhanced contrast, while the phase distributions appear blurred (Fig. 8d, f). In Fig. 8g the normalized focus value function obtained by spectral weighted analysis is plotted as a function of propagation distance Δz. The focus value passes through a global minimum for a propagation distance of $\Delta z = \Delta z_{AF} = 0.1$ cm which corresponds to the focus point in amplitude and phase distributions of the sample.

The black vertical lines in the unwrapped phase distributions in Fig. 8f show that diffraction patterns induced by defocusing may lead to phase singularities. These artifacts can cause phase unwrapping errors that inhibit further data evaluation. This scenario illustrates the urgent need for digital holographic autofocusing for automated DHM data processing during long-term observations of cultured cells with high numerical aperture microscope objectives.

Figure 9 illustrates the advantage of holographic autofocusing during long-term live cell observation (>50 h) on the example of human brain microvascular endothelia cells (HBMECs) in a Petri dish upon exposure to a toxin. The experiments were performed with a DHM setup as depicted in Fig. 2 using a 40× microscope objective (NA = 0.6) with a 532 nm laser illumination. Figure 9a–f shows representative amplitude and phase contrast images of the sample at two different time points. At the beginning of the experiment at $t = 0$, the HBMECs were imaged in focus of the hologram plane (Fig. 9a, d). Due to a mechanical instability of the experimental setup, the cells appeared defocused in the reconstructed amplitude and phase distributions after $t = 31.5$ h without numerical focus correction (Fig. 9b, e).

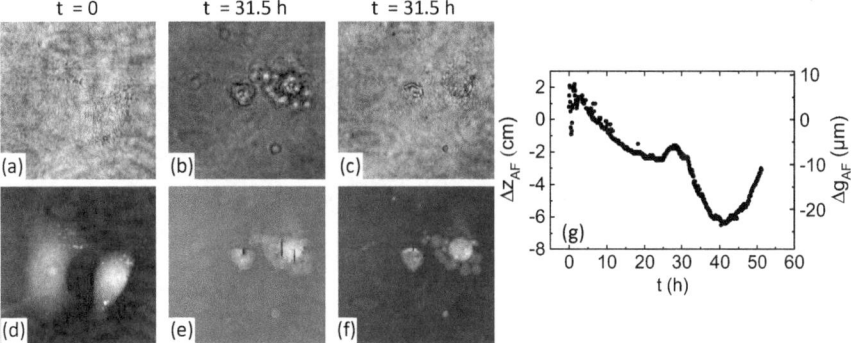

Fig. 9 Automated holographic refocusing during long-term imaging of living cell cultures. Reconstructed amplitude (**a**)–(**c**) and unwrapped phase (**d**)–(**f**) distributions retrieved from long-term investigations on living endothelia cells (HBMECs) in a Petri dish; (**a**), (**d**): $t = 0$; (**b**), (**e**): $t = 31.5$ h without autofocusing; (**c**), (**f**): $t = 31.5$ h with autofocusing; (**g**) time dependency of autofocus position z_{AF} and corresponding change of the axial object position Δg_{AF} (adapted from [76])

The resulting images after numerical refocusing are shown in Fig. 9c, f. Now the cells are in focus and subcellular structures are clearly resolved. The toxin-induced onset of apoptosis is clearly visible in Fig. 9f as apoptotic vesicles bud off the cell body. Figure 9g shows the focal position that was detected by numerical autofocusing for the whole measurement period ($t_{max} = 52$ h) as a function of time. The nonlinear drift illustrates the need for a permanent focus control during long-term investigations. Moreover, the autofocusing routine provides quantitative data about the object's axial position that may be helpful to improve the stability of the experimental setup or to identify the sources of instability.

2.6 Evaluation of Quantitative DHM Phase Images by Image Segmentation

Quantitative phase images of living cells are efficiently evaluated by image segmentation. The procedure is illustrated in Fig. 10 for a mixed cell culture that contains different pancreatic tumor cell types (PaTu 8988 T, PaTu 8988 S [22, 89]) with individual morphologies. Quantitative DHM phase images (Fig. 10b) are first calculated from a series of digital off-axis holograms (Fig. 10a) and are then normalized (Fig. 10c) and corrected for background in homogeneities (Fig. 10d) as described in [90]. Afterward common numerical algorithms for image segmentation are applied (Fig. 10e) as, for instance, provided by the free software *Cell Profiler* (http://www.cellprofiler.org, [91]) which allows using customized pipelines for automated image processing. In the example summarized in Fig. 10, larger objects like the conglomerates of PaTu 8988 S cells were identified first. Then, the identified

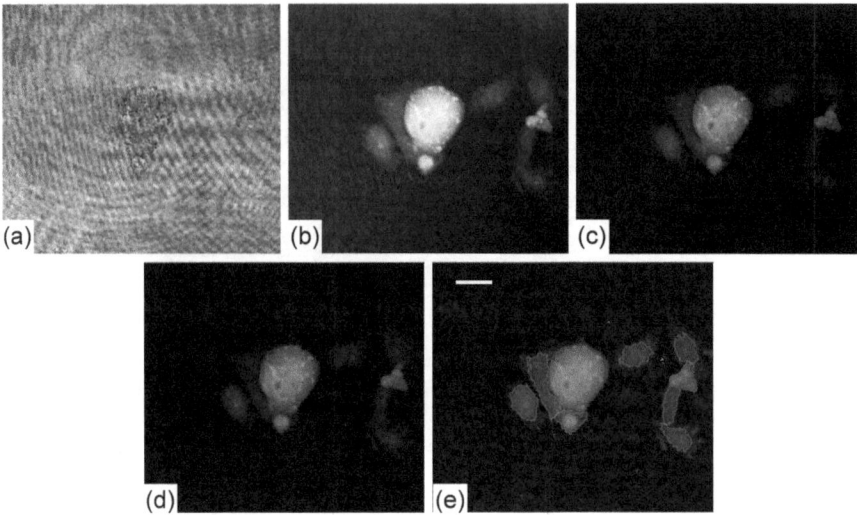

Fig. 10 Evaluation of quantitative phase images by image segmentation. (**a**) Digital off-axis hologram; (**b**) reconstructed quantitative phase image of a mixed cell culture with different tumor cell types (PaTu 8988 T, PaTu 888 S); (**c**) normalized quantitative phase image; (**d**) quantitative phase image after background correction; (**e**) background-corrected quantitative phase distribution after image segmentation. Red outlines in (**e**) indicate conglomerates of PaTu 8988 S cells, while yellow outlines mark PaTu 8988 T cells. The scale bar corresponds to 40 μm (adapted from [58])

objects were masked out in a second step, and the remaining areas of the quantitative phase images were analyzed for smaller objects, here PaTu 8988 T cells.

From the segmented images, the projected surface area S_c for each cell type and the average phase contrast $\Delta\bar{\varphi}$ are determined. This data may be used to quantify rates of cell proliferation by calculating the cellular dry mass (DM) for any time point of the experiment [92, 93]

$$\mathrm{DM} = \frac{10\lambda}{2\pi\gamma}\Delta\bar{\varphi}S_c \tag{6}$$

that correlates with various cellular phenotypes (see [94] and references therein). In Eq. (6) the symbol λ denotes the illumination wavelength, while γ is a constant known as the *specific refraction increment* (in m³/kg) that is related to the intracellular composition [95]. In addition to dry mass, the mean cell thickness $\bar{d}_{\mathrm{cell}}$ can be calculated from the average phase contrast $\Delta\bar{\varphi}$ by using Eq. (5). $\bar{d}_{\mathrm{cell}}$ may serve as a measure and to indicate dynamic changes of the global cell morphology [19, 22]:

$$\bar{d}_{\mathrm{cell}} = \frac{\Delta\bar{\varphi}}{2\pi}\lambda(n_{\mathrm{cell}} - n_{\mathrm{medium}}) \tag{7}$$

The parameter $\bar{d}_{cell}$ in Eq. (7) depends on the integral cellular refractive index n_{cell} that is efficiently determined from separate measurements on suspended cells as described in [68, 96] and Sect. 2.8.

2.7 Automated Single Cell Tracking and Quantification of Cell Motility

Quantitative DHM phase images of cell cultures are also efficiently evaluated by automated object tracking [97]. First, quantitative phase images of single cells are reconstructed from time-lapse series of digital off-axis holograms as described in Sects. 2.2–2.5. Then, the phase distributions are low-pass-filtered reducing the influence of substructures of the specimen in the phase distributions and noise, for instance, due to parasitic interferences and coherent noise. Afterward, the pixel coordinates of the maximum phase contrast are determined within a region of interest (ROI) in which the cell under study is located. Automated tracking of dynamic displacements in time-lapse sequences is then performed by successive recentering of the ROI to the coordinates of the preceding maximal phase value. By calibration of the imaging scale, the resulting lateral displacement trajectories of the cells in pixel coordinates are converted to metric units. Figure 11a illustrates the evaluation of quantitative DHM phase images by automated tracking of $n = 10$ preselected single breast cancer cells (MDA-MB-231). The resulting cell migration trajectories are shown in Fig. 11b.

To express cell motility with a single quantitative descriptor, the cell trajectories in Fig. 11 are analyzed with respect to the temporal dependency of the *mean square displacements* (MSD) [98]:

$$\text{MSD}(t) = \left\langle [x(t_0 + t) - x(t_0)]^2 \right\rangle_t + \left\langle [y(t_0 + t) - y(t_0)]^2 \right\rangle_t \tag{8}$$

In Eq. (8) the parameters x and y are the position coordinates of the cell under study, while t_0 denotes the starting time of the tracking experiment. Angular brackets denote temporal averaging. Figure 11c illustrates the mean square displacement (MSD) for each single cell in Fig. 11b (orange curves) as a function of time t together with the average MSD value for all cells (black curve).

The detection error for the lateral cell position in quantitative phase images is specific for the individual measurement setup and the cell type under study. It is estimated to range at about 10% of the average lateral cell diameter [97]. For those cell cycle phases in which the cells adhere tightly on the substrate, the nucleoli induce the maximum phase contrast and therefore dominate the determination of lateral coordinates. As the nucleoli are located in the nucleus, the resulting x, y values serve as a good approximation of the cell center. When the cells round up during

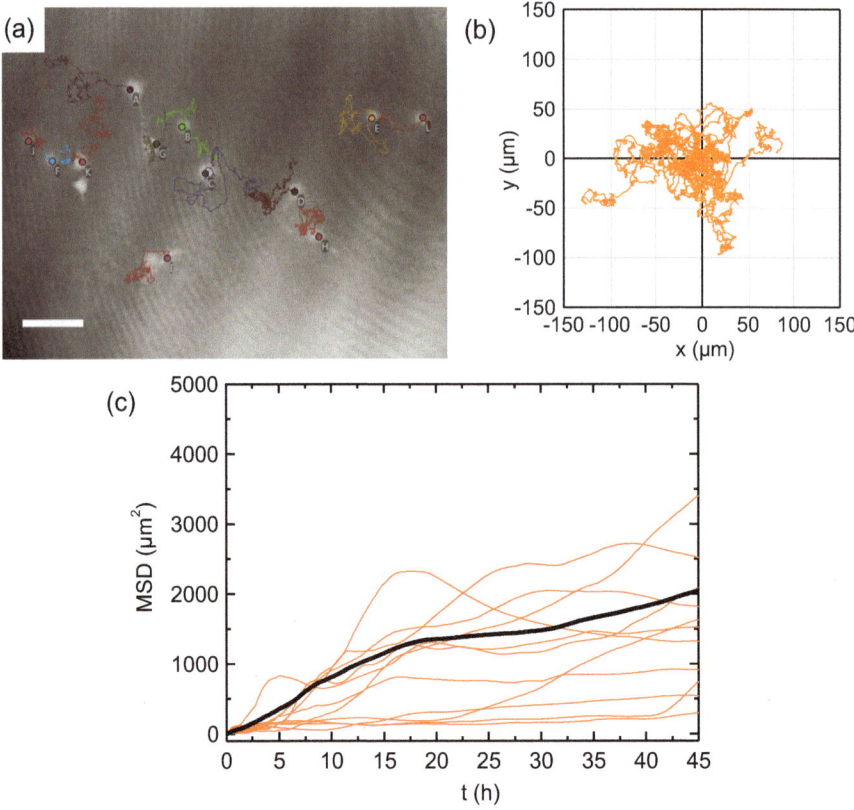

Fig. 11 (**a**) Quantitative evaluation of DHM phase images by automated tracking of $n = 10$ selected single cells (12Z) using the coordinates of the maximum phase contrast as the cells' center of gravity; (**b**) resulting plot of the cell migration trajectories in (**a**); (**c**) quantification of cell motility by calculating the mean square displacement (MSD) for each single cell (orange curves) and averaged MSD (black curve). Scale bar: 100 μm

cytokinesis, the maximum phase contrast is well defined by the center of the resulting spherical structure. Thus, also for this period of the cell cycle, the detection of the cell center is reliable. For those cell cycle phases during which the cells adhere on the substrate with flat morphology, the described tracking procedure algorithm is more efficient than the competing evaluation of DHM phase contrast images by image segmentation algorithms as described in Sect. 2.6. When the cells are capable of spreading out on the substrate, the thickness of the outer peripheral regions of the cell is often less than 1–2 μm. Consequently, the cells show only very little contrast along their perimeter in the DHM phase contrast images (see Fig. 11a) relative to the background which may affect the robustness of an edge detection-based determination of the lateral cell position. In conclusion, the evaluation of stacks of quantitative phase images of single cells by tracking their individual phase contrast maximum represents an efficient method to quantify cell migration and motility.

2.8 Refractive Index, Volume, and Dry Mass Determination of Suspended Cells

The cellular refractive index is an important parameter related to various cell phenotypes and processes [99]. In DHM phase contrast imaging of living cells, the refractive index determines the "visibility" of cells and subcellular structures on the one hand and represents the main limitation in accuracy for the thickness determination of transparent samples [22]. Knowing the cellular refractive index is also important for experiments using optical tweezers and related optical manipulation approaches as it dictates the resulting optical forces [100, 101]. Although measurements with high accuracy are accessible in principle, refractive index determination of adherent cells with quantitative phase imaging methods, as described in [22, 24, 26, 102], is time-consuming or requires special experimental equipment. DHM methods to determine the integral refractive index of suspended cells have been developed [68, 96, 103], which can be carried out with a minimum of sample preparation. Based on the option to numerically refocus suspended cells that are located in different focal planes, simultaneous imaging is possible and increases the effective data acquisition rates in DHM. Here, quantification of the integral refractive index of suspended cells is illustrated by a method in which the model of a circle is fitted to the two-dimensional DHM phase contrast data. For suspended spherical cells in focus, located at $x = x_0$ and $y = y_0$, with radius R_{cell}, the cell thickness $d_{cell}(x,y)$ amounts to:

$$
d_{cell}(x, y) = \begin{cases} 2 \cdot \sqrt{R_{cell}^2 - (x - x_0)^2 - (y - y_0)^2} & \text{for} \quad (x - x_0)^2 + (y - y_0)^2 \leq R_{cell}^2 \\ 0 & \text{for} \quad (x - x_0)^2 + (y - y_0)^2 > R_{cell}^2 \end{cases}
$$

$$(9)$$

Inserting Eq. (9) in Eq. (4) yields:

$$
\Delta\varphi_{cell}(x,y) = \begin{cases} \dfrac{4\pi}{\lambda} \cdot \sqrt{R_{cell}^2 - (x-x_0)^2 - (y-y_0)^2} \cdot (n_{cell} - n_{medium}) & \text{for } (x-x_0)^2 + (y-y_0)^2 \leq R_{cell}^2 \\ 0 & \text{for } (x-x_0)^2 + (y-y_0)^2 > R_{cell}^2 \end{cases}
$$

$$(10)$$

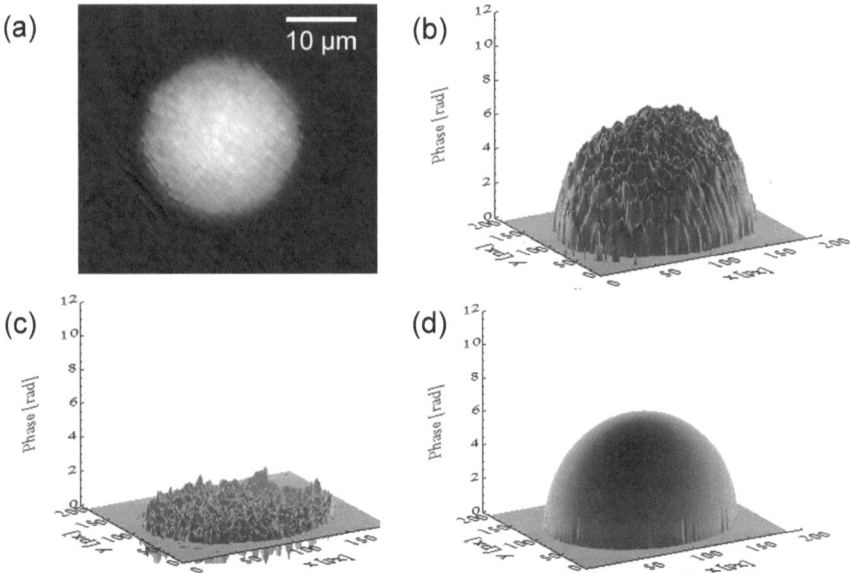

Fig. 12 Refractive index and radius determination of a suspended fibrosarcoma cell (HT1080). (**a**) Quantitative phase image of a suspended HT1080 cell; (**b**) rendered pseudo-three-dimensional plot of the phase data in (**a**); (**c**) deviation of the measurement data in (**a**) from the two-dimensional fit of the sphere function in (**d**) (adapted from [105])

with the unknown parameters n_{cell}, R_{cell}, x_0, and y_0. In order to obtain the parameters n_{cell}, R_{cell}, x_0, and y_0, Eq. (7) is fitted iteratively with the Gauß-Newton-Method [96, 104] to the measured phase data of spherical cells in suspension.

Figure 12 illustrates the evaluation process of the phase data by the example of a single fibrosarcoma cell (HT1080) in suspension. Figure 12a shows the DHM phase contrast image of the cell, coded to 256 gray levels. Figure 12b depicts a rendered pseudo-3D plot of the data in Fig. 12a after background correction and the application of a threshold value. Figure 12d shows the result of fitting Eq. (7) to the data in Fig. 12a. The absolute difference (residual) between experimentally recorded data and fit (Fig. 12c) underlines the suitability of the applied sphere model.

Figure 13 shows refractive index n_{cell} and cell radius R_{cell} values of $n = 152$ suspended human fibrosarcoma cells (HT1080). The cells were detached from the bottom of a Petri dish by trypsin treatment, now floating in cell culture medium. For each cell the parameters n_{cell} and R_{cell} were determined as described above with $n_{medium} = 1.339 \pm 0.001$. Only cells with spherical shape were selected for data analysis. They were observed in the main fraction of the recorded holographic phase contrast images. Figure 13a depicts the integral cellular refractive index n_{cell} in dependence of the cell radius R_{cell}. The plot shows that n_{cell} linearly decreases with increasing R_{cell}. This decrease of the refractive index with increasing cell radius and, thus, cell volume is caused by the cellular water content (see Sect. 3.2).

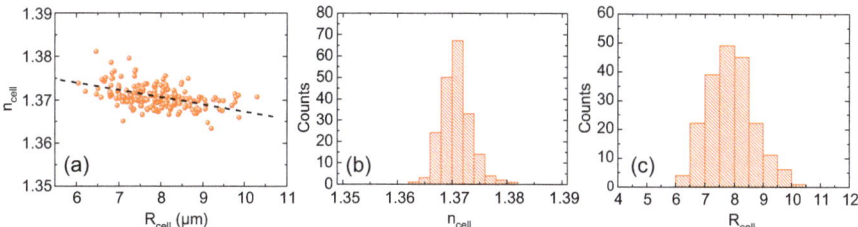

Fig. 13 Refractive index and radius of $n = 152$ suspended human fibrosarcoma cells (HT1080). (**a**) Integral cellular refractive index n_{cell} vs. cell radius R_{cell}, (**b**) histogram plot of n_{cell}, (**c**) histogram plot of R_{cell}

Figure 13b, c presents the histogram plots for the refractive index and radius. Both data sets show a mainly Gaussian distribution.

The results in Figs. 12 and 13 illustrate that DHM is an efficient and reliable way to determine the integral refractive index of living suspended single cells with mainly spherical morphology. From a cell with radius R_{cell}, the cell volume V is easily calculated:

$$V = \frac{4}{3} \pi R_{cell}{}^3 \tag{11}$$

Two-dimensional fitting of Eq. (10) to the measured phase data of spherical cells provides a higher accuracy for the cell radius (and cell volume) than edge detection-based algorithms.

Moreover, with Eqs. (6) and (7), the term $S_c = \pi R_{cell}^2$, and the average cell thickness $\bar{d} = V/S_c$ extracted from values for the cellular refractive index and the cell volume as well as the refractive index increment γ, i t $\dot{s}$ possible to calculate the cellular dry mass:

$$DM = 10 \cdot \frac{V}{\gamma} \cdot (n_{cell} - n_{medium}) \tag{12}$$

3   Selected Applications in Cell-Based Assays

This section demonstrates the potential of label-free multi-focus quantitative DHM phase contrast imaging to monitor cell-based assays in in vitro. DHM is first shown to allow for a precise characterization of suspended cells and their intracellular structures by cell-type dependent, absolute biophysical parameters. The latter are related to the osmotic pressure within the cells' microenvironment, give insides into the global intracellular composition and the response of the cells to non-isotonic stimulations, genetic modification, or exposure to drugs. When applied to adherent

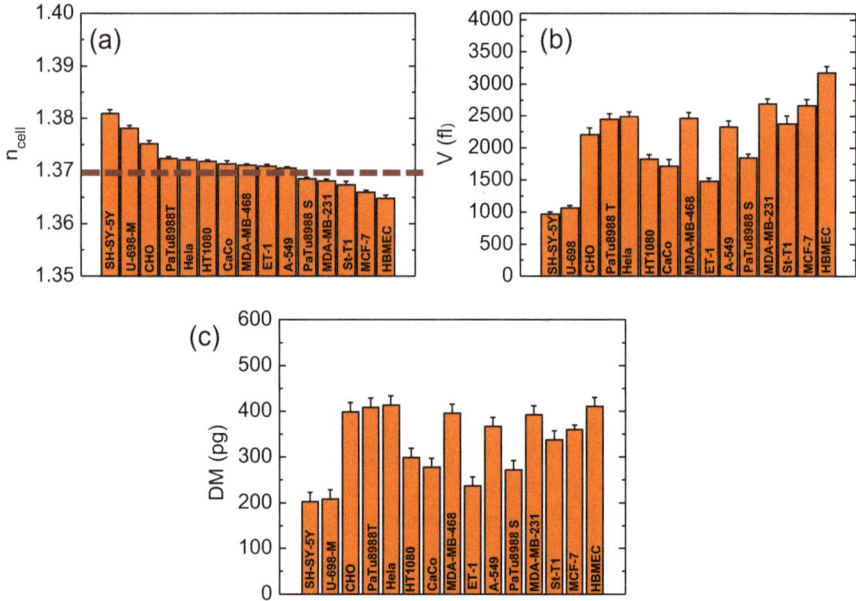

Fig. 14 (**a**) Average refractive indices n_{cell}, (**b**) volumes V, and (**c**) dry masses DM as recorded for different cell types in suspension ($n = 100–200$ individual cells were investigated for each cell type. Data are mean $\pm$ standard error)

cell cultures, DHM allows quantifying cell shape dynamics and motility by efficient automated tracking of migrating cells, dynamic cell thickness monitoring, thickness-based phenotyping, multimodal analysis of mixed cell cultures, and quantitative analysis of in vitro wound healing assays. Finally, the use of DHM to analyze the impact of genetic modifications or cytokines as well as the toxicity of pathogens and nanomaterials is demonstrated. Future prospects for enhanced quantitative imaging techniques are discussed at the very end of the chapter.

3.1 Characterization of Suspended Single Cells and Subcellular Organelles

The methods described in Sect. 2.8 allow for the characterization of living suspended cells by absolute biophysical parameters like refractive index, volume, or dry mass without extensive calibration procedures. Figure 14 shows the average cellular refractive index, the average cell volume, and the average dry mass for different cell types including fibroblasts, endothelial cells, epithelial cells, and different cancer cell types. All data was extracted from holograms acquired with an experimental setup as shown in Fig. 2 using laser light with a wavelength of $\lambda = 532$ nm.

After numerical reconstruction of the quantitative phase images as described in Sect. 2.2, the refractive indices n_{cell} and the cell radii R_{cell} were determined by fitting Eq. (10) to the phase data of spherical, suspended cells. Cell volume V and dry mass DM were then calculated according to Sect. 2.8.

Figure 14a summarizes that the average *cellular refractive index* n_{cell} ranges between 1.365 and 1.38 with an average value of 1.37 across all cell types under test (see dashed horizontal line in Fig. 14a). However, comparison of n_{cell} with the simultaneously recorded parameters *cell volume V* and *dry mass* DM reveals that n_{cell} is not correlated with the other two but individual for the various cell types (Fig. 14b, c). The *integral cellular refractive index* and the *cell volume* are both sensitive to the osmolality of the extracellular environment. In contrast, the *dry mass* is not affected by the osmolality and can serve as an independent control indicator for the total amount of intracellular solutes when the cell response to non-isotonic conditions is studied.

Besides the integral characterization of suspended cells, the methods described in Sect. 2.8 are also suitable for the characterization of intracellular organelles. Figure 15 shows results for the refractive index, volume, and dry mass of suspended HT1080 fibrosarcoma cells in comparison to data from isolated nuclei and nucleoli that were isolated from the same cell type, as described in [106] and

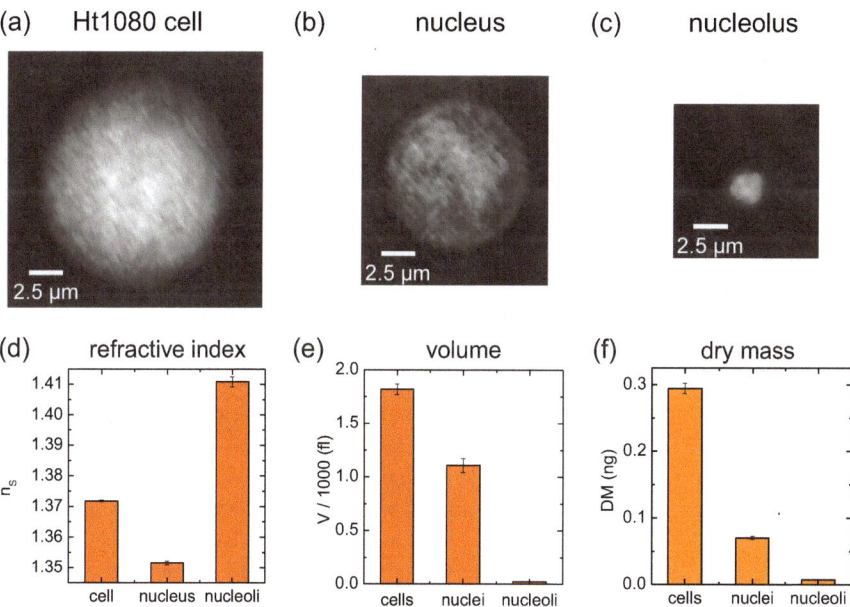

Fig. 15 Refractive index, volume, and dry mass of cells, nuclei, and nucleoli as recorded for fibrosarcoma cells. (**a–c**) Representative quantitative DHM phase images of an HT1080 fibrosarcoma cell, a nucleus of an HT1080 cell, and a nucleolus of an HT1080 cell. (**d–f**) Corresponding refractive indices (n_s), volumes (V), and dry mass (DM). In the measurements $n = 50$–100 samples of each specimen were analyzed. Data are mean $\pm$ standard error

[107]. The average refractive index of the nuclei is smaller than the integral cellular refractive index that is found for whole cells. Due to their inherent high density, the refractive index of the nucleoli amounts to about 1.41. So it is significantly higher than the average cellular refractive index. Both findings are in agreement with the results from investigations that were obtained with similar quantitative DHM phase imaging techniques [108–110] and tomographic phase microscopy (see, e.g., [24, 111, 112]).

The results in Figs. 14 and 15 demonstrate that DHM allows a highly precise characterization of suspended cells and their spherical subcellular organelles by absolute biophysical parameters. Please note that the uncertainty for the determination of cell volume and the dry mass is found in the range of a few femtoliters and picograms while the refractive index has been detected with a precision up to 10^{-3}–10^{-4} [68, 90, 96, 110]. In the following Sects. 3.2 to 3.4 and in Sect. 3.11, it is demonstrated how these biophysical parameters recorded from suspended cells are used for label-free monitoring of the intracellular water content, the response of cell volume to genetic modifications, as well as the impact of drugs and toxic substances.

3.2 Monitoring of the Intracellular Water Content

The analysis of changes in the volume and intracellular solute concentrations of biological cells represents an important topic in many research areas of life sciences. Label-free imaging with DHM can contribute to several of these topics by precise quantification of the absolute cell volume and changes of the intracellular solute concentration. Firstly, this is illustrated by results from a comparative study in which cell volumes and refractive indices of suspended living rat inner medullary collecting duct (IMCD) cells and human pancreatic tumor cells (PaTu 8988 T) are compared when bathed in buffer solutions of different tonicity [113]. As a second example, we show for adherent fibrosarcoma cells (HT1080) that internalized microspheres are useful in connection with quantitative DHM phase imaging to detect dynamic changes of the cytoplasmic solute concentration during regulatory cell volume changes [114].

Figure 16 shows representative quantitative DHM phase images of detached IMCD cells (encoded to 256 gray levels) at different osmolalities: 1,200 mOsmol/kg (Fig. 16a), 600 mOsmol/kg (Fig. 16b), and 200 mOsmol/kg (Fig. 16c). The gray levels of all images are normalized to the maximum phase contrast that was detected for the phase distribution in the left column of Fig. 16a. The corresponding gray level-encoded pseudo-3D plots of phase images are depicted in the right column of Fig. 16a–c. The micrographs show that the cell diameter increases with decreasing osmolality. In contrast, the phase contrast decreases as the integral cellular refractive index decreases with increasing water influx. Figure 16d–e present the outcome of a quantitative analysis of cellular refractive index and cell volume changes upon exposure of the cells to different extracellular tonicities. The mean cellular refractive

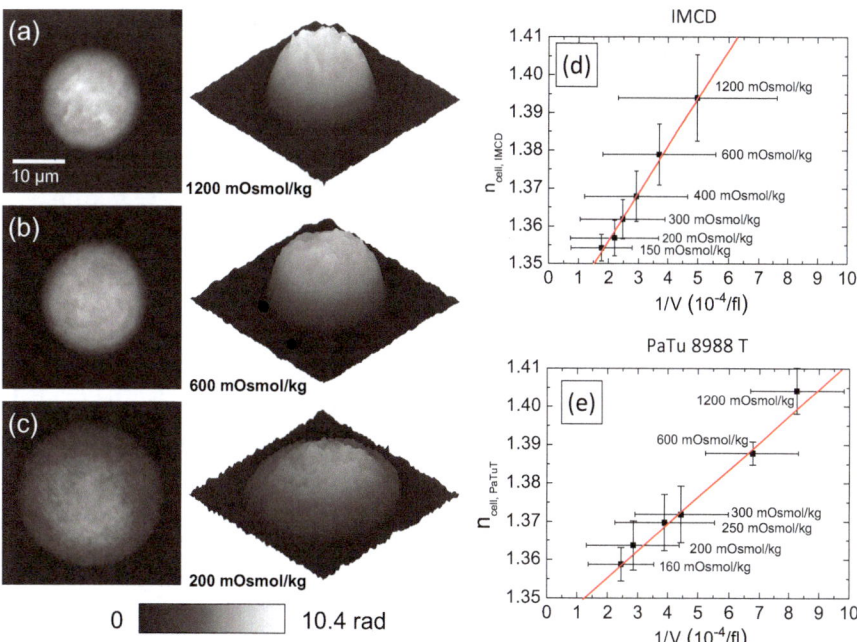

Fig. 16 Cell volume and refractive index in buffer solutions with different osmolality. Representative DHM phase contrast images of trypsinized IMCD cells coded to 256 gray levels (left columns of (**a–c**)) and corresponding pseudo-3D plots (right columns of (**a–c**)). (**a**) 1,200 mOsmol/kg, (**b**) 600 mOsmol/kg, (**c**) 200 mOsmol/kg; (**d**) refractive index $n_{cell, \, IMCD}$ of IMCD cells vs. reciprocal cell volume; (**d**) refractive index $n_{cell, \, PaTuT}$ of pancreatic tumor cells vs. reciprocal cell volume. Data are mean ± standard deviation (adapted from [113])

indices of IMCD and PaTu 8988 T cells are plotted as a function of the inverse cell volume. Fitting the data linearly (solid line in Fig. 16d–e) illustrates that for both IMCD cells and PaTu 8988 T cells, n_{cell} is proportional to $1/V$, while the different slopes mirror the individual responses of both cell types to osmotic stimulation. The cellular refractive index reflects the concentration of cytoplasmic components like proteins, solutes, and organelles [95]. A refractive index decrease indicates intracellular solute dilution.

The linear relationship between n_{cell} and $1/V$ supports the conclusion that refractive index changes are mainly caused by water uptake while the amount of intracellular substances remains constant. Accordingly, the observed changes in cellular volume are due to different water contents of the cells and not due to import or export of solutes.

It is also possible to detect dynamic changes of the refractive index of the cytoplasm of living *adherent* cells. Figure 17 illustrates DHM-based sensing of refractive index changes of adherent single cells in a microfluidic device. Silica microspheres with known diameter and refractive index, which have been internalized by these cells by means of phagocytosis, are used as optical reference probes to

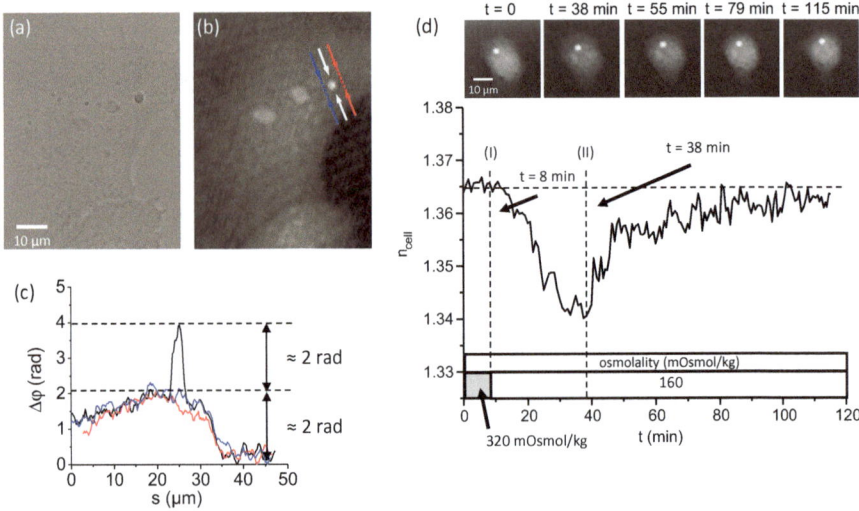

Fig. 17 Utilization of internalized silica microspheres as optical probes for monitoring of refractive index changes of the cytoplasm inside living adherent cells. (**a**)–(**c**) Verification of the internalization of silica microspheres in an adherent pancreatic tumor cell (PaTu 8988 T) by quantitative DHM phase imaging. (**a**) White light image of an adherent cell with an incorporated microsphere. (**b**) Quantitative phase image corresponding to (**a**). (**c**) Cross sections through the phase contrast image in (**b**), through and near the microspheres [see arrows in (**b**)]. (**d**) Temporal response of the cellular refractive index n_{cell} to a decrease of the osmolality (evaluation of $n = 180$ digital holograms, $\Delta t = 40$ s); upper panel, representative quantitative DHM phase images of a HT1080 cell with an incorporated microsphere at indicated time; lower panel, temporal dependency of n_{cell}; at $t = 8$ min (II) the osmolality of the cell culture medium is decreased from 320 to 160 mOsmol/kg; at $t = 38$ min (II) the flow is stopped (adapted from [114])

determine the refractive index of the cytoplasm from single quantitative phase images [114]. Figure 17a, b depicts a white light image (a) and a DHM quantitative phase image (b) of a pancreatic tumor cell (PaTu8988T) with internalized microsphere. The evaluation of the phase contrast that is caused by the microsphere with known refractive index and diameter (see cross sections through the phase contrast image in Fig. 17b, through and near the microspheres that are marked with arrows) allows the determination of the refractive index of the cytoplasm by fitting of Eq. (10) to the recorded data. The reliability of this approach as a tool to monitor refractive index changes dynamically is demonstrated by recording the refractive index response of an adherent fibrosarcoma cell to non-isotonic stimulation in a microfluidic perfusion chamber [114]. Therefore, a HT1080 cell with an internalized microsphere ($\varnothing = 3.44$ μm; refractive index, 1.433 [114]) was exposed to laminar flow in a perfusion channel for $t = 8$ min, while digital off-axis holograms were recorded continuously every 40 s. Then osmolality was decreased from 320 to 160 mOsmol/kg. At $t = 38$ min the flow was stopped, and the cell was further observed until $t = 120$ min. The upper panel of Fig. 17d shows representative DHM phase contrast images and the temporal dependency of n_{cell} as extracted from the images. Cell swelling and slight detachment of the cell are observed until $t \approx 38$ min. However, for $t > 40$ min, the cell starts to shrink again. The shrinking process is accompanied by an increase of n_{cell} that is detected with the microsphere. At $t = 120$ min the cell has almost completely recovered its initial refractive index, and there are no significant morphology changes in the DHM phase images besides a slight detachment resulting from the swelling process. The re-establishment of the refractive index – even though the osmolality was kept at 160 mOsmol/kg – reports on a *regulatory volume decrease* (RVD) [115] as a cellular mechanism to fight non-isotonic conditions. Accordingly, the method is sufficiently sensitive to detect refractive index changes during cell volume regulations and may serve as a tool to study them in more detail. The results in Figs. 16 and 17 open up future applications of DHM quantitative phase imaging for phenotyping cultured cells with respect to their water homeostasis and regulatory cell volume processes.

3.3   Monitoring the Influence of Cytokines

Simultaneous measurements of cell volume, refractive index, and dry mass are also of great interest when it comes to characterizing the response of cells to potential drug candidates. However, the simultaneous recording of these parameters from individual adherently growing cells requires special experimental setups, for instance, combining DHM with fluidics, using specific cell culture media [102] and multiwavelength [116] or tomographic phase microscopy techniques as reported in [24, 26, 111, 117]. The analysis is a lot simpler and more efficient for suspended cells as illustrated by results obtained from DHM studies on suspended colon carcinoma cells (Caco-2) that were stimulated with either epidermal growth factor (EGF) or mitomycin c. The latter is a cytokine that prevents mitosis [90]. Quantitative DHM phase contrast images of suspended EGF- and mitomycin c-treated single cells as well as untreated control cells were analyzed as described in Sect. 2.8 to quantify refractive index, cell volume, and dry mass. A total of $n = 89$ suspended cells were studied in each experiment.

Figure 18a–c presents quantitative DHM phase images of suspended single Caco-2 cells (encoded to 256 gray levels) with typical size and refractive index. Mean cell radii and volumes are summarized in Fig. 18d–f by means of false color-coded pseudo-3D representations of the quantitative phase images. EGF-treated Caco-2 cells (Fig. 18b, e) show a significant bigger mean cell radius compared to unstimulated Caco-2 cells (Fig. 18a, d), while the mitomycin c treatment increased the average cell radius even further (Fig. 18e, f). The cell volume changed accordingly. Whereas the volume of EGF-stimulated cells was almost doubled compared to untreated control cells, the average volume of mitomycin c-treated cells was even quadrupled (Fig. 18i). After stimulation with EGF, the cells' refractive index

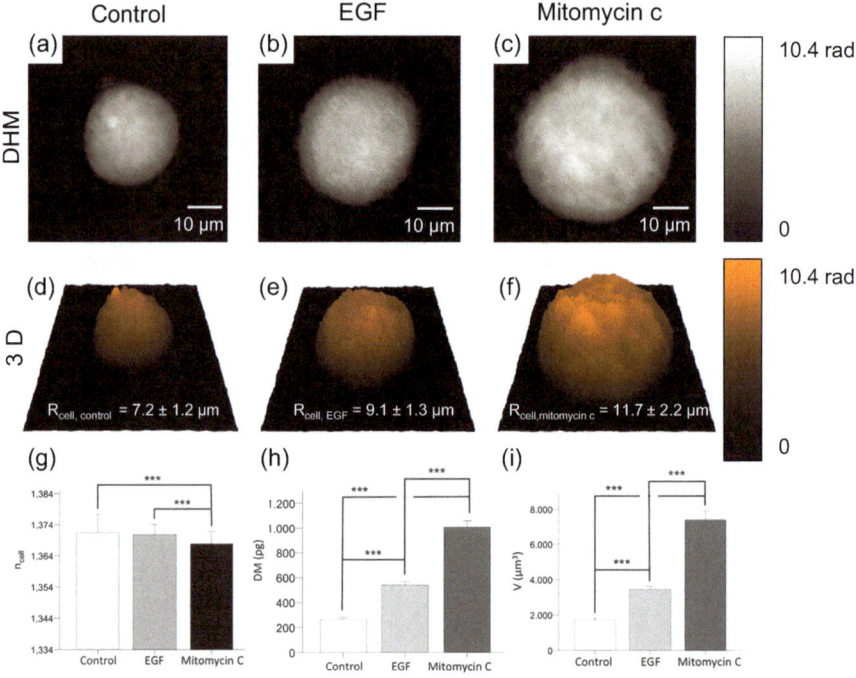

Fig. 18 Refractive index, dry mass, and cell volume of suspended colon carcinoma cells (Caco-2) after exposure to EGF or mitomycin c. (**a–c**): Representative quantitative DHM phase images (encoded to 256 gray levels). (**a**) Untreated control cells, (**b**) after exposure to epidermal growth factor (EGF), (**c**) after exposure to mitomycin c. (**d–f**) The mean cellular radius R_{cell} as determined from quantitative phase images was slightly increased after EGF exposure and markedly increased after mitomycin c treatment. (**g**) The refractive index n_{cell} of mitomycin c-treated cells was found to be significantly smaller compared to EGF-treated and untreated cells. (**h, i**) Dry mass DM and cell volume V of EGF-treated Caco-2 cells were significantly increased relative to the untreated cells. Mitomycin c induced an even stronger increase in DM and V. Data given as mean $\pm$ standard error; each experiment included $n = 89$ cells, ***$p < 0.001$ (adapted from [90])

remained similar to that of unstimulated control cells (Fig. 18g). In contrast, mito- mycin c treatment induced a significant reduction of the refractive index relative to control. The cellular dry mass increases significantly upon EGF and mitomycin c exposure relative to unstimulated control cells (Fig. 18h) in good agreement with the findings on cell volume. The dry mass increase after mitomycin c treatment is significantly higher compared to EGF-treated cells. In conclusion, the results in Fig. 18 demonstrate that DHM is capable of quantifying reliably morphological cell characteristics like volume, density, and intracellular solute content as well as alterations of these parameters upon treatment with cytokines.

3.4   Analysis of Cell Volume and Refractive Index Changes After Transfection with MicroRNA

MicroRNAs are pivotal posttranscriptional regulators of gene expression and play significant roles in tumorigenesis and tumor progression. In [118] the role of microRNA miR-142-3p in breast cancer was investigated. Within this study DHM was used to quantify the impact of miRNA on the breast cancer cell line MDA-MB-468. Suspended MDA-MB-468 cells were transiently transfected using Dharmafect reagent and 10 nM negative control miRNA #1, miR-142-3p precursor, or 20 nM anti-miR-142-3p. The DHM analysis was performed 72 h after transfection. Digital off-axis holograms of selected MDA-MB-468 cells after transfection and control cells in DMEM (osmolality of 320 mOsmol/kg) were recorded with a setup as shown in Fig. 2. Cell volumes and refractive indices were determined and are described in Sect. 2.8. Figure 19a, c shows typical quantitative DHM phase contrast images of transfected MDA-MB-468 cells and control cells in suspension (coded to 256 gray levels). The dashed circles illustrate the cell volume increase as a consequence of transfection. The pseudo-3D representations (Fig. 19b, d) of the phase images (Fig. 19a, c) show the corresponding impact on cell thickness. Figure 19e illustrates that the cell volume V of $n = 251$ miR-142-3p transfected cells is significantly decreased in comparison to the volume of $n = 172$ control cells. In contrast, the mean cellular refractive index (RI) that reflects the integral density of the cells (Fig. 19) remains constant. In conclusion, the results in Fig. 19 demonstrate that volume and refractive index changes of genetically modified cells are reliably accessible by DHM.

3.5   Quantification of Cell Motility as an Indicator for Invasiveness in Endometriosis Research

In recent years, microRNAs, for instance, small noncoding RNA molecules of approximately 21 nucleotides, have grown in significance for endometriosis research. In the framework of a wider study, quantitative DHM phase microscopy was used to investigate the influence of miR-200b on the motility of endometrial stroma cells (ST-T1b) [119]. ST-T1b cells were cultured for 24 h under basal conditions, transfected with miR-200b, and transferred to Petri dishes 24 h post transfection. Subconfluent transfected cells and their untreated correspondents (control) were observed by quantitative DHM phase contrast with the setup shown in Fig. 2 at 37°C for 48 h in HEPES-buffered cell culture medium. The time delay between two subsequent holograms was 3 min. From this series of quantitative phase images, the mean square displacement and the maximum distance the cells migrated from their initial position were calculated and evaluated (cf. Sect. 2.7). Figure 20a, b shows typical migration tracks of control cells ($n = 30$) and miR-200b-transfected ($n = 30$) ST-T1b cells, respectively.

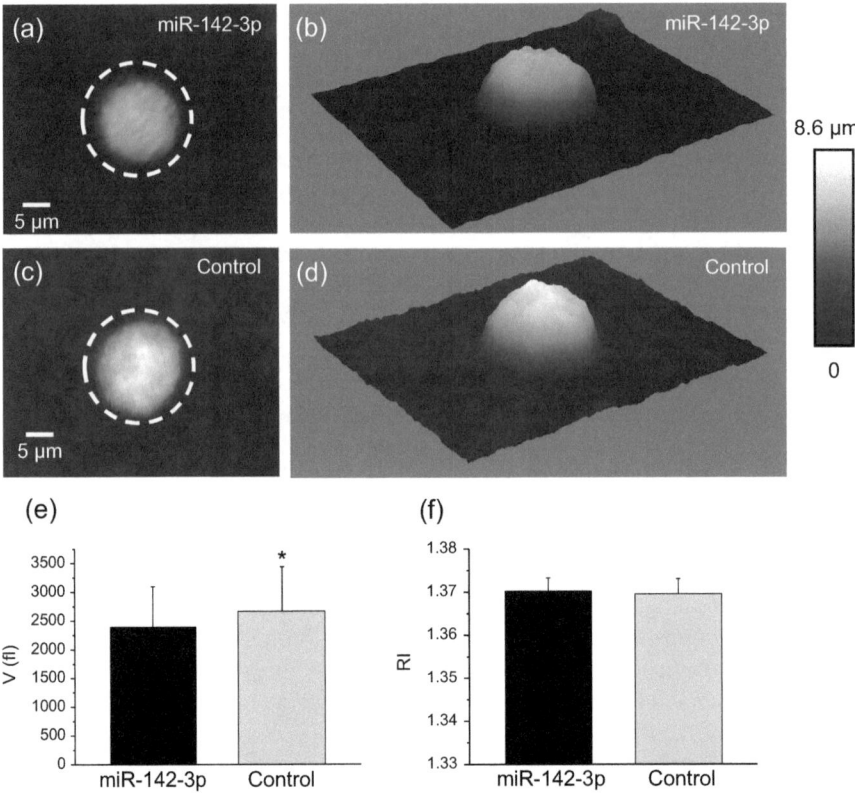

Fig. 19 Volume and refractive index of microRNA-transfected breast cancer cells (MDA-MB-468) compared to control cells. Evaluation of quantitative DHM phase contrast images reveals a decrease in cell volume after miR-142-3p transfection, while refractive index changes are not statistically significant. (**a, c**) Representative quantitative DHM phase contrast images of transfected MDA-MB-468 cells and control cells in suspension (coded to 256 gray levels); the dashed circles illustrate the cell volume increase. (**b, d**) Pseudo-3D representations of the phase images in (**a**), (**b**) demonstrate the corresponding impact cell thickness. (**e**) Volumes V of $n = 251$ miR-142-3p-transfected cells are significantly decreased in comparison to the volume of $n = 172$ control cells. (**f**) Mean cellular refractive index (RI). Data are mean $\pm$ SEM, $*p < 0.05$ (adapted from [118])

Figure 20c presents the mean square displacement (calculated by Eq. (8)) of control (green) and miR-200b-transfected (red) ST-T1b cells. Each curve represents the mean of 8–12 individually tracked cells of three independently conducted experiments. Figure 20d illustrates a quantitative analysis of cell migration experiments using the maximal migration distance from the starting point as indicator. It reveals a reduced migration distance in miR-200b-transfected cells compared with controls. In conclusion, the results in Fig. 20 illustrate that ST-T1b cells show a significantly decreased motility after miR-200b transfection that is precisely monitored by DHM.

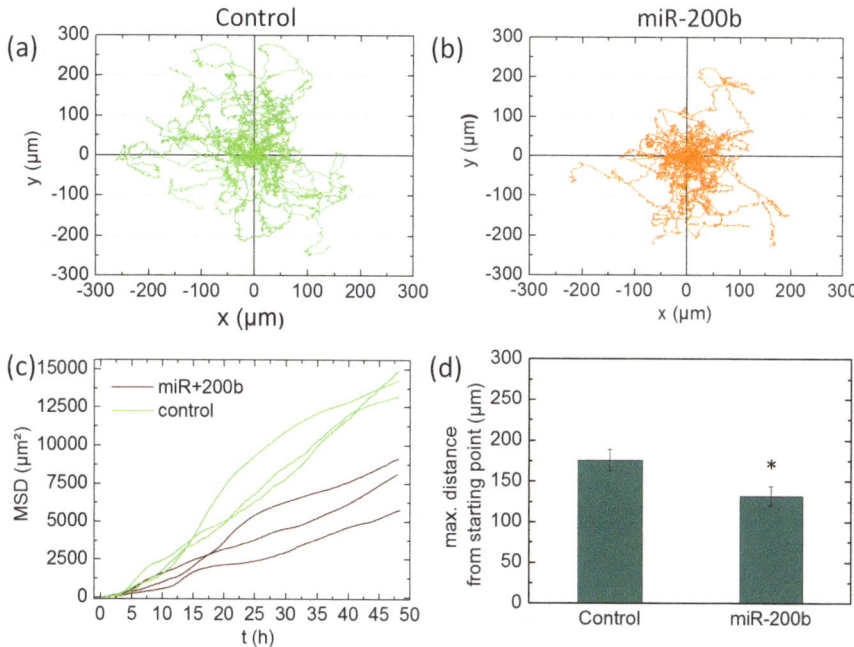

Fig. 20 Impact of microRNA (miR-200b) on endometrial stroma cell motility. For evaluation of motility, ST-T1b cells were tracked in sequences of quantitative DHM phase contrast images. Increased levels of miR-200b result in decreased cell motility. (**a**, **b**) Typical migration tracks of control ($n = 30$) and miR-200b-transfected ($n = 30$) ST-T1b cells. (**c**) Mean square displacement of control (green) and miR-200b-transfected (red) ST-T1b cells. Each curve represents the mean of 8–12 individual cells observed in one out of three independent experiments. (**d**) The maximum migration distance from the starting point reveals a reduced migration distance in miR-200b-transfected cells compared to controls (* $= p < 0.016$, $n = 30$, error bars $=$ SEM) (adapted from [119])

3.6 Imaging of Cytokinesis After Downregulation of Survivin in Ewing's Sarcoma Cells

Survivin (Birc5) is a radiation-inducible protein that belongs to the *inhibitor of apoptosis* (IAP) family in Ewing's sarcoma, and its downregulation sensitizes cells toward irradiation [120]. In [120] digital holographic phase contrast was used to quantify proliferation of STA-ET-1 cells under control conditions (random siRNA) and after transfection with Birc5 siRNA. Transfection was performed using Lipofectamine, and cells were observed with a DHM setup as depicted in Fig. 2.

In the resulting quantitative DHM phase images, the area covered by the cells was quantified by image segmentation as described in Sect. 2.6. Figure 21 shows representative results. Proliferation slowed down in surviving knockdown cells (Fig. 21a, b)

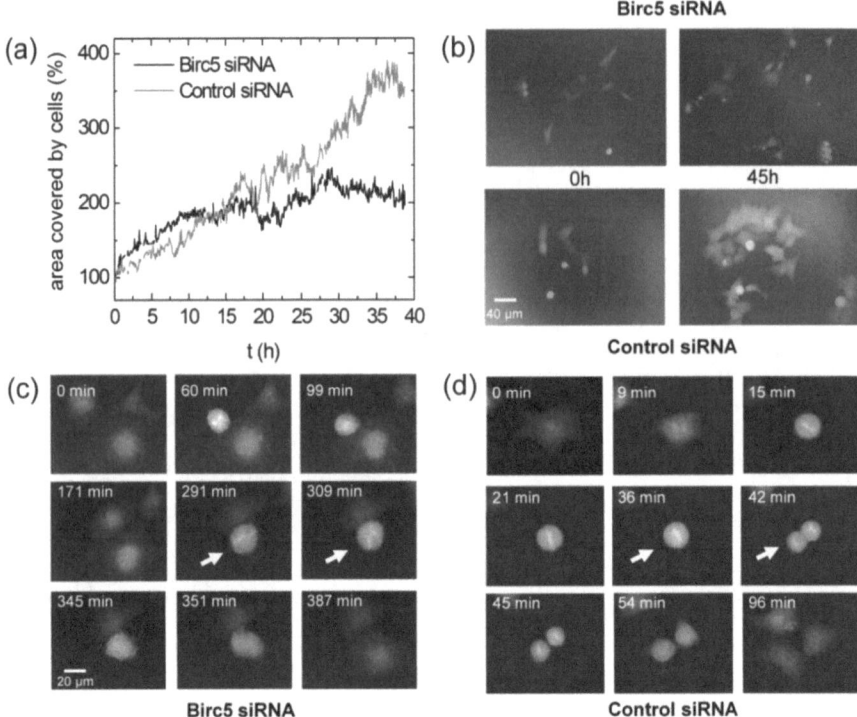

Fig. 21 Label-free DHM analysis of STA-ET-1 cells after Birc5 knockdown. Cell proliferation was quantified by determining the area covered by cells for the control siRNA- and Birc5 siRNA-treated population. Relative to the area covered by cells at the beginning of the experiment (100%), the area covered by control cells was approx. 400%, while the Birc5 siRNA-treated cells showed a maximum of 200% coverage after 40 h (**a**). Quantitative DHM phase contrast images of cells at the beginning of the experiment (0 h) and at the end (45 h) illustrate the observed differences (**b**). Knockdown of survivin led to an aberrant morphology with a less regular cell structure. Polyploid cells resulted from incorrect segregation of chromosomes (white arrow), and many cells no longer divided (**c**). Cells treated with random control siRNA cells showed a proper arrangement of chromosomes at the equatorial axis with correct segregation of chromosomes during mitosis (white arrows) and the development of two well-shaped daughter cells (**d**). The time points of analysis are given in each figure (adapted from [120])

along with strong morphological changes and aberrant cytokinesis as marked in Fig. 21c compared to control cells in Fig. 21d. In conclusion, the results in Fig. 21 demonstrate that functional properties (proliferation, chromosome segregation) as well as structural features (cell morphology, chromosome structure) after genetic modification are simultaneously accessible by DHM for label-free quantification.

3.7   Label-Free Phenotyping by Cell Thickness

DHM also allows for label-free phenotyping of adherently growing cells in terms of cell thickness. This is illustrated by results from a comparative study on pancreas tumor cells. *PaTu8988T* cells were compared with genetically modified *PaTu8988T pLXIN-E-Cadherin* cells expressing the epithelial cell adhesion protein E-Cadherin [22]. For the experiments, holograms of adherent *PaTu8988T* cells and *PaTu8988T pLXIN-E-Cadherin* cells were recorded with a transmission DHM setup as depicted in Fig. 2. Thickness d_{cell} of either cell type was determined from quantitative phase images by application of Eq. (5) using an integral refractive index $n_{cell} = 1.38$ [22]. Figure 22 depicts typical results obtained from *PaTu8988T* and two *PaTu8988T pLXIN-E-Cadherin* cells. Figure 22a, c shows the reconstructed holographic amplitude images, whereas Fig. 22b, d depicts the unwrapped phase data including a gray level legend for the phase and the corresponding cell thickness. Figure 22f, h shows rendered pseudo-3D representations of the cell bodies in comparison to typical scanning electron micrographs (SEM) (Fig. 22e, g) of both cell types. For both cell types, cell thicknesses as determined from DHM are found in good agreement with the cells' appearance in SEM images.

In Fig. 22i, line profiles of phase shift and cell thickness along the white lines in Fig. 22b, d are compared for both cell lines. The two cell lines are clearly different with respect to phase difference $\Delta\varphi_{cell}$ as well as their calculated thickness. For the *PaTu8988T* cell, the maximum thickness amounts to $d_1 = (23 \pm 1)$ μm, while the *PaTu8988TpLXIN-E-Cadherin* cells are significantly flatter with a maximum thickness of $d_2 = (7 \pm 1)$ μm. Assuming complete adhesion of all cells under study upon the growth substrate, cell thicknesses as provided by DHM are supported by the independent SEM images and may serve as a quantitative descriptor of cell morphology.

3.8   Multimodal Growth and Morphology Analysis of Mixed Cell Cultures

A minimally invasive but quantitative monitoring of different cell types in a single culture is of particular interest to analyze the impact of drugs, pathogens, or toxins on different cell species under identical experimental conditions and to analyze interactions between different cell species. Thus, we addressed the feasibility of DHM for label-free quantitative observation of mixed cell cultures in [58]. Moreover, we explored if quantitative phase images provide sufficient information to distinguish between different cell types and to extract cell-specific parameters in mixed populations. Human pancreatic ductal adenocarcinoma cell lines *PaTu8988S* and *PaTu8988T* [22, 89] were used in the experiments. *PaTu8988S* represents a highly differentiated carcinoma, while *PaTu8988T* cells are poorly differentiated with a high metastatic potential. The morphologies of these two cell lines that originate

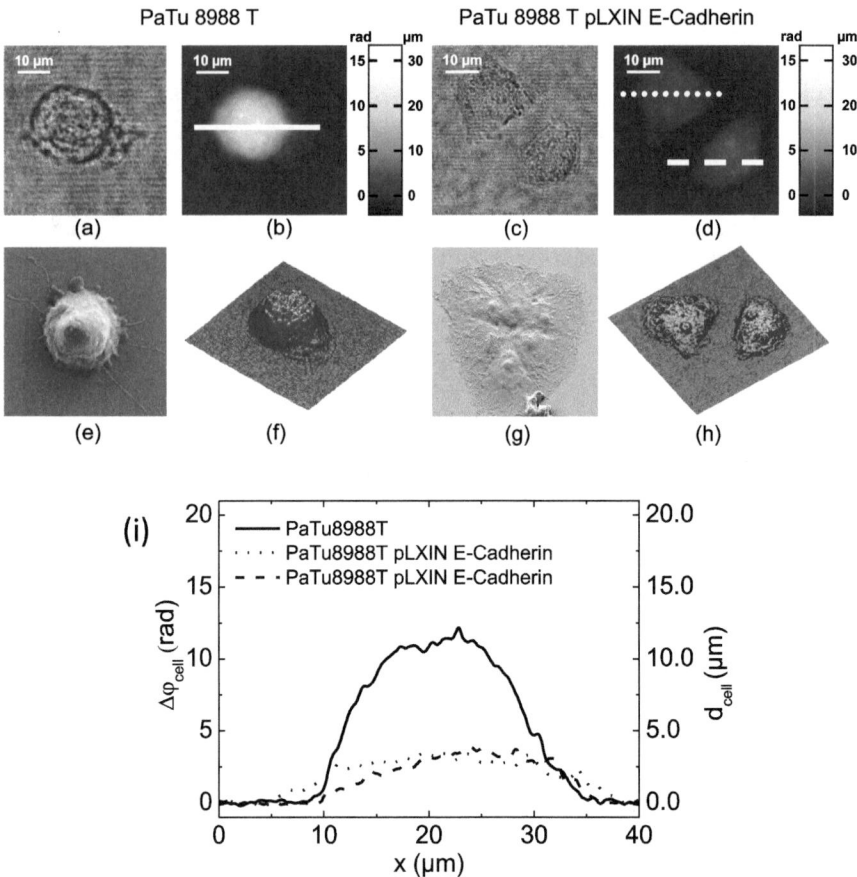

Fig. 22 Label-free cell phenotyping by comparing the thickness of two pancreatic tumor cell lines. (**a**), (**c**) Holographic amplitude images of one *PaTu8988T* and two genetically modified *PaTu8988T-pLXIN-E-Cadherin* cells; (**b**), (**d**) corresponding unwrapped quantitative phase contrast images; (**e**), (**g**) SEM images for comparison; (**f**), (**g**) rendered pseudo-3D plots as obtained from (**b**) and (**d**). (**i**) Line profiles of phase and cell thickness for *PaTu8988T* and *PaTu8988T-pLXIN-E-Cadherin* cells along the section lines in (**b**) and (**d**) (adapted from [22])

from the same tumor are different and have been studied previously by scanning electron microscopy (SEM) and DHM (Sect. 3.7 and [22]). *PaTu8988S* cells built thick insular conglomerates with tight cell-cell contacts, while *PaTu8988T* cells adhere upon their growth surface expressing only few intercellular contacts. Furthermore, *PaTu8988S* and *PaTu8988T* cells are grown in the same cell culture medium. For the experiments, *PaTu8988T* and *PaTu8988S* were mixed and cultured subconfluently in Petri dishes. Quantitative DHM phase contrast imaging was performed with a setup as depicted in Fig. 2 ($\lambda = 532$ nm). The resulting series of digital off-axis holograms were evaluated as described in Sect. 2.2, and the resulting quantitative phase images were analyzed by image segmentation (Sect. 2.6).

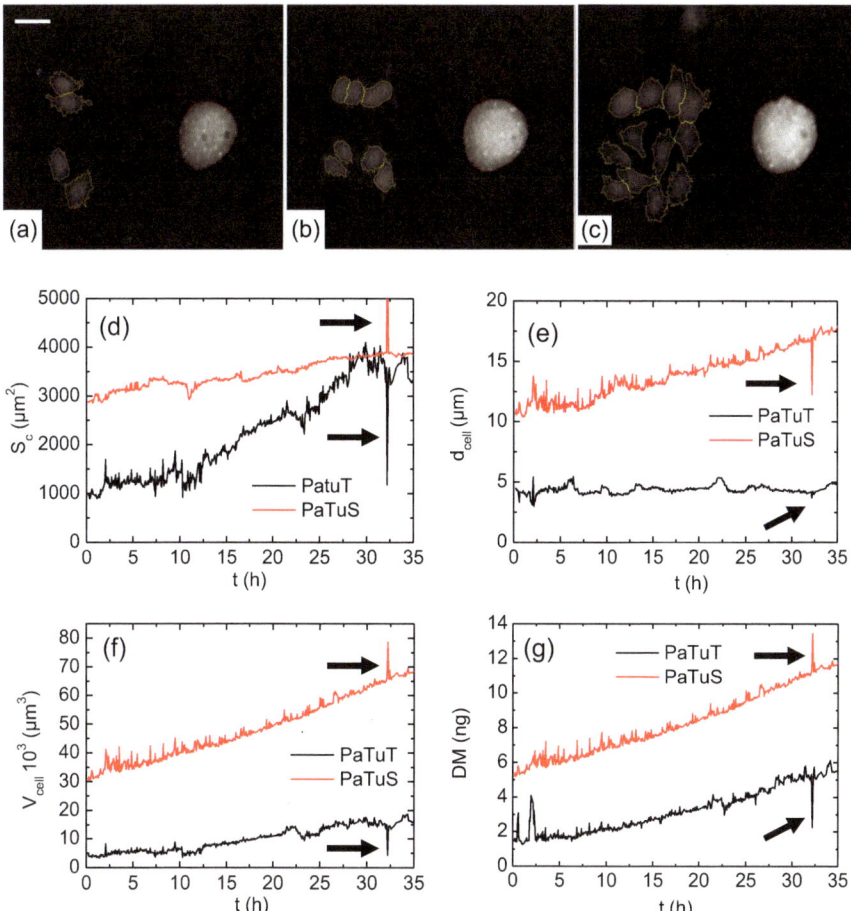

Fig. 23 Time-resolved growth and morphology analysis of *PaTu8988T* and *PaTu8988S* cells grown in co-culture. (**a**)–(**c**) Quantitative phase images after 2 h, 21 h, and 32 h of co-culture: *PaTu8988T* cells are indicated by yellow and *PaTu8988S* by red outlines, respectively. (**d**) Time course of the area S_c occupied by the different cell types; (**e**) mean cell thickness calculated for each cell type from the averaged phase contrast in the area S_c; (**f**) cell volume V_{cell} calculated by multiplying the data in (**d**) and (**e**); (**g**) cellular dry mass DM calculated for each cell type from the average phase contrast in the area S_c. Black curves indicate data assigned to *PaTu8988T* cells, while red curves represent data from *PaTu8988S* cells. Peaks with opposite amplitudes (see arrows) in the plots indicate errors in the correct identification of the different cell types. The scale bar corresponds to 40 μm (adapted from [58])

Figure 23a–c illustrates the quantitative evaluation of phase images by segmentation of representative images for *PaTu8988T* cells (yellow outlines) and *PaTu8988S* cells (red outlines) at different time points of the experiment. Due to their individual cell morphology, *PaTu8988T* and *PaTu8988S* cells are easy to

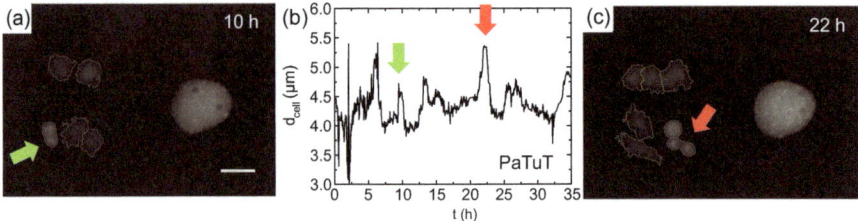

Fig. 24 Analysis of cell proliferation by in-depth analysis of the time course of the average cell thickness of PaTu 8988 T cells. The local thickness maxima which are marked in (**b**) with arrows correlate with cell division events as documented by quantitative DHM phase contrast images after 10 h and 22 h in (**a**) and (**c**). The scale bar corresponds to 40 μm (adapted from [58])

identify. Figure 23d–f shows the individual time courses of area S_c, thickness d_{cell}, volume V_{cell}, and dry mass DM as determined for each cell type by image segmentation (black curves, *PaTu8988T* cells; red curves, *PaTu8988S* cells). Figure 23d compares the area S_c occupied by each cell type along the experiment. The data demonstrates that *PaTu8988S* cells show a faster increase of the area that is covered by this cell type compared to the *PaTu8988T* cells. Figure 23e shows the time course of the corresponding mean cell thickness d_{cell} within the area S_c. The average thickness of *PaTu8988T* cells remains constant at about 4.9 μm for the entire observation time, while the thickness of *PaTu8988S* cells increases almost linearly from 11 μm up to 17 μm in the same time. Figure 23f presents the corresponding cell volume V_{cell} for each cell type as a function of time which is calculated by multiplying the data in Fig. 23d, e for each time point. Figure 23g compares the dry mass DM for each species along the observation time. Cell volume V_{cell} and dry mass DM increase slightly nonlinear. This is in agreement with the expectation of an exponential increase of the cell number during proliferation. Peaks with opposite amplitudes (see arrows) in the plots of Fig. 23d–g indicate errors in the correct identification of the different cell types. Taken together, the data summarized in Fig. 23 provides a wealth of information about the individual morphologies and the growth rates of *PaTu8988T* and *PaTu8988S* cells in mixed co-cultures.

In-depth analysis of the time course of cell thickness as shown in Fig. 23e provides additional information about cell doublings. Figure 24b shows a magnified plot of the *PaTu8988T* data from Fig. 23e. The local thickness maxima which are marked with arrows correlate favorably with cell division events as documented after 10 h (Fig. 24a) and 22 h (Fig. 24c) by DHM phase contrast images.

In conclusion, Figs. 23 and 24 illustrate that quantitative phase imaging with DHM allows the continuous analysis of different cell types even when they are grown in co-culture as long as they have significantly different cell morphologies. This is achieved by adapted segmentation of quantitative DHM phase contrast images. From the segmented phase images, cell type-specific parameters like the area covered by the different cell types and their average thickness, volume, and dry mass are accessible. However, cell types with very similar morphologies require more sophisticated image segmentation algorithms. Nevertheless, current and future developments of DHM clearly foreshadow applications of quantitative phase imaging for label-free investigations addressing the influence of drugs and pathogens on different cell types or a continuous analysis of cell-cell interactions in co-culture.

3.9 Multiparameter Monitoring of In Vitro Wound Healing Assays

DHM has been used as a novel, label-free method to monitor wound healing assays in vitro quantitatively and continuously. The performance of the device is illustrated for human fibrosarcoma cells (HT1080). In these experiments, wound healing assays were conducted and observed in special Petri dishes as described with details in [90]. Imaging with quantitative DHM phase contrast was performed with a setup as depicted in Fig. 2 ($\lambda = 532$ nm). Time-lapse series of digital off-axis holograms were recorded for 24 h with a time delay of 3 min between two consecutive holograms. Images were reconstructed as described in Sect. 2.2. The resulting quantitative phase images were analyzed by image segmentation (see Sect. 2.6 and [90]) in order to determine the area covered by cells, the cellular dry mass, and the cell thickness. In these assays a linear, cell-free lesion is experimentally introduced or *scratched* into an otherwise confluent cell layer. Accordingly, this assay is also called *scratch assay*. The lesion defines the in vitro wound. Cells from the periphery migrate toward the center of the lesion and thereby close the wound. The rate of wound closure is quantified from the time course of the cell-free or cell-covered area. Figure 25 summarizes the evaluation of an in vitro wound healing assay with quantitative digital holographic phase contrast by image segmentation [35]. Figure 25a shows typical quantitative phase images of HT1080 cells after segmentation at times indicated in the upper left of each image. Figure 25b (left) shows the time course of the area S_c that is covered by cells as determined from the images in Fig. 25a. The increase in S_c indicates wound closure with time. The corresponding evolution of the cellular dry mass (DM) is shown on the right of Fig. 25b. Figure 25c shows the averaged line profiles of phase contrast $\Delta\varphi$ and the corresponding cell thickness d as obtained from horizontal sections of the phase images in Fig. 25a. These profiles allow studying this in vitro wound healing process quantitatively. In conclusion, the results summarized in Fig. 25 underline that DHM allows accurate, label-free, and continuous multimodal monitoring of wound healing assays in vitro that may pave the way for new applications in related migration assays.

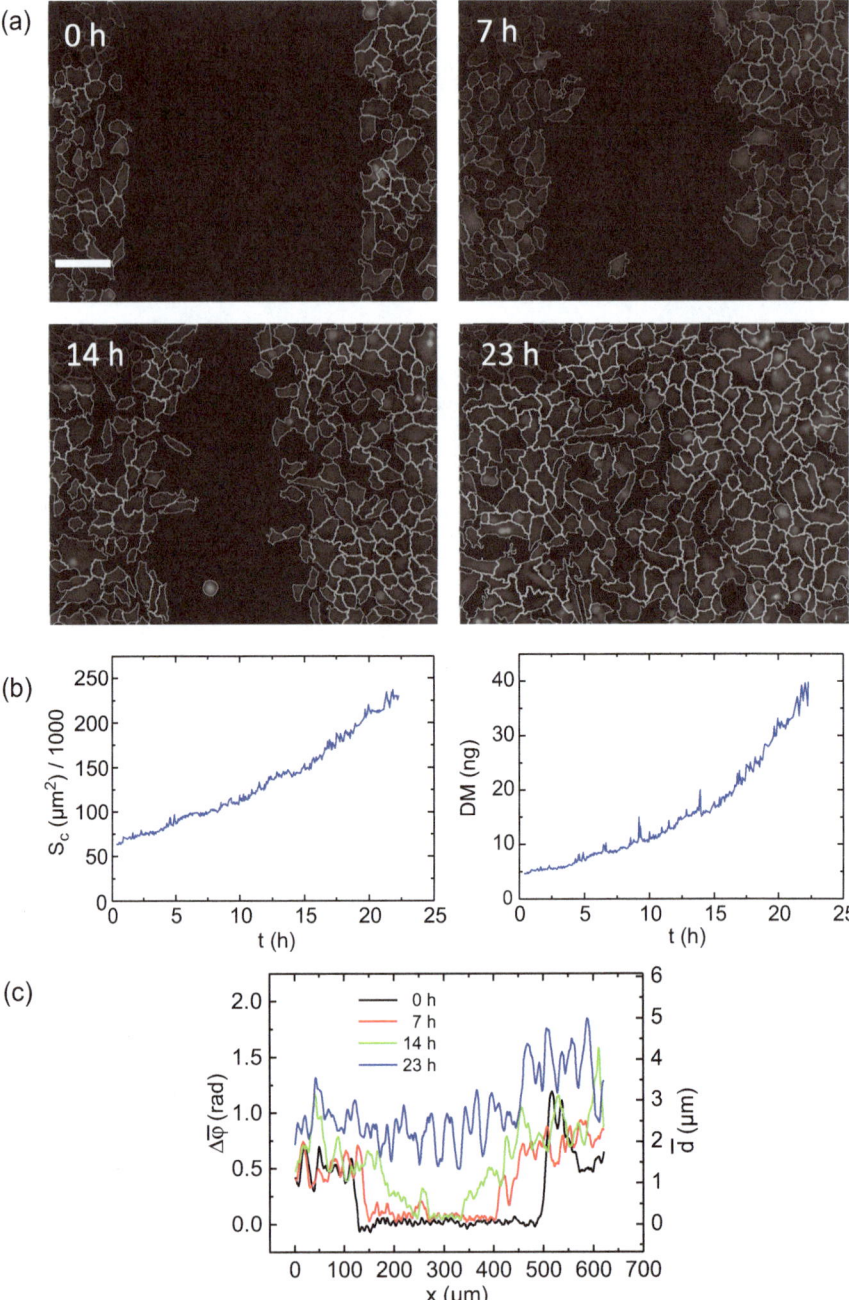

Fig. 25 Multimodal analysis of in vitro wound healing assays. (**a**) Quantitative phase images of human fibrosarcoma cells (HT1080) at indicated time after segmentation. (**b**) Cell migration or wound healing was quantified from the time course of S_c which defines the area covered by cells

3.10   Label-Free In Vitro Toxicity Testing of Pathogens

Common in vitro toxicity testing of drugs, chemicals, or nanomaterials involves the analysis of phenotypic changes of the cells under study like activation of stress responses, changes in metabolic activity or proliferation, eventually impacts on viability. Most approaches to quantify these phenotypes are so-called end point assays that only read the cell response at one predefined time point. The assays typically determine enzyme activity or protein expression by colorimetric or fluorescence-based optical readouts. These standard procedures have several disadvantages. For example, the process under study has to be stopped at a distinct time point to perform the assay without any information on the time course of the cell response. Moreover, only one parameter is typically measured, and the readout requires several time-consuming incubations and washing steps. In contrast, DHM provides a continuous, label-free, and quantitative monitoring of toxicity-induced changes of various cell phenotypes. This is illustrated by results from human colon carcinoma cells (HCT-8) that were exposed to toxic vesicles (LB226692 OMVs) containing 580 ng/ml shiga toxin 2a (Stx2a) or cell culture medium as control [121]. The experiments were performed using an incubator-integrated microscope as shown in Fig. 2. Digital holograms of the cells were recorded continuously every 3 min for 48 h.

To analyze cell growth and morphology changes, the area occupied by the cells and the cell-induced average phase contrast were determined by image segmentation providing the cellular dry mass as described in Sect. 2.6. Figure 26 summarizes representative results. Control cells show regular cell proliferation resulting in an increase of the cell-covered area, an increase in cell density, and an increase in cell layer thickness within the 48 h of the experiment (Fig. 26a, left panel). Dry mass of the control cell patch increased threefold within 48 h, whereas cells treated with toxic vesicles showed no growth. This is reflected by the rather stationary time course of the cellular dry mass in Fig. 26c. Moreover, single cells treated with toxic vesicles left the cell aggregate, and five out of eight cells under study underwent apoptosis (after 22 h, 29 h, 37 h, 40 h, and 47.5 h, respectively). This led to the disintegration of the cell patch (right panel of Fig. 26a, 48 h). Quantitative phase imaging with DHM confirmed the OMV LB226692-mediated apoptosis of HCT-8 cells [121].

In conclusion, the results in Fig. 26 demonstrate the suitability of DHM to monitor in vitro toxicity testing as well as to quantify cell proliferation and cell death. In contrast to other established in vitro toxicity assays, DHM analysis requires no additional staining or colorimetric/fluorescence-based readouts.

Fig. 25 (continued) (left) and the corresponding evolution of the cellular dry mass (DM). The latter was obtained from quantitative evaluation of the phase data (right). (c) Average phase contrast ($\Delta\varphi$) and corresponding cell thickness (d) line profiles obtained from horizontal section of the phase images in (a). The scale bar in (a) corresponds to 100 μm (adapted from [35])

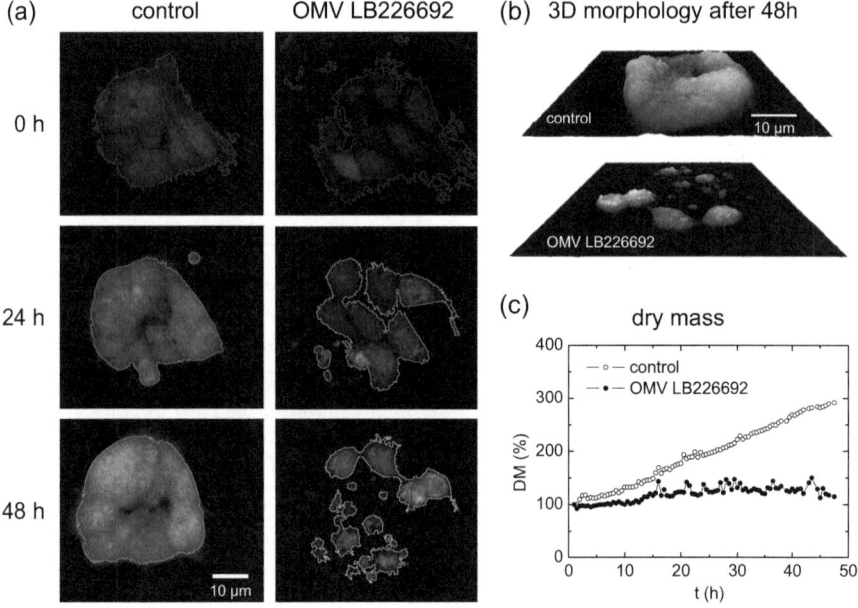

Fig. 26 DHM monitoring of cell morphology changes caused by toxic vesicles. (**a**) Quantitative DHM phase images of human colon carcinoma cells (HCT-8) after segmentation (green outlines) for 0 h, 24 h, and 48 h of incubation with cell culture medium (control) or toxic vesicles (OMV LB226692) containing 580 ng/ml of shiga toxin 2a. (**b**) Pseudo-3D representation of the quantitative phase images after 48 h of exposure. (**c**) Time-dependent increase of cellular dry mass normalized to the respective dry mass at t_0 (cf. Sect. 2.6 for details of the calculation) (adapted from [121])

3.11 Analysis of Cell-Nanomaterial Interactions

In vitro cytotoxicity assessment of engineered nanomaterials commonly involves the measurement of different endpoints like the formation of reactive oxygen species, impacts on metabolic activity, or cell death [122]. Usually, these parameters are determined by colorimetric or fluorescence-based optical readouts of enzymatically converted substrates that are, however, often affected by the nanomaterials themselves [7]. Therefore, more precise and reliable nanomaterial toxicity testing strategies that allow high nanomaterial doses to determine low effect levels (LOEL) even for rather nontoxic materials are highly desirable.

The potential of DHM to monitor cytotoxic effects of nanomaterials has been demonstrated by quantifying the impact of spherical silver nanoparticles (NM 300) on macrophages [123]. Therefore, quantitative DHM phase images of suspended cells (Fig. 27) and adherent single cells (Fig. 28) were recorded with an experimental setup as the one shown in Fig. 2 ($\lambda = 532$ nm). The images were analyzed for the

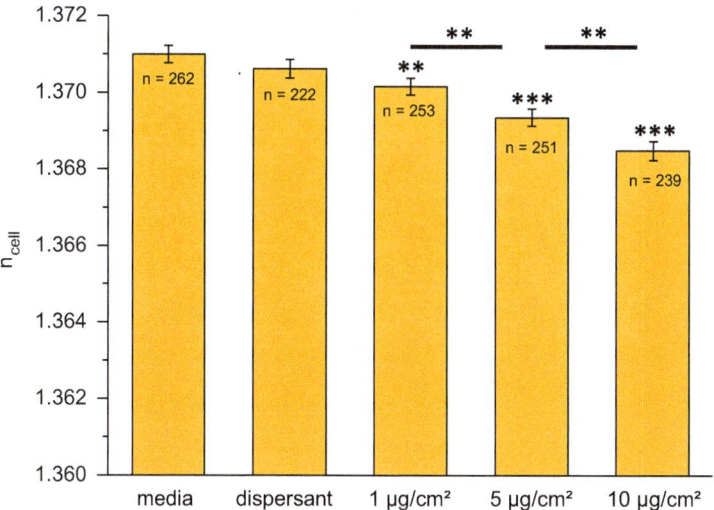

Fig. 27 Refractive index n_{cell} of RAW264.7 macrophages after exposure to increasing doses of a spherical silver nanomaterial (NM 300). The refractive index decreases dose dependently after 2 h of incubation with NM 300. For the dispersant no significant influence was detected (n denotes the number of cells under study; data are mean $\pm$ SEM, $**p < 0.01$, $***p < 0.001$, adapted from [123])

integral refractive index, the volume, the occupied surface area, and the dry mass of the cells under study applying the procedures described in Sects. 2.6 and 2.8. Figure 27 shows the average integral refractive index n_{cell} of RAW264.7 macrophages that was determined from suspended cells after exposure to different silver particle (NM300) concentrations [123]. The parameter n_{cell} decreases with increasing dose indicating a decreased cell density and protein content. The cell volume, which has been measured simultaneously, was not affected significantly (data not shown).

Via automated image segmentation, the surface area S_c occupied by adherently grown RAW264.7 cells (Fig. 28a, b) is determined during nanomaterial exposure.

Combining S_c readings with the average phase contrast data, the dry mass DM of the adherent cells is calculated providing a direct indicator for the time course of proliferation. The slopes of S_c and DM decrease upon exposure to the nanomaterial relative to the medium and dispersant control (Fig. 28c, d).

In conclusion, data summarized in Figs. 27 and 28 proves DHM to be a valuable tool to study cell-nanomaterial interactions. DHM overcomes some of the limiting problems in current nanomaterial characterization and toxicity testing as it detects simultaneously different cellular phenotypes independent from labels that might interfere with the nanomaterials themselves. DHM is very well suited to allow quantifying LOELs of nanomaterials with rather weak toxicity.

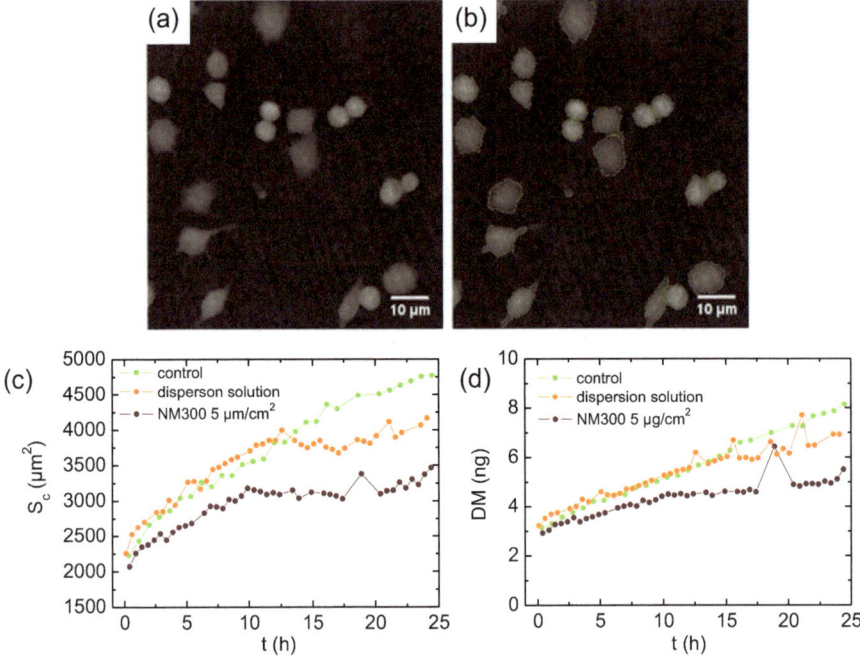

Fig. 28 Time-lapse observation of RAW264.7 macrophages after incubation with silver nanoparticles (NM 300) compared to control cells and cells exposed to the dispersant. (**a**) Quantitative DHM phase images, (**b**) segmented quantitative phase images; (**c, d**) time courses of the area S_c occupied by the adherent RAW264.7 cells and their corresponding dry mass DM after incubation with a dose of 5 µg/ml NM 300 (adapted from [123])

3.12 Challenges and Future Prospects of Quantitative DHM Phase Imaging

The application examples described in Sects. 3.1–3.11 demonstrate DHM as a powerful multifunctional imaging tool suitable for label-free in vitro quantification of cell morphology, growth, and motility and their changes after genetic modifications and/or exposure to drugs, pathogens, or nanomaterials. Nevertheless several challenges remain that require further developments and improvements of the technology as well as advanced novel algorithms for improved evaluation of quantitative phase contrast images. The experimental results in Sect. 3.5–3.11 underline that high-resolution quantitative phase imaging is essential for robust and reliable live cell analysis. However, major drawbacks of using laser light in DHM are noise due to coherence-induced scattering effects and parasitic reflections that cause disturbing interferences in the optical setup. These disturbances significantly affect the reconstructed quantitative phase images and, thus, restrict the measurement accuracy of optical path length changes and subsequent parameter extraction. Various approaches for the reduction of such effects have been proposed that include, for example, use of partially coherent light sources [75, 124–129], multiwavelength techniques [130–132], or a modulated object illumination [60, 124, 133]. Modulation of object illumination has been found particularly useful for quantitative phase imaging with DHM setups that are based on common research microscopes (see Sect. 2.1 and [60]).

The importance of high-quality quantitative phase imaging in DHM-based live cell analysis is illustrated by results recorded for human fibrosarcoma cells (HT-1080) that were observed in a time-lapse experiment with a Michelson interferometer setup (cf. Fig. 3) attached to a common cell culture microscope. For enhanced quantitative phase imaging, sequences of off-axis holograms of the cells were recorded, while the illumination was modulated with an electrically focus tunable lens integrated into the illumination path [60]. From the resulting hologram series, quantitative DHM phase images were reconstructed (cf. Sects. 2.2 and 2.3) and subsequently averaged. Figure 29a, b shows typical single ($n = 1$) and averaged ($n = 15$) quantitative phase images of HT-1080 cells. To illustrate the performance and the impact of noise reduction, the images were also segmented using the software *Cell Profiler* (cf. Sect. 2.6). The averaged quantitative phase images (Fig. 29b, d) provide a considerably enhanced resolution of intracellular structures like nucleoli or vacuoles and more clearly visualize thin extrusions and cell borders (arrows in Fig. 29c, d) which led to improved image segmentation.

Figure 29e, f shows images that were obtained from the phase information in Fig. 29a, b by calculating the first derivative in horizontal direction following [134, 135] in order to simulate images similar to differential interference contrast (DIC) [44]. The individual image qualities reflect the reduction of coherent disturbances in the images in Fig. 29a, b. In summary, the results in Fig. 29 demonstrate the importance of high-performance quantitative phase imaging for a precise parameter extraction especially from very thin cell specimens.

Studies on mixed cell cultures in Sect. 3.8 highlighted that the differentiation of cell types with very similar morphologies is challenging. In addition to improved methods for improved quantitative phase contrast imaging as demonstrated in Fig. 29, novel evaluation procedures like the extraction of generalized parameter sets [136], advanced algorithms for cell shape differentiation [137–139], or machine learning [140] may contribute to an improved identification of different cell types or advanced phenotyping. Future technology developments may also address simplification and acceleration of the image acquisition process, e.g., by integration of automated microscope stages and methods of flow cytometry. Moreover, improved algorithms for data extraction will help to record a statistically convincing amount of measurement data with reduced effort and time.

Additional advances in label-free quantitative live cell imaging with DHM may arise from recent developments in tomographic phase microscopy (TPM). For TPM, the sample is either rotated (see [112, 141]) or illuminated from different directions as described in [24, 26]. The recorded sets of quantitative phase images provide

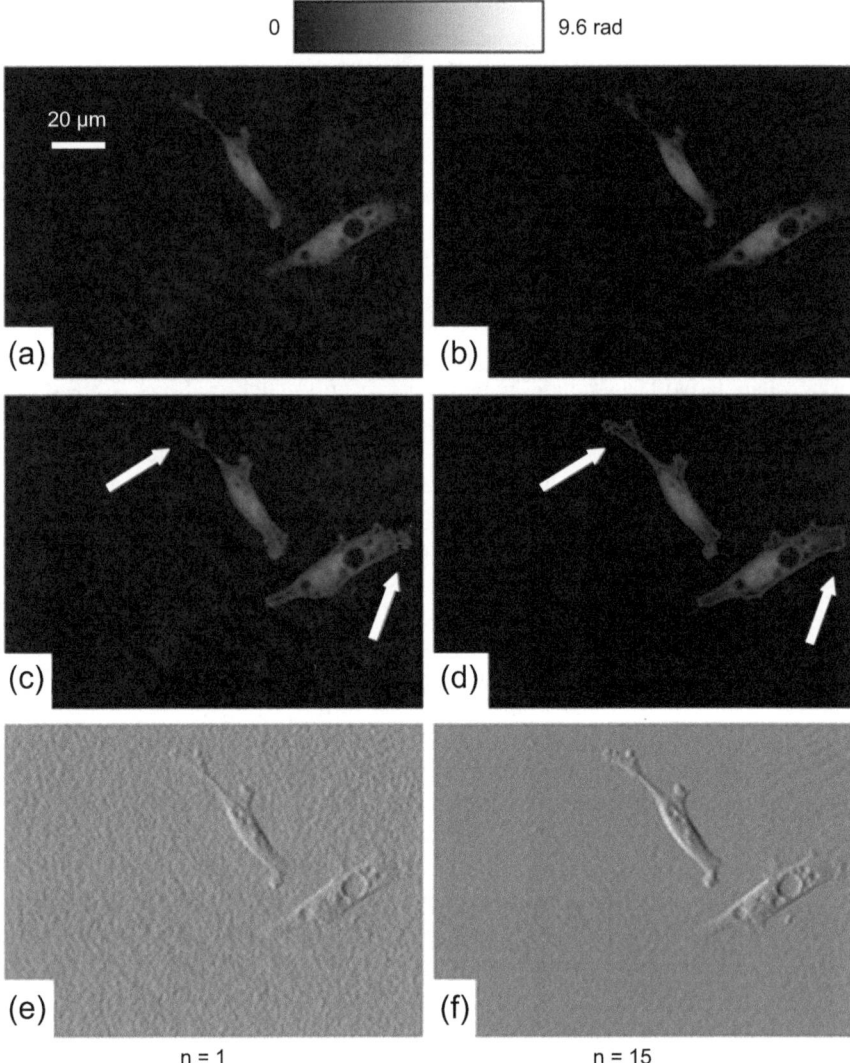

Fig. 29 Enhanced quantitative phase imaging by modulation of the object wave illumination. (**a, b**) Typical quantitative phase images from reconstructed holograms of human fibrosarcoma cells (HT1080); (**c, d**) quantitative phase images after segmentation; (**e, f**) simulated differential interference contrast images calculated by the first derivative of the images in (**a**) and (**b**) in horizontal direction. Left column: phase images reconstructed from a single hologram ($n = 1$). Right column: averaged phase images reconstructed from $n = 15$ holograms that were acquired during modulated illumination with an electrically focus tunable lens. Arrows in (**c**), (**d**) mark differences in image segmentation yielding different results of the surface area occupied by adherent cells (adapted from [60])

high-resolution 3D information. Recently, it has been shown that TPM in combination with algorithms for complex deconvolution of the holographic wave fields provides even nanoscopic resolution ([111] and references therein as well as www.nanolive.com and www.tomocube.com).

4 Conclusions

In summary, DHM allows high-resolution label-free quantitative in vitro imaging of suspended single cells and adherent cell cultures. It has been demonstrated by selected examples that DHM is capable of quantifying minimally invasive and reliably cell growth, motility, and features of cell morphology like volume, dry mass, or intracellular density after genetic modifications or exposure to drugs, pathogens, and nanomaterials. Compared to bright-field microscopy, Zernike's phase microscopy, and differential interference contrast (DIC), DHM provides quantitative phase contrast with optional numerical focus correction (multi-focus imaging) that allows extracting absolute biophysical parameters from the recorded images. The latter are valuable complementary data to biochemical spectroscopic or microscopic studies. Although DHM-based methods as presented here do not achieve a similar spatial resolution as scanning force microscopy (SFM) or scanning electron microscopy (SEM), they overcome some particular limitations of these techniques, for example, the slow scanning process or the requirement of fixed cells in vacuum. As DHM only requires low light intensities for object illumination, the technique minimizes the interference of the measurement with the sample and offers minimally invasive long-term time-lapse observations of living cells. In combination with common research microscopes, DHM enables a contactless, robust, and fast data acquisition in the regular environment of biomedical laboratories. Future challenges for technology developments are simplification, automation, and acceleration of image acquisition and data extraction in order to collect statistically solid amounts of experimental data with reduced efforts and time. But already today DHM is a multifunctional approach capable of producing novel insights into various areas of the life sciences including cell biology with obvious applications in cancer research as well as in drug and toxicity testing.

References

1. Ntziachristos V (2006) Fluorescence molecular imaging. Ann Biomed Eng 8:1–33
2. Goldys EM (2009) Fluorescence applications in biotechnology and life sciences. Wiley-Blackwell, Hoboken
3. Schermelleh L, Heintzmann RB, Leonhardt H (2010) A guide to super-resolution fluorescence microscopy. J Cell Biol 190(2):165–175
4. Spence MT, Johnson ID (2010) The molecular probes handbook: a guide to fluorescent probes and labeling technologies. Live Technologies Corporation, Carlsbad

5. Monici M (2005) Cell and tissue autofluorescence research and diagnostic applications. Biotechnol Annu Rev 11:227–256
6. Chang C, Sud D, Mycek M (2007) Fluorescence lifetime imaging microscopy. Methods Cell Biol 81:495
7. Kroll A, Dierker C, Rommel C, Hahn D, Wohlleben W, Schulze-Isfort C, Göbbert C, Voetz M, Hardinghaus F, Schnekenburger J (2011) Cytotoxicity screening of 23 engineered nanomaterials using a test matrix of ten cell lines and three different assays. Part Fibre Toxicol 8(1):1
8. Shimomura O (2005) The discovery of aequorin and green fluorescent protein. J Microsc 217 (1):3–15
9. Felgner PL, Gadenk TR, Holm M, Roman R, Chan HW, Wenz M, Northrop JP, Ringold GM, Danielsen M (1987) Lipofection: a highly efficient, lipid-mediated DNA-transfection procedure. Proc Natl Acad Sci U S A 84(21):7413–7417
10. Tsampoula X, Taguchi K, Cizmar T, Garces-Chavez V, Ma N, Mohanty S, Mohanty K, Gunn-Moore F, Dholakia A (2008) Fibre based cellular transfection. Opt Express 16 (21):17007–17013
11. Drexler W (2004) Ultrahigh-resolution optical coherence tomography. J Biomed Opt 9 (1):47–74
12. Fercher AF (2010) Optical coherence tomography–development, principles, applications. Z Med Phys 20(4):251–276
13. Seddon AB (2013) Mid-infrared (IR) – a hot topic: the potential for using mid-IR light for non-invasive early detection of skin cancer in vivo. Phys Stat Solid (b) 250(5):1020–1027
14. Hughes C, Baker MJ (2016) Can mid-infrared biomedical spectroscopy of cells, fluids and tissue aid improvements in cancer survival? A patient paradigm. Analyst 141(2):467–475
15. Rodriguez LG, Lockett SJ, Holtom GR (2006) Coherent anti-stokes Raman scattering microscopy: a biological review. Cytometry A 69(8):779–791
16. Cuche E, Marquet P, Depeursinge C (1999) Simultaneous amplitude-contrast and quantitative phase-contrast microscopy by numerical reconstruction of Fresnel off-axis holograms. Appl Optics 38(34):6994–7001
17. Carl D, Kemper B, Wernicke G, von Bally G (2004) Parameter-optimized digital holographic microscope for high-resolution living-cell analysis. Appl Optics 43(36):6536–6544
18. Popescu G, Deflores LP, Vaughan JC, Badizadegan K, Iwai H, Dasari RR, Feld MS (2004) Fourier phase microscopy for investigation of biological structures and dynamics. Opt Lett 29 (21):2503–2505
19. Marquet P, Rappaz B, Magistretti PJ, Cuche E, Emery Y, Colomb T, Depeursinge C (2005) Digital holographic microscopy: a noninvasive contrast imaging technique allowing quantitative visualization of living cells with subwavelength axial accuracy. Opt Lett 30(5):468–470
20. Mann CJ, Yu L, Lo CM, Kim MK (2005) High-resolution quantitative phase-contrast microscopy by digital holography. Opt Express 13(22):8693–8698
21. Ikeda T, Popescu G, Dasari RR, Feld MS (2005) Hilbert phase microscopy for investigating fast dynamics in transparent systems. Opt Lett 30(10):1165–1167
22. Kemper B, Carl D, Schnekenburger J, Bredebusch I, Schäfer M, Domschke W, von Bally G (2006) Investigation of living pancreas tumor cells by digital holographic microscopy. J Biomed Opt 11(3):34005
23. Popescu G, Ikeda T, Dasari RR, Feld MS (2006) Diffraction phase microscopy for quantifying cell structure and dynamics. Opt Lett 31(6):775–777
24. Choi W, Fang-Yen C, Badizadegan K, Oh S, Lue N, Dasari RR, Feld MS (2007) Tomographic phase microscopy. Nat Methods 4:717–719
25. Kemper B, von Bally G (2008) Digital holographic microscopy for live cell applications and technical inspection. Appl Optics 47(4):A52–A61
26. Debailleul M, Georges V, Simon B, Morin R, Haeberlé O (2009) High-resolution three-dimensional tomographic diffractive microscopy of transparent inorganic and biological samples. Opt Lett 34(1):79–81

27. Kozacki T, Krajewski R, Kujawińska M (2009) Reconstruction of refractive-index distribution in off-axis digital holography optical diffraction tomographic system. Opt Express 17 (16):13758–13767

28. Shaked NT, Rinehart MT, Wax A (2009) Dual-interference-channel quantitative-phase microscopy of live cell dynamics. Opt Lett 34(6):767–769

29. Bon P, Maucort G, Wattellier B, Monneret S (2009) Quadriwave lateral shearing interferometry for quantitative phase microscopy of living cells. Opt Express 17(15):13080–13094

30. Jang J, Bae CY, Park JK, Ye JC (2010) Self-reference quantitative phase microscopy for microfluidic devices. Opt Lett 35(4):514–516

31. Wang Z, Millet L, Mir M, Ding H, Unarunotai S, Rogers J, Gilette MU, Popescu G (2011) Spatial light interference microscopy (SLIM). Opt Express 19(2):1016–1026

32. Frank J, Matrisch J, Horstmann J, Altmeyer S, Wernicke G (2011) Refractive index determination of transparent samples by noniterative phase retrieval. Appl Optics 50(4):427–433

33. Wang Z, Tangella K, Balla A, Popescu G (2011) Tissue refractive index as marker of disease. J Biomed Opt 16(11):116017

34. Phillips KG, Velasco CR, Kolatkar A, Luttgen M, Bethel K, Duggan B, Kuhn P, McCarthy OJ (2012) Optical quantification of cellular mass, volume, and density of circulating tumor cells identified in an ovarian cancer patient. Front Oncol 2:72

35. Bettenworth D, Lenz P, Krausewitz P, Brückner M, Ketelhut S, von Bally G, Domagk D, Kemper B (2013) Quantification of inflammation in colonic tissue sections and wound healing in vitro with digital holographic microscopy. SPIE Proc 8797:879702

36. Marquet P, Depeursinge C, Magistretti PJ (2014) Review of quantitative phase-digital holographic microscopy: promising novel imaging technique to resolve neuronal network activity and identify cellular biomarkers of psychiatric disorders. Neurophotonics 1(2):020901

37. Jenkins MH, Gaylord TK (2015) Quantitative phase microscopy via optimized inversion of the phase optical transfer function. Appl Optics 54(28):8566–8579

38. Barankov R, Baritaux JC, Mertz J (2015) High-resolution 3D phase imaging using a partitioned detection aperture: a wave-optic analysis. J Opt Soc Am A 32(11):2123–2135

39. Kreis T (1996) In: Osten W (ed) Holographic interferometry: principles and methods, vol 1. Akademie-Verlag, Berlin

40. Beek M, Hentschel W (2000) Laser metrology – a diagnostic tool in automotive industry. Opt Lasers Eng 34:101–120

41. Ostrovsky YI, Shchepinov VP, Yakovlev VV (2013) Holographic interferometry in experimental mechanics. Wiley, New York

42. Cuche E, Bevilacqua F, Depeursinge C (1999) Digital holography for quantitative phase-contrast imaging. Opt Lett 24(5):291–293

43. Zernike F (1955) How I discovered phase contrast. Science 121(3141):345–349

44. Nomarski G (1955) Differential microinterferometer with polarized waves. J Phys Radium 16 (9):9S–11S

45. Gabor D (1948) A new microscopic principle. Nature 161(4098):777–778

46. Leith EN, Upatnieks J (1962) Reconstructed wavefronts and communication theory. J Opt Soc Am 52(10):1123–1130

47. Leith EN, Upatnieks J (1963) Wavefront reconstruction with continuous-tone objects. J Opt Soc Am 53(12):1377–1381

48. Schnars U, Jüptner WP (2002) Digital recording and numerical reconstruction of holograms. Meas Sci Technol 13(9):R85

49. Lee K, Kim K, Jung J, Heo JH, Cho S, Lee S, Chang G, Jo YJ, Park H, Park YK (2013) Quantitative phase imaging techniques for the study of cell pathophysiology: from principles to applications. Sensors 13(4):4170–4191

50. Kim MK (2010) Principles and techniques of digital holographic microscopy. SPIE Rev 1:018005

51. Popescu G (2011) Quantitative phase imaging of cells and tissues. McGraw Hill Professional, New York

52. Kemper B, Carl D, Höink A, von Bally G, Bredebusch I, Schnekenburger J (2006) Modular digital holographic microscopy system for marker free quantitative phase contrast imaging of living cells. SPIE Proc 6191:61910T
53. Rommel CE, Dierker C, Schmidt L, Przibilla S, von Bally G, Kemper B, Schnekenburger J (2010) Contrast-enhanced digital holographic imaging of cellular structures by manipulating the intracellular refractive index. J Biomed Opt 15(4):041509
54. Kemper B, Langehanenberg P, Höink A, von Bally G, Wottowah F, Schinkinger G, Guck J, Käs J, Bredebusch I, Schnekenburger J, Schütze K (2010) Monitoring of laser micromanipulated optically trapped cells by digital holographic microscopy. J Biophotonics 3(7):425–431
55. Esseling M, Kemper B, Antkowiak M, Stevenson DJ, Chaudet L, Neil MA, French PW, von Bally G, Dholakia K, Deny C (2012) Multimodal biophotonic workstation for live cell analysis. J Biophotonics 5(1):9–13
56. Barroso Á, Woerdemann M, Vollmer A, von Bally G, Kemper B, Denz C (2013) Three-dimensional exploration and mechano-biophysical analysis of the inner structure of living cells. Small 9:885–893
57. Odenthal-Schnittler M, Schnittler HJ, Kemper B (2016) Online quantitative phase imaging of vascular endothelial cells under fluid shear stress utilizing digital holographic microscopy. SPIE Proc 9718:97180U
58. Kemper B, Wibbeling J, Ketelhut S (2014) Analysis of mixed cell cultures with quantitative digital holographic phase microscopy. SPIE Proc 9129:91290W
59. Kemper B, Vollmer A, Rommel CE, Schnekenburger J, von Bally G (2011) Simplified approach for quantitative digital holographic phase contrast imaging of living cells. J Biomed Opt 16(2):026014
60. Schubert R, Vollmer A, Ketelhut S, Kemper B (2014) Enhanced quantitative phase imaging in self-interference digital holographic microscopy using an electrically focus tunable lens. Biomed Opt Express 5(12):4213–4222
61. Poon TC (ed) (2006) Digital holography and three-dimensional display. Springer, Boston
62. Yaroslavsky L (2004) Digital holography and digital image processing: principles, methods, algorithms. Kluwer Academic Publishers, Boston
63. Kreis T (2005) Handbook of holographic interferometry: optical and digital methods. Wiley-VCH, Weinheim
64. Kim MK, Yu L, Mann CJ (2006) Interference techniques in digital holography. J Opt A8:518–523
65. Goodman JW (1996) Introduction to Fourier optics. McGraw-Hill, New York
66. Colomb T, Montfort F, Depeursinge C (2008) Small reconstruction distance in convolution formalism. Digital holography and three-dimensional imaging. OSA Technical Digest, Optical Society of America, St. Petersburg
67. De Nicola S, Finizio A, Pierattini G, Ferraro P, Alfieri D (2005) Angular spectrum method with correction of anamorphism for numerical reconstruction of digital holograms on tilted planes. Opt Express 13:9935–9940
68. Kemper B, Kosmeier S, Langenhanenberg P, von Bally G, Bredebusch I, Domschke W, Schnekenburger J (2007) Integral refractive index determination of living suspension cells by multifocus digital holographic phase contrast microscopy. J Biomed Opt 12:054009
69. Shaked NT, Zhu Y, Bursac N, Wax A (2010) Reflective interferometric chamber for quantitative phase imaging of biological sample dynamics. J Biomed Opt 15(3):030503
70. Klokkers J, Langenhanenberg P, Kemper B, Kosmeier S, von Bally G, Riethmüller C, Wunder F, Sindic A, Pavenstädt H, Schlatter E, Edemir B (2009) Atrial natriuretic peptide and nitric oxide signaling antagonizes vasopressin-mediated water permeability in inner medullary collecting duct cells. Am J Physiol Renal Physiol 297(3):693–703
71. Creath K (1993) Temporal phase measurement methods. In: Robinson D, Reid G (eds) Interferogram analysis. Institute of Physics Publishing, Bristol, pp 94–140
72. Creath K (1994) Phase-shifting holographic interferometry. In: Rastogi RK (ed) Holographic interferometry. Springer, Berlin, pp 109–150

73. Liebling M, Blu T, Unser M (2004) Complex-wave retrieval from a single off-axis hologram. J Opt Soc Am A 21(3):367–377

74. Kemper B, Kandualla J, Dirksen D, von Bally G (2003) Optimization of spatial phase shifting in endoscopic electronic-speckle-pattern-interferometry. Opt Commun 217:151–160

75. Remmersmann C, Stürwald S, Kemper B, Langenhanenberg P, von Bally G (2009) Phase noise optimization in temporal phase-shifting digital holography with partial coherence light sources and its application in quantitative cell imaging. Appl Optics 48:1463–1472

76. Langehanenberg P, von Bally G, Kemper B (2011) Autofocusing in digital holographic microscopy. 3D. Research 2(1):1–11

77. Marquet P, Rappaz B, Charrière F, Emery Y, Depeursinge C, Magistretti P (2007) Analysis of cellular structure and dynamics with digital holographic microscopy. SPIE Proc 6633:66330F

78. Takeda M, Ina H, Kobayashi S (1982) Fourier-transform method of fringe-pattern analysis for computer-based topography and interferometry. J Opt Soc Am 72:156–160

79. Kreis T (1986) Digital holographic interference-phase measurement using the fourier transform method. J Opt Soc Am A 3:847–855

80. Rasheed S, Nelson-Rees WA, Toth EM, Amstein P, Gardner MB (1974) Characterization of a newly derived human sarcoma cell line (HT-1080). Cancer 33:1027–1033

81. Kemper B, Langenhanenberg P, Kosmeier S, Schlichthaber F, Remmersmann C, von Bally G, Rommel C, Dierker C, Schnekenburger J (2013) Digital holographic microscopy: quantitative phase imaging and applications in live cell analysis. Handbook of coherent-domain optical methods. Springer, Berlin, pp 215–257

82. Dubois F, Schockaert C, Callens N, Yourassowsky C (2006) Focus plane detection criteria in digital holography microscopy by amplitude analysis. Opt Express 14:5895–5908

83. Langehanenberg P, Kemper B, Dirksen D, von Bally G (2008) Autofocusing in digital holographic phase contrast microscopy on pure phase objects for live cell imaging. Appl Optics 47:D176–D182

84. Groen FC, Young IT, Ligthart G (1985) A comparison of different focus functions for use in autofocus algorithms. Cytometry A 6:81–91

85. Sun Y, Duthaler S, Nelson BJ (2004) Autofocusing in computer microscopy: selecting the optimal focus algorithm. Microsc Res Tech 65:139–149

86. Firestone L, Cook K, Culp K, Talsania N, Preston Jr K (1991) Comparison of autofocus methods for automated microscopy. Cytometry 12:195–206

87. Bravo-Zanoguera M, von Massenbach B, Kellner AL, Price JH (1998) High-performance autofocus circuit for biological microscopy. Rev Sci Instrum 69:3966–3977

88. Langehanenberg P, Kemper B, von Bally G (2007) Autofocus algorithms for digital-holographic microscopy. SPIE Proc 6633:66330E

89. Elsässer HP, Lehr U, Kern HF (1992) Establishment and characterisation of two cell lines with different grade of differentiation derived from one primary human pancreatic adenocarcinoma. Virchows Arch B 61(1):295–306

90. Bettenworth D, Lenz P, Krausewitz P, Brückner M, Ketelhut S, Domagk D, Kemper B (2014) Quantitative stain-free and continuous multimodal monitoring of wound healing in vitro with digital holographic microscopy. PLoS One 9(9):07317

91. Carpenter AE, Jones TR, Lamprecht MR, Clarke C, Kang IH, Friman O, Guertin DA, Chang JH, Lindquist RA, Moffat J, Golland P, Sabatini DM (2006) Cell profiler: image analysis software for identifying and quantifying cell phenotypes. Genome Biol 7(10):R100

92. Popescu G, Park Y, Lue N, Best-Popescu C, Deflores L, Dasari RR, Feld MS, Badizadegan K (2008) Optical imaging of cell mass and growth dynamics. Am J Physiol Cell Physiol 295(2): C538–C544

93. Rappaz B, Canno E, Colomb T, Kühn J, Depeursinge C, Simanis V, Magistretti PJ, Marquet P (2009) Noninvasive characterization of the fission yeast cell cycle by monitoring dry mass with digital holographic microscopy. J Biomed Opt 14(3):034049

94. Zangle TA, Teitell MA (2014) Live-cell mass profiling: an emerging approach in quantitative biophysics. Nat Methods 11(12):1221–1228

95. Barer R (1952) Interference microscopy and mass determination. Nature 169:366–367
96. Kosmeier S, Kemper B, Langenhanenberg P, Bredebusch I, Schnekenburger J, Bauwens A, von Bally G (2008) Determination of the integral refractive index of cells in suspension by digital holographic phase contrast microscopy. SPIE Proc 6991:699110
97. Kemper B, Bauwens A, Vollmer A, Ketelhut S, Langenhanenberg P, Muthing J, Karch H, von Bally G (2010) Label-free quantitative cell division monitoring of endothelial cells by digital holographic microscopy. J Biomed Opt 15(3):036009
98. Sridharan S, Mir M, Popescu G (2011) Simultaneous optical measurements of cell motility and growth. Biomed Opt Express 2(10):2815–2820
99. Liu PY, Chin LK, Ser W, Chen HF, Hsieh CM, Lee CH, Sung KB, Avi TC, Yap PH, Liedberg B, Wang K, Bourouina T, Leprince-Wang Y (2016) Cell refractive index for cell biology and disease diagnosis: past, present and future. Lab Chip 16(4):634–644
100. Ashkin A (1997) Optical trapping and manipulation of neutral particles using lasers. Proc Natl Acad Sci U S A 94:4853–4860
101. Guck J, Ananthakrishnan R, Moon TJ, Cunningham CC, Käs J (2000) Optical deformability of soft biological dielectrics. Phys Rev Lett 84(23):5451
102. Rappaz B, Marquet P, Cuche E, Emery Y, Depeursinge C, Magistretti PJ (2005) Measurement of the integral refractive index and dynamic cell morphometry of living cells with digital holographic microscopy. Opt Express 13(23):9361–9373
103. Kemmler M, Fratz M, Giel DM, Saum N, Brandenburg A, Hoffmann C (2007) Noninvasive time-dependent cytometry monitoring by digital holography. J Biomed Opt 12(6):64002
104. Björk A (1996) Numerical methods for least squares problems. SIAM, Philadelphia
105. Kemper B, Dartmann S, Schlichthaber F, Vollmer A, Ketelhut S, von Bally G (2012) Self interference digital holographic microscopy for live cell imaging. SPIE Proc:842709
106. Rosner M, Schipany K, Hengstschläger M (2013) Merging high-quality biochemical fractionation with a refined flow cytometry approach to monitor nucleocytoplasmic protein expression throughout the unperturbed mammalian cell cycle. Nat Protoc 8(3):602–626
107. Vandelaer M, Thiry M, Goessens G (1996) Isolation of nucleoli from ELT cells: a quick new method that preserves morphological integrity and high transcriptional activity. Exp Cell Res 228(1):125–131
108. Chalut KJ, Ekpenyong AE, Clegg WL, Melhuish IC, Guck J (2012) Quantifying cellular differentiation by physical phenotype using digital holographic microscopy. Integr Biol 4 (3):280–284
109. Ekpenyong AE, Man SM, Achouri S, Bryant CE, Guck J, Chalut KJ (2013) Bacterial infection of macrophages induces decrease in refractive index. J Biophotonics 6(5):393–397
110. Schürmann M, Scholze J, Müller P, Guck J, Chan CJ (2016) Cell nuclei have lower refractive index and mass density than cytoplasm. J Biophotonics 9(10):1068–1076
111. Cotte Y, Toy F, Jourdain P, Pavillon N, Boss D, Magistretti P, Marquet P, Depeursinge C (2013) Marker-free phase nanoscopy. Nat Photonics 7(2):113–117
112. Kuś A, Dudek M, Kemper B, Kuiawinski M, Vollmer A (2014) Tomographic phase microscopy of living three-dimensional cell cultures. J Biomed Opt 19(4):046009
113. Kemper B, Klokkers J, Przbilla S, Vollmer A, Ketelhut S, von Bally G, Pavenstädt HJ, Schlatter E, Edemir B (2012) Tonicity induced changes in volume and refractive index of suspended cells quantified with digital holographic microscopy. Photonics Lett Poland 4 (2):45–47
114. Przibilla S, Dartmann S, Vollmer A, Ketelhut S, Greve B, von Bally G, Kemper B (2012) Sensing dynamic cytoplasm refractive index changes of adherent cells with quantitative phase microscopy using incorporated microspheres as optical probes. J Biomed Opt 17(9):097001
115. Hoffmann EK, Lambert IH, Pedersen SF (2009) Physiology of cell volume regulation in vertebrates. Physiol Rev 89(1):193–277
116. Rappaz B, Charrière F, Depeursinge C, Magistretti PJ, Marquet P (2008) Simultaneous cell morphometry and refractive index measurement with dual-wavelength digital holographic microscopy and dye-enhanced dispersion of perfusion medium. Opt Lett 33(7):744–746

117. Debailleul M, Simon B, Georges V, Haeberle O, Lauer V (2008) Holographic microscopy and diffractive microtomography of transparent samples. Measur Sci Technol 19(7):074009

118. Schwickert A, Weghake E, Brüggemann K, Engbers A, Brinkmann BF, Kemper B, Seggewiß J, Stock C, Ebnet K, Kiesel L, Riethmüller C, Götte M (2015) microRNA miR-142-3p inhibits breast cancer cell invasiveness by synchronous targeting of WASL, integrin alpha V, and additional cytoskeletal elements. PLoS One 10(12):e0143993

119. Eggers JC, Martino V, Reinbold R, Schäfer SD, Kiesel L, Starzinski-Powitz A, Schüring AN, Kemper B, Greve B, Götte M (2016) microRNA miR-200b affects proliferation, invasiveness and stemness of endometriotic cells by targeting ZEB1, ZEB2 and KLF4. Reprod Biomed Online 32(4):434–445

120. Greve B, Sheikh-Mounessi F, Kemper B, Ernst I, Götte M, Eich HT (2012) Survivin, a target to modulate the radiosensitivity of Ewing's sarcoma. Strahlenther Onkol 188(11):1038–1047

121. Kunsmann L, Rüter C, Bauwens A, Greune L, Glüder M, Kemper B, Fruth A, Wai SN, He X, Lloubes R, Schmidt MA, Dobrindt U, Mellmann A, Karch H, Bielaszewska M (2015) Virulence from vesicles: Novel mechanisms of host cell injury by Escherichia coli O104:H4 outbreak strain. Sci Rep 5:13252

122. Farcal L, Torres Andón F, Di Christo L, Rotoli BM, Bussolati O, Bergamaschi E, Mech A, Hartmann NB, Rasmussen K, Riego-Sintes J, Ponti J, Kinsner-Ovaskainen A, Rossi F, Oomen A, Bos P, Chen R, Bai R, Chen C, Rocks L, Fulton N, Ross B, Hutchison G, Tran L, Mues S, Ossig R, Schnekenburger J, Campagnolo L, Vecchione L, Pietroiusti A, Fadeel B (2015) Comprehensive in vitro toxicity testing of a panel of representative oxide nanomaterials: first steps towards an intelligent testing strategy. PLoS One 10(5):e0127174

123. Mues S, Antunovic J, Ketelhut S, Kemper B, Schnekenburger J (2016) Novel optical approaches for label-free quantification of nano-cytotoxic effects. SPIE Proc 97190:97190J

124. Dubois F, Joannes L, Legros JC (1999) Improved three-dimensional imaging with a digital holography microscope with a source of partial spatial coherence. Appl Optics 38 (34):7085–7094

125. Kemper B, Stürwald S, Remmersmann C, Langenhanekamp P, von Bally G (2008) Characterisation of light emitting diodes (LEDs) for application in digital holographic microscopy for inspection of micro and nanostructured surfaces. Opt Lasers Eng 46:499–507

126. Langehanenberg P, von Bally G, Kemper B (2010) Application of partial coherent light in live cell imaging with digital holographic microscopy. J Mod Opt 57:709–717

127. Girshovitz P, Shaked NT (2013) Compact and portable low-coherence interferometer with off-axis geometry for quantitative phase microscopy and nanoscopy. Opt Express 21 (5):5701–5714

128. Singh AK, Faridian A, Gao P, Pedrini G, Osten W (2014) Quantitative phase imaging using a deep UV LED source. Opt Lett 39(12):3468–3471

129. Dohet-Eraly J, Yourassowsky C, Mallahi AE, Dubois F (2016) Quantitative assessment of noise reduction with partial spatial coherence illumination in digital holographic microscopy. Opt Lett 41(1):111–114

130. Kühn J, Charrière F, Colomb T, Cuche E, Montfort F, Emery Y, Marquet P, Depeursinge C (2008) Axial sub-nanometer accuracy in digital holographic microscopy. Measur Sci Technol 19:074007

131. Kosmeier S, Langenhanekamp P, Przbilla S, von Bally G, Kemper B (2010) Multi-wavelength digital holographic microscopy for high resolution inspection of surfaces and imaging of phase specimen. SPIE Proc 7718:77180T

132. Kosmeier S, Langenhanenberg P, von Bally G, Kemper B (2012) Reduction of parasitic interferences in digital holographic microscopy by numerically decreased coherence length. Appl Phys B 106(1):107–115

133. Choi Y, Yang TD, Lee KJ, Choi W (2011) Full-field and single-shot quantitative phase microscopy using dynamic speckle illumination. Opt Lett 36(13):2465–2467

134. Kemper B, Kosmeier S, Langenhanenberg P, Przibilla S, Remmersmann C, Stürwald S, von Bally G (2009) Application of 3D tracking, LED illumination and multi-wavelength

techniques for quantitative cell analysis in digital holographic microscopy. SPIE Proc 7184:71840R

135. Miccio L, Finizio A, Puglisi R, Balduzzi D, Galli A, Ferraro P (2011) Dynamic DIC by digital holography microscopy for enhancing phase-contrast visualization. Biomed Opt Express 2 (2):331–344

136. Girshovitz P, Shaked NT (2012) Generalized cell morphological parameters based on interferometric phase microscopy and their application to cell life cycle characterization. Biomed Opt Express 3(8):1757–1773

137. Liu R, Dey DK, Boss D, Marquet P, Javidi B (2011) Recognition and classification of red blood cells using digital holographic microscopy and data clustering with discriminant analysis. J Opt Soc Am A 28(6):1204–1210

138. Moon I, Javidi B, Yi F, Boss D, Marquet P (2012) Automated statistical quantification of three-dimensional morphology and mean corpuscular hemoglobin of multiple red blood cells. Opt Express 20(9):10295–10309

139. Yi F, Moon I, Javidi B, Boss D, Marquet PP (2013) Automated segmentation of multiple red blood cells with digital holographic microscopy. J Biomed Opt 18(2):026006

140. Nguyen TH, Sridharan S, Marcias V, Balla AK, Do MN, Popescu G (2015) Prostate cancer diagnosis using quantitative phase imaging and machine learning algorithms. SPIE Proc 9336:933619

141. Charrière F, Marian A, Montfort F, Kuehn J, Colomb T, Cuche E, Marquet P, Depeursinge C (2006) Cell refractive index tomography by digital holographic microscopy. Opt Lett 31 (2):178–180

Index